WHEN LIFE NEARLY DIED

Thames & Hudson

WHEN LIFE NEARLY DIED

THE GREATEST MASS EXTINCTION OF ALL TIME

MICHAEL J. BENTON

48 illustrations

To Mary, Philippa and Donald
for their forbearance and inspiration, as ever

ACKNOWLEDGMENTS

In this new edition, I am grateful again to Colin Ridler of Thames & Hudson for managing the process, and for his encouragement throughout. Fifteen years ago, he first suggested that I should write this book, and then he discreetly guided me through various false starts. Thanks to him, and to his team, for their thorough and extensive work on the book as it went through the publication process. The entire text of the first edition was read by Paul Pearson and Richard Twitchett at the University of Bristol, James Lovelock at the University of Oxford, and Paul Wignall at the University of Leeds. I value the time they spent reading the draft manuscript, their wise suggestions about style and content, but especially for picking up some glaring misunderstandings and errors of fact. Thanks also to archivists and librarians at the Geological Society of London, the British Library and the Natural History Museum, for finding ancient letters and notebooks by Murchison, Owen and the others. Finally, the appearance of the book has been enhanced immeasurably by the artwork, and I thank John Sibbick for his exquisite craftsmanship informed by a real knowledge of the ancient beasts he draws.

Frontispiece. A sabre-toothed reptile, the gorgonopsian *Lycaenops* from the Late Permian of South Africa. This reptile, together with nearly all its contemporaries, was wiped out 252 million years ago by the biggest mass extinction of all time.

First published in hardcover in the United States of America in 2003 by Thames & Hudson Inc., 500 Fifth Avenue, New York, New York 10110

thamesandhudsonusa.com

First paperback edition 2005

Reprinted with revisions 2008, 2015

Library of Congress Catalog Card Number 2014952827
ISBN 978-0-500-29193-1

Printed and bound in China by Toppan Leefung Printing Limited

CONTENTS

PROLOGUE

When I was a student of palaeontology in the late 1970s, I remember being puzzled by some extraordinary differences of view. These were not minor disagreements – that sort of thing is common enough, and such disputes are usually resolved eventually by the discovery of some new evidence that settles the matter one way or the other. This dispute was over the biggest mass extinction of all time: some palaeontologists accepted that, around 250 million years ago, all of life on Earth came very close to complete annihilation, while another group argued that nothing at all had really happened. How could there be any debate about such a fundamental question? The fossil record of the history of life may be patchy and full of holes, but surely there should be no doubt about such an event? Either it had happened or it hadn't.

As I delved deeper into the question, I was convinced that there had indeed been a mass extinction of huge magnitude about 250 million years ago, at the end of the Permian period. But why had some palaeontologists denied it? The extinction deniers were not creationists, flat-earthers or members of some other fringe group. They knew their fossils, and yet their reading of the record seemed to tell them that the end of the Permian passed with only the merest blip, the smallest disturbance, and really it was nothing to be concerned about.

The story has moved on now. Where there was lack of clarity in 1970, or even 1990, there are now two strongly argued catastrophic models for the end-Permian mass extinction. One ties the event to massive volcanic eruptions, producing thousands of cubic kilometres of lava and poisoning the atmosphere. The other links it to impact – dramatic new evidence published in 2001 suggests that a huge meteorite hit the Earth and caused global destruction. Such an extraterrestrial proposition would have been treated as wild scare-mongering by my geology professors 35 years ago, and yet today it is debated seriously.

The end-Permian mass extinction is very much a live issue. But it is not merely a question of interest to palaeontologists, the scientists who study fossils and the history of life, and to juvenile dinosaur fanatics. This is not an arcane debate that might yield only a few amusing historical or philosophical insights. There is immediate relevance today. It has often been claimed that we are living through a mass extinction at the moment, where the main agent of destruction is human activity. Whatever the causes and outcomes of the current ecological crisis, the only analogues we have for realistic comparison are extinction events in the fossil record. They can give indications of what might, or might not, happen in the future.

Death of the dinosaurs

In the late 1970s, much more was written about the death of the dinosaurs, the mass extinction that happened 66 million years ago, at the end of the Cretaceous period. But even for this heavily studied event, there was huge disagreement. The common view was that the dinosaurs had dwindled to extinction gradually over perhaps 5 million years. Certainly, from time to time, rather wild proposals had been put forward – that there had been a huge solar flare, a massive meteorite impact or some other extraterrestrial catastrophe – but these were all apparently nonsense.

Geologists in 1970 knew that the Earth was subject to huge forces that moved the continents at a slow pace, but massive impacts? Not likely. Then, in 1980, came one of the most astonishing publications of my lifetime. A group of physicists and geologists in California announced evidence that the Earth had indeed been hit by a huge meteorite at around that time in the distant past (then dated to 65 million years ago), and that the consequences of this impact had wiped out much of life, including the dinosaurs. Their idea was met by instant ridicule and derision by most geologists and palaeontologists. And yet it is broadly accepted now.

This marks one of the biggest shifts in scientific opinion of recent decades. From being regarded as pariahs, the *catastrophists*, geologists who point to larger-than-normal crises in the geological past, have won the argument, in terms of extinctions of life in the past at least. In retrospect now, it is extraordinary to see how mainstream geologists denied the reality of catastrophes for so long. Their stance dates back to the 1830s, and early debates in geology which were won by the *uniformitarians*, scientists who argued that everything in the past could be explained by reference to modern slow-moving processes. As a result, it became pretty well impossible to discuss mass extinctions as we now understand them, without being branded a madman.

Now that it is considered acceptable to talk about sudden extinction events in the past, the end-Cretaceous mass extinction has been more closely studied than any other. Dozens of scientific publications on the subject appear each year, and these are accompanied by hundreds of treatments in children's books, news items and web sites. This is partly because of the appeal of the dinosaurs: everyone wants to know why they died out. But also, the end-Cretaceous event was the last of several mass extinctions, and so it is easier to study. It's a general principle of geology and palaeontology that the quality of information declines the further back in time one goes. An event that happened 66 million years ago is easier to date accurately – and associated fossils are more abundant – than an event that happened 252 million years ago.

Near-annihilation

The end-Permian mass extinction may be less well known than the end-Cretaceous, but it was by far the biggest mass extinction of all time. Perhaps as few as 10% of species survived the end of the Permian, whereas 50% survived the end of the Cretaceous. Fifty percent extinction was clearly significant enough, and it was associated with devastating environmental upheaval. The catalogue of crises 66 million years ago makes terrifying reading: the impact of a vast meteorite in the Caribbean, vulcanism in India, major environmental deterioration, dust clouds, blacking-out of the sun, freezing cold, acid rain, and dinosaurs going extinct on all sides. But there is an enormous difference between 50% survival through the end-Cretaceous crisis and only 10% survival through the end-Permian.

The key difference between 50% survival and 10% survival is the diversity of founders available for the reflowering of life after the catastrophe. Fifty per cent of species should offer a reasonable cross-section of the range of organisms that existed before the event, and the chance that the ecology of habitats on land and in the sea could eventually recover to something like a balanced system. But survival of only 10% of species means that many major groups of plants, animals and microbes have probably gone forever.

Life can be thought of best as a great tree, originating with a single founding species many billions of years ago and continually branching and expanding upwards through time as new species arise. Here and there twigs of the tree die off as species become extinct, but the overall shape of the tree expands ever upward. During a mass extinction, vast swathes of the tree are cut short, as if attacked by crazed, axe-wielding madmen. Whole branches and twigs are brutally removed. The ragged remnants of the tree, though, will reshape themselves and grow back to full luxuriance. At the end of the Permian, however, the slashing of the tree of life was vicious and sustained. Entire regions of its diverse branches were cut and hacked off. After the crisis, only 10 in 100 of the branches remained, a pathetic remnant. There was certainly no guarantee that these would be sufficient to survive in the long term. After such a severe attack, the great tree of life, with over 3000 million years of history behind it at the time, might have withered away and died completely.

The fact that you are reading this book is evidence that this did not happen, of course. The 10% of surviving species after the end-Permian crisis clearly were sufficient to allow the diversity of life to recover. But it was an astonishingly slow process, and the recovery phase lasted much longer than any other known post-extinction recovery phase.

But what kind of environmental crisis would be sufficient to wreak such havoc? Remember that palaeontologists here are not looking for some process that might have killed a few large or marginal animals, the Permian equivalents of dodos, tigers or pandas. They are looking for a process, or combination of processes, that would kill 90% of species of plants and animals in the sea and on land, as well as, presumably, 90% of microscopic organisms in all habitats.

Impact

The *New York Times* for 23 February 2001 reported 'Meteor crash led to extinctions in era before dinosaurs'. *The Times* of London for the same day said, 'Asteroid collision left the world almost lifeless'. The *Daily Mail* went further in announcing 'The great dying. This week one of the

Earth's greatest mysteries was solved: What happened 250 million years ago to wipe out life on the globe and usher in the dinosaurs.' A team of scientists from the University of Washington, NASA and other institutions, claimed they had found unequivocal proof that the end-Permian mass extinction had been caused by the impact of a meteorite. But is the problem really solved?

Certainly, the news reports were confident enough. The researchers estimated that an asteroid, or giant meteorite, some 6 to 12 kilometres across, had hit the Earth, and that this had led to a series of environmental shocks that in turn caused the extinctions in the sea and on land. Luann Becker, the leading scientist of the group, was reported as saying, 'To knock out 90 percent of organisms, you've got to attack them on more than one front'.

The significance of this event was highlighted on many websites. One of them[1] gave some graphic comments about what followed the impact:

> Earthquakes and volcanoes would have rattled the planet. . . . Lava poured out in volume – enough to cover the planet 10 feet (3 metres) deep. The oceans dropped 820 feet (250 metres).
>
> Worse, the combined effects of the object vaporizing on impact, along with all the vulcanism, poisoned what was left of the seas and choked the air with ash and deadly gases. Sunlight may have disappeared for months.
>
> The researchers say that the volcanic activity was likely going on before the impact, but was then fuelled into a frenzy. The one-two punch, it seems, may be what's needed to precipitate the worst extinctions. Poreda [one of the contributing scientists] called the whole scenario a 'blast from a double-barrelled shotgun'.

The picture, then, seemed clear, in February 2001. The biggest mass extinction all time, the end-Permian, had been caused by a sequence of events of unimaginable ghastliness, triggered by an extraterrestrial impact. And that wasn't all.

Following from this scenario, a NASA scientist was quoted as saying, 'This suggests that the evolution of life on Earth is strongly coupled with our cosmic environment'. He thought that 'the Earth's biosphere is regularly disrupted, every 100 million years or so, by giant impacts that would render human life impossible.' So when will the next one happen?

It's not just wild-eyed eccentrics who scan the skies, fearing our impending annihilation as a result of the impact of another such asteroid.

Politicians and scientific advisers in many countries are planning for this eventuality. Money is being spent to fund astronomers who search for asteroids that might approach the Earth. Politicians warn that we must develop technologies that would allow us to blast any asteroid that looked dangerous. Fortunately, the man from NASA assures us that 'there's no need to panic, and no need right now to spend money to defend Earth against any potential species-ending impact.' He goes on to explain that, if major impacts happen about once every 100 million years, then we might be able to hold on for a few more decades, or even centuries, without too much problem.

This theory of impact was rapidly rejected by most researchers in 2001, though further evidence for impact, and even a crater, was published in 2004. The alternative model is that massive eruptions of lava, sustained over half-a-million years or more, caused catastrophic environmental deterioration – poison gas, heating, freezing, stripping of soils and plants from the landscape, eruption of gases from their frozen locations deep in the oceans and mass de-oxygenation. They debate the synergistic effects of such a break-down in earth systems, when normal feedback processes are overwhelmed. We have oceanic burps and runaway greenhouses. The volcanic model has been developed step-by-step through the 1990s, and it has progressively picked up adherents and enthusiasts. So who is right – was the biggest mass extinction of all time triggered by volcanic eruptions or by meteorite impact? This book sets out to provide the answer.

The ripples from ancient mass extinctions spread a long way. The end-Permian catastrophe is clearly no longer a minor sideline in the history of the Earth, as I was taught by my geology professors. Perhaps nothing so familiar as the dinosaurs was killed off, but the event was profoundly more important. Scientific findings about the events of 252 million years ago can inform our understanding of modern extinction threats in a direct way.

When life nearly died

My personal involvement in exploring the end-Permian mass extinction took me to Russia several times in the 1990s. Shortly after *glasnost* and *perestroika* we had the opportunity to mount expeditions to the Ural Mountains, the borderland between Europe and Asia, where some of the best Permian rocks are to be seen. We were looking at the successions of ancient environments there, and the evolution of amphibians and reptiles on land. In a way, too, we were recreating history. The Permian period was named in 1841 after the city of Perm on the western edge of the Ural Mountains, and it was named by a roving Scotsman. Sir Roderick Impey Murchison, one of the most distinguished geologists of his day, was engaged in a rapid and gruelling tour of Russia during which he made the first studies of the classic Permian, and his early role is a critical part of our story.

In this book, we shall explore several themes. We meet Sir Richard Owen, and some other Victorian worthies, who were the first to explore the fossil reptiles of the past, including the dinosaurs. These were early days for palaeontology, and, with new discoveries coming in from all corners of the world, they had to grapple with the astonishing concept of huge antediluvian reptiles, much larger than any modern lizard or crocodile. At the same time, geologists were disentangling the sequence of the rocks. They were laying the foundations of our understanding of geological time and dating. This story includes a journey through the gentlemen's clubs of London in the late 1830s, Murchison's stately progress through Russia in 1840 and 1841, and his evidence for naming the Permian period.

Then, with the evidence in place, we shall explore in Chapter 3 why many, if not most, geologists and palaeontologists denied the reality of a substantial crisis at the end of the Permian. In parallel with this denial was the problem of understanding the demise of the dinosaurs, and other extinctions. Some extraordinary ideas were put forward right up to the 1970s.

The phase of denial of the end-Permian mass extinction has ended only relatively recently. Chapters 4 and 5 explore the two major developments that forced this shift of opinion: the rediscovery of catastrophism – 'neocatastrophism' – and the acceptance that the Earth has undergone unexpected and huge upheavals in the past. The change of opinion was crystallized in that single remarkable publication in 1980 that offered a robust presentation of evidence that the dinosaurs had died out as a result of a massive meteorite impact. Through this time of adjustment, from about 1950 to 1990, progress was slow: many geologists found it immensely hard to accept such wild and preposterous kinds of processes in Earth history.

Then in Chapter 6 we explore the history of life, and how it has diversified from a single species to many millions today. Several mass extinctions, and many smaller extinction events, have punctuated the evolution of life from time to time. What has been the role of extinction, and are there any general or predictable patterns?

Next, we look at the end-Permian mass extinction in detail. To do this, we follow geologists on their recent explorations of the evidence in China, Russia, Pakistan, Italy, Greenland and other parts of the world, as they strive to pin down the exact timing of the event (Chapter 7). Their work has also revealed some startling new evidence about the precise sequence of environmental changes, before, during and after the crisis. The latest research on the mass extinctions in the sea (Chapter 8) reveals a great deal of detail about the magnitude of the event, which groups died out and which survived, whether the extinctions were ecologically selective or not, and whether they happened equally in the tropics and at the poles.

Dramatic new discoveries also tell us much about what was happening on land at the end of the Permian. Intensely detailed studies of ancient soils, river sands and lake muds have revealed the changes in plant life and in the amphibians and reptiles that fed on the plants, and on each other (Chapter 9). These studies also show what the post-apocalyptic scene was like. After the mass dyings, the Earth was a cold, gloomy place, and the few surviving plants and animals experienced a strange, empty world. Our work in Russia (Chapter 10) contributes new evidence concerning the extinctions of life on land, and in particular the extinctions of large reptiles.

Drawing all the threads together – the evidence of ancient environments read from the rocks, the evidence for major climatic changes read from chemicals within those rocks and the detailed record of appearances and disappearances of microbes, plants and animals on land and in the sea – leads to a cohesive picture of just what happened at the end of the Permian (Chapter 11). The debate about meteorite impact vs. volcanic eruption continues apace, and I highlight the strengths and weaknesses of each theory. In the end, there does seem to be a clear winner, one which provides a convincing model of the sequence of killings. This is a model that could not have been known, and certainly would not have been believed, 30 years ago. Indeed, it's a model that could not have been offered even five years ago. But it is a model that is attracting increasing approval from earth scientists as they examine the new evidence.

But what does this mean for humans today? Potentially a great deal, especially in understanding the greenhouse effect and current threats to biodiversity. The growing research links between palaeontological and geological studies of mass extinctions of the past, and the current ecological crisis on the Earth are explored in the last chapter. Scientists, conservationists and politicians are looking ever more closely at the evidence from the past. In planning for the future, perhaps the best evidence comes from the past.

Since 2003 ...

Inevitably, many new discoveries have been made in the twelve years since this book was first published. In this new edition, I have updated the text throughout. Of course, much of the background and historical detail remains unchanged – this has still been a story in which varied scientists were unwilling to admit that extinctions had happened, and where we have seen several revolutions in ways of thinking even within living memory.

I have also had the pleasure of continuing research along several lines relating to the great extinction. We carried out further fieldwork on the Permo-Triassic boundary in Russia in 2004, 2006, and 2009, and this has shown that the extinction of amphibians and reptiles was indeed sudden, but also that the ecological recovery after the event lasted for at least 15 million years, far longer than previously thought. This work also allowed us to reconstruct the habitats and environmental settings of some of the extraordinary fossil tetrapods, and how those linked to climatic changes towards greater aridity across the Permo-Triassic boundary.

Most importantly, I have extended my experiences to South China, one of the most productive areas in recent years for improving understanding of the great mass extinction. There, many kilometres of thickness of rocks not only span the crucial extinction horizons, but also document the recovery of life during the turbulent post-extinction period. Studying the nature of the recovery of life, and determining whether this was a slow or fast process, has occupied many researchers over the past decade, and additional materials in this edition concern this new theme.

Does extinction matter?

It is easy to take highly polarized views on the current ecological situation. On the one hand, the doom-mongers declare that everything is lost and that life will become entirely extinct within a few hundred years. Their calculations suggest that we are losing 2000 species per day, and that our every advance in civilization is murderous. Humans should instantly cut their population sizes and return to self-subsistence, stone-age modes of life.

The opposing view looks at the past history of human population growth and to agricultural advances, and assumes complacently that nothing has really gone wrong. Who really cares about the odd dodo, panda or elephant? Clearly, in some way, it is natural to expect that such an advanced species as *Homo sapiens* will take over the world, and use its great brain to discover ever more efficient ways of producing food and curing disease.

Such standpoints are too extreme for most people. It is clear that human populations cannot be cut back to pre-industrial levels, nor would many people choose to live in rude huts and subsist on brown rice. Equally, the world was not created simply for human benefit. This idea is somewhat akin to the creationist viewpoint that horses have conveniently curved backs so that they may carry human riders. The complacent assumption that ever more food can be squeezed out of a limited area of agricultural land around the world is absurd. Advancing populations and farming are daily destroying vast tracts of habitat. But how can one draw a realistic line between the doom-mongers and the growth-enthusiasts?

Look to the past. Extinctions have happened before, and life has recovered in different ways. Comparison of present crises with documented ancient examples at least allows scientists and policy-makers to work with real facts and figures. Questions can be posed: 'If 10% of species are wiped out worldwide, what happens?', 'If half the species on one continent are destroyed, do the remainder repopulate the area, or is the continent invaded from elsewhere?', 'If half of life disappears, how long does the recovery phase last?'. The step-by-step dissection of ancient mass extinctions can provide some of the answers.

The geological timescale. Mass extinctions and minor extinctions are shown by arrows. The 'big five' are shown with heavier arrows.

MASS EXTINCTIONS

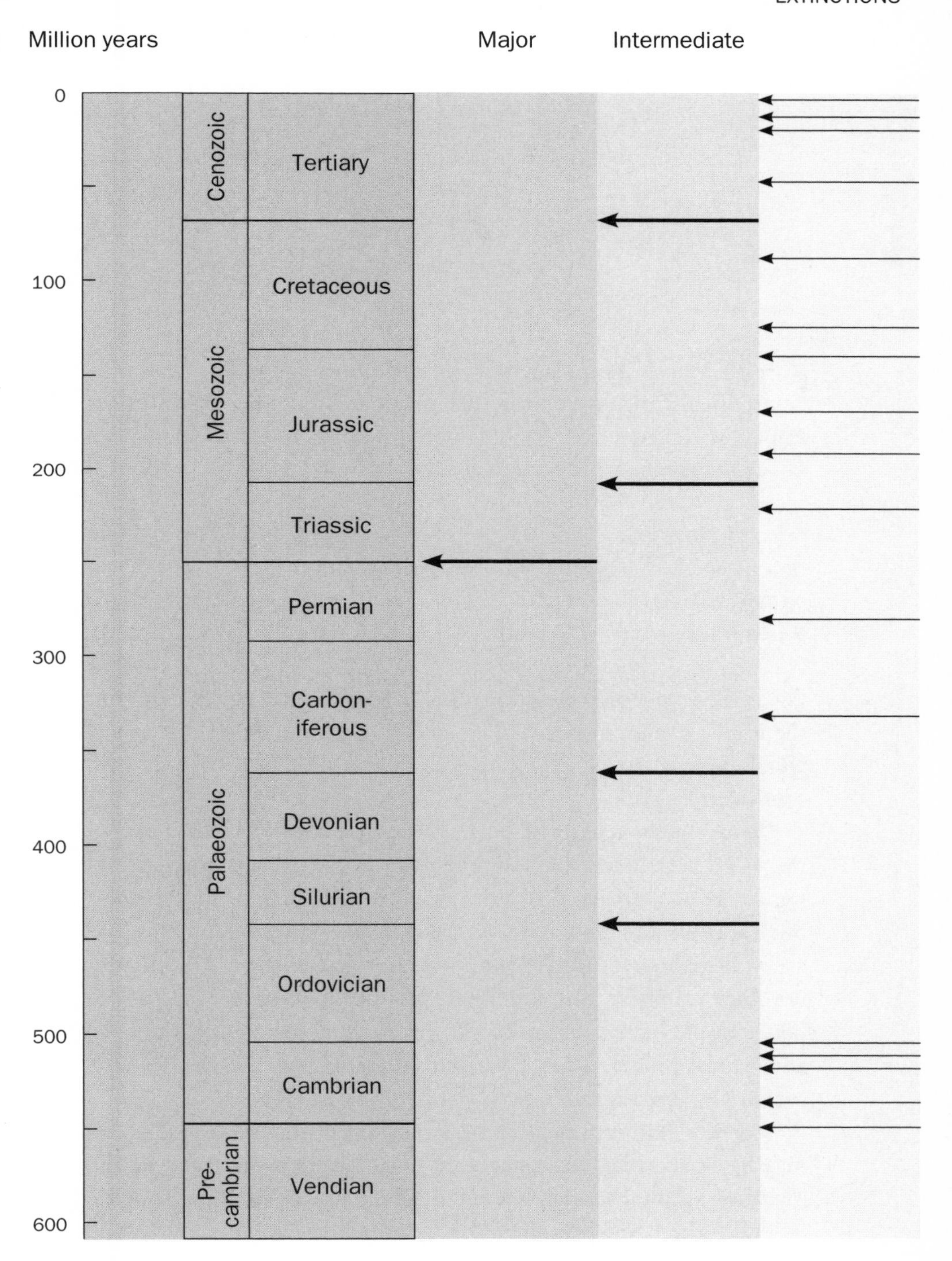
MINOR
EXTINCTIONS
Million years
Major
Intermediate
0
100
200
300
400
500
600
Cenozoic
Mesozoic
Palaeozoic
Pre-
cambrian
Tertiary
Cretaceous
Jurassic
Triassic
Permian
Carbon-
iferous
Devonian
Silurian
Ordovician
Cambrian
Vendian

1
ANTEDILUVIAN SAURIA

On 5 March 1845, Professor Richard Owen wrote an account of some bones of ancient reptiles that had been brought back by Sir Roderick Impey Murchison from his long peregrinations in Russia in 1840 and 1841. Owen determined one section of a backbone as

> belonging to the Crocodilian division of Sauria by the strong, short, rib-like processes from the sides of the two anchylosed sacral vertebrae,— a modification . . . introduced to give a firm 'point d'appui' to the hinder extremities of those higher Sauria which occasionally walk on dry land.[1]

Owen then went on to describe other bones he had been sent. He compared them all to modern crocodiles, and to a reptile called *Thecodontosaurus* which had been named in 1836 from the Triassic rocks of Bristol in southwest England.

What does this passage mean? All that Owen was in fact saying was that the sacral vertebrae, that is the elements of the backbone that lie in the hip region, showed strong ribs at the side to which the hip bones could attach. Such a strong attachment is seen today in crocodiles and other reptiles that are well adapted for walking on land. He was stressing the fact that this was not some primitive kind of animal that lived most of the time in water and hence would have no need for a strong skeletal framework for the legs. An advanced reptile such as the Russian crocodile indicates that it comes from rocks of Triassic age, or later.

Owen's determination of the Russian reptiles, arcane as his language may be, was critical evidence used by Roderick Murchison in his interpretation of the Permian System of rocks in Russia. Murchison had just invented the name 'Permian' for the fossiliferous sediments around the city of Perm on the west side of the Ural Mountains, and, by implication, for all rocks of the same age from other parts of the world. In addition, for Owen, these specimens offered further information towards a fuller understanding of the early history of the reptiles.

It is important to place Owen's views in context, and examine where they came from. In the absence of adequate specimens, and with extensive confusion between beasts from what we now recognize as Permian and Triassic rocks, it is no wonder that savants of early Victorian times could not conceive of a vast end-Permian extinction event. Let us look briefly at what we know now of amphibian and reptile evolution, and then piece together what Owen knew in 1845.

Tetrapods: batrachian and saurian

Modern amphibians and reptiles are easy to tell apart.[2] The amphibians, such as frogs and salamanders, generally lay their eggs in water, and from these hatch tadpoles. The tadpoles develop as entirely aquatic creatures – essentially as fishes – before they metamorphose into the adult form which lives partly on land and still partly in the water. Indeed, the name amphibian means 'life on both sides' (from the Greek *amphi*, 'on both sides', and *bios*, 'life'), in other words, life both in the water and on land.

Modern reptiles, particularly lizards and snakes, live their lives entirely on the land and are adapted to life in dry conditions, with horny waterproof scales over their bodies. They lay eggs with white calcareous shells (like a hen's egg) or with a leathery coat – in either case, the egg is waterproof and the young reptile develops safely inside before hatching straight into life on land.

Naturalists have long realized that amphibians are essentially intermediate between fishes and reptiles. Today, of course, we see this as an expression of the evolution of the respective groups – fishes came first, then amphibians evolved from fishes when they took their first faltering steps on to the land, and finally the reptiles evolved from the amphibians by dispensing with life in the water altogether. Owen was famously not an evolutionist – he was later one of Charles Darwin's most forceful critics – although he could not deny the apparent sequence of forms.

But, in the 1840s, the fossils were rare. Equally, the dividing line between the fossil amphibians and reptiles was hard to draw. Owen, and others at the time, frequently referred all the early tetrapods – that is, four-footed creatures – to Reptilia. The older, amphibian-like ones were sometimes called Batrachia, or Batrachian Reptilia, and those that could be compared more closely with modern lizards or crocodiles were sometimes called Sauria.

Tetrapods of the coal forests and Permian deserts

We now know that the first tetrapods came ashore in the Late Devonian, some 370 million years ago, and that these basal tetrapods – amphibians – radiated into a variety of forms in the subsequent Carboniferous period. The Carboniferous amphibians included land-living forms with four sturdy limbs, as well as many largely aquatic species, some of which swam with their paddle-like limbs, and others of which had lost their limbs entirely. They ranged in size from just a few centimetres to 3- and 4-metre long monsters.

The Carboniferous is famous for its coal forests – great tangled masses of vegetation growing luxuriantly in the tropical conditions that were experienced all over Europe and North America at the time (both continents straddled the Equator during the Carboniferous). Crawling through the low vegetation were myriad millipedes, centipedes, spiders and the like. Higher in the trees were insects, some of them remarkable for their huge size – there were dragonflies as large as small gulls in those days. The amphibians feasted.

Among the Carboniferous amphibians were a few rather more terrestrialized forms, animals that had waterproof skins and which laid eggs with shells. At first, these basal reptiles did not make much of a mark, since the amphibians, with food and damp habitats in abundance, flourished. But the great tropical forests of Europe and North America began to dry out towards the end of the Carboniferous and into the subsequent Permian period. Most of the amphibians died out, and only some smaller aquatic forms survived in the reduced watercourses.

Now it was the turn of the basal reptiles to flourish. During the Permian, the group called the synapsids, or mammal-like reptiles, rose to dominance, especially in the Late Permian. Ecosystems became complex, with synapsids of all shapes and sizes feeding on the plants, insects and other invertebrates, and each other. By the end of the Permian, some ecosystems were as complex as any today. Top-level animals had evolved – massive herbivores the size of rhinos, and sabre-toothed carnivores that could pierce the thick hides of these huge plant-eaters. Among the smaller reptiles, in addition to the synapsids, were also representatives of other groups, including forms distantly ancestral to turtles, and to crocodiles and lizards.

But these animals abruptly disappeared, wiped out by the mysterious end-Permian crisis. The complex ecosystems collapsed and all the richness of Late Permian life was destroyed. What came next was one of the most extraordinary times in the history of the Earth.

When pigs ruled the Earth

A few years ago, I had a rather plaintive phone call from a researcher who worked for a small, independent film-maker near London. They had completed a one-hour documentary about the end-Permian mass extinction, but the programme had just been turned down by the commissioning television company. I asked why, and was told it was because there was too much about lemurs in the film. This surprised me, since lemurs have absolutely nothing to do with the end-Permian mass extinction – indeed, the first lemur fossils date back only a million years or so, and their distant relatives, some of the first primates, are at most 60 million years old.

It turned out that the film-makers had spent all their budget flying to Madagascar and filming the rare lemurs and other primitive primates of that island. I asked them how much footage they had of Permo-Triassic rocks in Russia and South Africa – the sequences that span the geological boundary and that contain clues that might point to the causes of the event. I detected a gulp at the other end of the line. The researchers had clearly failed in their job, and the film-makers had panicked and thought, 'if we can't fill the hour with dinosaurs, let's go for the monkeys'. They did, however, have animatronic rubber dicynodonts.

The dicynodonts were one of those key synapsid groups that had been hugely dominant in the Late Permian. From poles to equator, the dicynodonts were the major plant-eaters, some of them reaching the size of hippos. It was well known that they had been virtually wiped out by the end-Permian mass extinction, and only one form had survived: a medium-sized dicynodont called *Lystrosaurus*. Subsequently, from *Lystrosaurus*, the dicynodonts re-flowered in the Triassic, radiating back to something like their former diversity.

The film-makers had at least heard of *Lystrosaurus*, this extraordinary reptile that had survived the end-Permian cataclysms. All the impacts, eruptions, freezing, poisoning and other catastrophic events had not discouraged *Lystrosaurus*, which not only survived, but repopulated the Earth from Antarctica to China, from Argentina to Russia. Surely *Lystrosaurus* must have been the original bionic organism, possessing superpowers that its friends and relatives lacked?

To rescue their film, I was interviewed at the animatronics studio in London. As the model *Lystrosaurus* gulped and rolled its eyes beside me, I tried to explain that it was in fact a very ordinary animal. It had no special survival qualities that the other animals lacked. It was simply lucky. 'Why', they kept asking, 'did *Lystrosaurus* survive, and nothing else?' A clear adaptive statement was required. I explained that *Lystrosaurus* was a survivor of a sort, but it was not particularly fast, fearsome or intelligent. It was really something like a Triassic pig in appearance, and even perhaps in habits, since it was probably a generalist, without specific adaptations in its diet, living requirements or mode of locomotion.

The point is that good fortune is a characteristic of mass extinctions. The survivors are more lucky than specially adapted. The most advanced, intelligent, fast-breeding animal species may be wiped out by the chance calamity of an extinction event when it is obliged to face challenges that have never been encountered before. Evolution works to hone the fine details of the adaptations of organisms against commonly encountered problems, such as droughts, floods, predators and diseases, but rare events that happen perhaps once every few million years just cannot be accommodated. The phenomenon is what the palaeontologist David Raup memorably described as 'bad luck, not bad genes'.

So, *Lystrosaurus* was not a specially tough survivor, a perfect animal to weather the storms and found a major new dynasty. It was rather nondescript, about 1.5 metres long, with a bulky, blimp-like body and rather inadequate hindquarters (Fig. 1). Its legs were short and its head was heavy, armed with a small snout and horn-edged jaws that lacked all but the canine teeth. *Lystrosaurus* was a plant-eater that evidently sliced tough stems and chopped them into manageable fragments by a circular backwards-and-forwards rotatory jaw action. All of this was nicely shown by

1 Lystrosaurus, *survivor of the end-Permian mass extinction.*

the animatronic model which, by the operation of a control panel, could be made to snuffle or drool, while technicians added suitable snorts and howls.

A few weeks later, when the film had been put together, it was re-titled 'When pigs ruled the Earth', the title appearing in all the pre-publicity handouts and notices. The *Sunday Times* ran a full-page story, accompanied by lurid colour illustrations, about how pigs had been hugely successful in the Triassic, and how they had later evolved into the array of other mammals we know and love today. On the day after the film was shown, I encountered withering looks of pity from colleagues and students alike. But at least *Lystrosaurus* became known to a wider audience.

Reptilian renaissance in the Triassic

The earliest Triassic was indeed a bleak, impoverished time. Gone were the specialized plant- and flesh-eaters of the Late Permian. All that remained were literally one or two tetrapod species worldwide, including *Lystrosaurus*, and these slowly populated the Earth, eventually giving rise to a second flowering of the reptiles. But the old days were over. New animals came on the scene, including the dinosaurs. Some synapsids reradiated in the early part of the Triassic, and they diversified to a certain extent. But ecosystems did not reach the levels of complexity achieved in the Late Permian for millions of years. And the synapsids were unable, for a long time, to regain their former ascendancy.

In fact, the end-Permian mass extinction triggered an evolutionary dance. The synapsids largely withdrew, and their place was taken by the great reptilian group called the diapsids ('two arches', referring to their two cheek bones behind the eye socket), represented today by crocodiles, lizards, snakes and birds, and deriving from a distant, Carboniferous ancestor.

During the Carboniferous and Permian, the diapsids had been a minor element of most faunas, only a small lizard-like creature here and there, never large, and rarely more than 2 or 3% of the total numbers of animals. Some admittedly took to gliding in the latest Permian, but they were hardly a major group. In the Early Triassic some diapsids, in particular the group called the archosaurs ('ruling reptiles') took over the carnivore niches. They preyed on the re-evolving synapsid plant-eaters. During the first 20 million years of the Triassic, the basal archosaurs diversified slowly, and eventually included huge predators, some of them up to 5 metres long. The first dinosaurs appeared about this time, and diversified in the Late Triassic; rather small, bipedal forms at first, they soon reached huge size. Mammals were around during the entire age of dinosaurs, small descendants of the formerly dominant synapsids, but the dinosaurs ruled the Earth for the next 165 million years, until their extinction 66 million years ago. Then, and after a long wait in the wings, the synapsid descendants, the mammals, finally moved back to dominate the Earth, a position they had last held in the Late Permian. The synapsid-diapsid-synapsid cycle had gone full-circle.

This is our present understanding, founded on extensive collecting and study of specimens from around the world since the 1840s. But Richard Owen knew very little of this. He had information from only scattered fossil remains, and indeed, many of the principles of his work had only just been established. In particular, the new science of comparative anatomy, which was Owen's forte, was only around 20 years old when he was studying the Russian saurian fossils. No wonder that neither he, nor any of his colleagues, could even conceive of the magnitude of the mass extinction event that they had just begun to document.

The nature of fossils

I remember collecting my first fossils at the age of eight. I was brought up in Aberdeen, in northeast Scotland, a city where all the older buildings are constructed from the native silver-coloured granite, a resilient rock that looks as fresh today as it did when it was cut 100 or 200 years ago. Granite is an igneous rock, formed deep in the Earth's crust from molten magma, and there was no chance of ever finding fossils in that.

On family holidays in Yorkshire, though, I was able to explore more fruitful rocks, in particular the marine Jurassic rocks of the coast around Saltburn, Staithes and Whitby. There it is easy to collect fossils – clams, brachiopods, coiled ammonites, rare corals and even isolated bones of marine reptiles and fishes. Easiest to find were specimens of *Gryphaea*, formerly called 'devil's toenails' since they look like the calloused, horny toenails of some great dragon. A *Gryphaea* specimen fits neatly into the palm of a child's hand – a deeply curved shell, shaped something like half a doughnut, but with a flat shell covering the top. *Gryphaea*, an oyster, lived with its curved lower shell half-buried in the sea-bed mud, and its upper, flat shell able to flap open and shut, rather like the lid of a kitchen bin. I collected 160 of these shells on my first fossil-hunting trip, and insisted that we carry every one of them back with us to our hotel, and then on the train home to Aberdeen.

It was obvious to me then that these fossils were the remains of animals that had once been alive. Of course, today we read that fact in all the books about fossils, and any child who is interested in dinosaurs will know that the shells, bones and impressions buried in the rocks can be brought back to life by palaeontologists. But this has not always been self-evident.

In the Middle Ages, scientists discussed the nature of fossils long and hard. Surely, they argued, the fossil shells that they found high in the hills of Italy or France could not be the remains of real sea creatures, since the sea was miles away? Some of them might have been carried there by Roman soldiers, but not all of them. As in so many things, however, Leonardo da Vinci had the correct answer. He argued that if a fossil looks in every detail like a modern shell or bone, then it must clearly be the remains of some ancient organism related to the modern forms. The fossil shells got into the high hills since the seas had once covered those hills, or the hills had been lifted up after the rocks had been deposited, to form the mountains we see today.

The debate about the nature of fossils continued actively through the seventeenth century. The great savants in England and France presented weighty arguments on both sides. Some claimed that the fossil shells, sea urchins, sharks' teeth, mammoth bones and other fossils they found were indeed the remains of ancient organisms. Others argued that this was impossible: surely the so-called fossils had been placed in the rocks by the Almighty to test our faith. Since fishes and molluscs clearly could not live in rocks, the fossils were no more than inorganic productions, the result of the action of plastic forces deep within the Earth. To lend philosophical seriousness to their proposals, the opponents of fossils as the remains of ancient plants and animals often used the term *vis plastica* for 'plastic force', the Latin version presumably sounding more authoritative.

But as the fossils piled up, this miraculous view had to be rejected. However, if all the fossils represented plants and animals from long ago, this opened up two further difficult questions: first, where had all these plants and animals gone, and second, how long ago had they lived?

The Yorkshire crocodile and the Ohio *incognitum*

By 1750, most naturalists accepted the organic nature of fossils, but they still denied the possibility of extinction. This may seem paradoxical, but any Christian of the time could not readily accept this notion, since extinction, the loss of one or more species, would imply an error of some kind in God's plan. But if they were not extinct, where now were all these strange plants and animals of the past as represented by the fossils, but not known to be alive anywhere?

One way out of the dilemma was to suggest either that the fossil creatures had lived before the biblical Great Flood, or that they perhaps still existed in some unexplored part of the world, such as the far reaches of North America, Australia, India, or any of the other lands that were just then being explored by Europeans.

In 1758, a beautiful fossil of a crocodile was reported from the foot of a cliff on the Yorkshire coast, near Whitby – indeed close to the site of my early fossil-hunting forays, as it happens. The crocodile skeleton, laid out as black fossil bones, all in their correct relations, was embedded in Lias limestones, recognized now as part of the Jurassic. The specimen was described in two separate reports to the *Philosophical Transactions of the Royal Society of London*, by William Chapman and a Mr Wooler, whose first name has been lost to history. Wooler argued that the fossil was identical 'in every respect' to the modern Indian crocodile from the River Ganges. As to its position, at the foot of the cliff and buried under 180 feet (55 metres) of rock, there could be no doubt[3] that the animal itself must have been antediluvian, and that it could not have been buried or brought there any otherwise than by the force of waters of the universal deluge. The different *strata* above this skeleton never could have been broken through at any time, in order to bury it to so great a depth as 180 feet; and consequently it must have been lodged there, if not before, at least at the time when those *strata* were formed.

This was all very well for a fossil animal that apparently looked like a living form, but what about something that seemed to have no obvious living counterparts?

Such was the mastodon.[4] The first specimens were found on the banks of the Ohio River in 1739 by a French Canadian army officer, Baron Charles de Longueuil, who sent the bones and teeth back to Paris. There, a great furore broke out, as the leading naturalists of the day debated whether the mastodon could be extinct or not. Further specimens were collected by British settlers, and they were discussed in London, but, by 1769, the issue was pretty much settled.

The intervention of the great anatomist and physician, William Hunter (1718–83), was critical. He compared the jaw bone of the American mastodon, which he termed the Ohio *incognitum*, with that of a modern elephant, and saw that they were identical, except for the teeth. So this must have been an elephant, but one that was not the same as either the African or Indian elephants of today. Hunter pursued his argument remorselessly. He had examined most of the specimens sent both to Paris and to London, and he therefore concluded with the comment that, 'though we may as philosophers regret it, as men we cannot but thank heaven that its whole generation is probably extinct'.

Georges Cuvier (1769–1832), the great French anatomist and geologist,[5] finally resolved the extinction issue, with a flurry of publications in 1795, at the beginning of his glittering career in Paris. He wrote accounts of the mastodon and of the Siberian mammoth, many specimens of which had been dug out of the frozen tundra, some even retaining hair and flesh. His third case study was the giant ground sloth *Megatherium*, a huge animal that lived formerly in the Americas, particularly South America.

Cuvier used a new scientific approach in these papers, one that came to be known as comparative anatomy. In this approach, the anatomist compares every fine detail of the skeleton and other structures of the body of different animals with a view to finding regularities and similarities of structure. By using his thorough knowledge of the anatomy of modern animals, Cuvier was able to demonstrate that these ancient animals were indeed similar to modern forms – to elephants and tree sloths, for example – but they were clearly also anatomically distinct. These were all large mammals, and unlikely to be missed if they were still alive, Cuvier argued, and his case was so ably demonstrated that the last doubters had to accept the reality of extinction.

This is ancient history now, but to Richard Owen the debates were just settling down. As a young man studying medicine and anatomy in Edinburgh in the 1820s, Owen heard constant reference to the great Cuvier. British anatomists looked to Paris for the best, most up-to-date thinking, and stories of Cuvier's abilities spread. It was said, for example, that he could identify any animal from a single bone, and he demonstrated his skill in public presentations of his discoveries. Owen set himself the task early in his life to try to emulate the great Frenchman. But that was a long time in the future. In 1845, as he pondered the unusual new reptile bones from the Russian Permian, Owen recalled some obscure references to such beasts having been found in the previous century, at the height of the mastodon-mammoth-extinction debates.

First finds in the Russian Copper Sandstones

The first finds of the famous Permian fossil reptiles from Russia came to light in the 1770s, but only in obscure ways.[6] The fossil bones were found in the Copper Sandstones, a belt of ore-rich rocks stretching for hundreds of kilometres along the western slope of the Ural Mountains. The specimens came from the extensive mining works for copper, especially from a small number of mines in what is now Orenburg Province, as well as in Bashkortostan and some from Perm Province. The Russian Academy of Sciences sent out scientific expeditions to the Ural Mountains from 1765 to 1805, but the naturalists did not always understand what they had found. One of them, P. I. Rychkov, mentioned the discovery of fossil reptile bones in the Copper Sandstones in his diary of 1770, but he took them to be the remains of ancient miners.

The Copper Sandstones reptile fossils continued to be collected sporadically through the late eighteenth century, mainly by mining engineers who had a personal interest in these curious remains. As the mines were so remote from fashionable St Petersburg, however, little was reported to the scientific societies there and hardly any formal study was undertaken. Distances in Russia are huge, and in those days the roads were appalling, so the fastest post-chaise would take at least four weeks to cover the 2500 kilometres from Perm or Ekaterinburg back to St Petersburg. To the French-speaking tsars and princes in their glittering court, the mineral workings of the Ural Mountains might as well have been on another planet. And almost nothing of these remarkable discoveries filtered out to the West until somewhat later. Cuvier had nothing to say about the Russian Permian reptiles because he had never heard of them, and he had died before Owen received some of the first specimens to reach the West, from Murchison. But Cuvier did play a large part in the interpretation of the first dinosaurs to be found and identified.

Giant saurians

Bones of what we now know to have been a large meat-eating dinosaur[7] came to light in Middle Jurassic rocks north of Oxford in about 1818, and were taken to William Buckland (1784–1856), Professor of Geology at the University of Oxford and Dean of Christ Church. At that time it was not uncommon for churchmen to combine their careers with science. Buckland was a brilliant mix of philosopher and eccentric. His house was full of fossils and he was sought after for his impressive lectures to undergraduates about geology and the life of the past. His dinner party invitations were received with a mixture of pleasure and nervousness: pleasure at the thought of his agreeable company and amusing conversation, but nervousness at what might grace the table. Buckland was famous for having eaten his way through the entire animal kingdom. He enjoyed most of his culinary experiments, but could not find an agreeable way to prepare house flies.

Buckland could not identify the huge fossil bones, and he showed them to Cuvier in Paris and to other experts. In the end, Buckland classified the animal as a giant reptile, probably a lizard, and he estimated it had been 40 feet (12 metres) long in life. After six years of consideration, Buckland finally published a description of the bones in 1824, and he stated that they came from a giant reptile which he named *Megalosaurus* ('big reptile'). This was the first dinosaur to be described formally.

At the same time, independently, Gideon Mantell (1790–1852), a country physician in Sussex, was amassing large collections of fossils. Mantell was as fanatical about geology and palaeontology as Buckland, but he lacked the family background and wealth to indulge his philosophical pursuits freely. His life was dogged by the tension between his passion for fossils and the need to earn a living and to achieve some kind of position in society. During a visit to a patient near Cuckfield in 1820 or 1821, so the story goes, his wife Mary, who had come with him, picked up some large teeth from a pile of road-builders' rubble. The teeth had evidently come from the Wealden beds, recognized now as Early Cretaceous in age.

Mantell realized the teeth belonged to some large plant-eating animal, but as he was not skilled in anatomy he sent the teeth to Cuvier, who assured him that the animal must have been a rhinoceros. Mantell then compared the teeth with those of other modern animals in the Hunterian Museum in London, and a student there, Samuel Stutchbury, showed him that they were like the teeth of a modern plant-eating lizard, the iguana, except they were much bigger. So, Mantell described the second dinosaur, named by him *Iguanodon* ('iguana tooth') in 1825, based both on the teeth and on some other bones he had found since.

William Buckland and Gideon Mantell had great difficulty in interpreting these early dinosaur bones since there were no modern animals to compare them with. In the end, they decided that *Megalosaurus* and *Iguanodon* were giant lizards. The third dinosaur to be named, *Hylaeosaurus*, came to light in southern England in the same Wealden beds that had produced *Iguanodon.* Mantell named it in 1833. Further dinosaurs were named in the 1830s, including two from the Triassic: *Thecodontosaurus* was named in 1836 by Henry Riley and Samuel Stutchbury from Bristol, southwest England, and *Plateosaurus* in 1837 by Hermann von Meyer from southern Germany. In 1842, Richard Owen himself named the sixth dinosaur, a giant plant-eater, again from the Middle Jurassic of Oxfordshire, *Cetiosaurus.* But still no one knew what these extraordinary giant beasts actually were.

Richard Owen, the 'British Cuvier'

By 1841, when Murchison brought back his valuable bones from the Permian of Russia, Richard Owen (1804–92) was rising rapidly in reputation.[8] Owen had arrived in London in 1825 after his medical training in Edinburgh, and was appointed as an assistant at the Royal College of Surgeons in 1826, with the task of sorting out their collections of anatomical specimens, both human and animal. He began to produce a series of catalogues in 1830, continuing with the task until 1860. In 1836 Owen was appointed Hunterian Professor of Anatomy at the Royal College of Surgeons. A reward for his labours, this also gave him the status he craved in order to make an impact in scientific circles in London.

Although he had worked mainly on the anatomy of living (or at least recently dead) animals up to 1836, his new status as the 'British Cuvier' made him the obvious choice to review the burgeoning materials relating to the giant fossil reptiles. Owen at the time was aware of his own intelligence, and he was intensely ambitious. With his high forehead and arresting eyes (Fig. 2), his capacity for fast and accurate work and his skills as a lecturer, he made a strong impression on his older colleagues.

Thus Owen, a scientific gentleman of the new breed – one who earned his living – was awarded one of the first scientific research grants for palaeontology. In 1838, the British Association for the Advancement of Science (the BA) offered him £200 to pay his expenses in compiling a survey of the British fossil reptiles. For this he travelled extensively throughout England, examining specimens in public museums, but many more in the collections of wealthy amateurs. He presented his report on the marine reptiles, the ichthyosaurs and plesiosaurs from the Jurassic rocks of the Dorset and Yorkshire coasts, and from the Cretaceous of Kent, at the BA meeting in 1839 in Birmingham.

The committee of the BA was so pleased with Owen's work that they promptly awarded him a further £200 to extend his survey to the terrestrial reptiles – the crocodiles, turtles and giant saurians. Owen set off again on his tours around England, travelling mostly by carriage but using the new

2 *Portrait of Richard Owen, the man who named the Dinosauria, as he was around 1840.*

railways where he could. He prepared a full and detailed account of all that he had seen during the course of two years of intense work, and he presented his paper at the next meeting of the BA, this time in Plymouth, on 2 August 1841. The talk lasted for two and a half hours, and the audience was reported in contemporary newspaper accounts to have received the peroration with satisfaction and approbation. In any case, much pleased with his brilliant overview, the BA awarded Owen a further sum of £250 to complete another report, and £250 towards the cost of engravings of the extinct reptiles.

Owen's work was straightforward as regards the fossil turtles and crocodiles: they could be compared readily with their living relatives. However, the giant saurians – *Megalosaurus*, *Iguanodon*, *Hylaeosaurus* and the rest – were much harder to explain. In the first version of his report, the basis for the talk in Plymouth, Owen simply tried to shoe-horn them all into existing groups. So some were overblown lizards, others were crocodiles, and yet others were something in between. But Owen was much more daring in the final report, which was published some months later, in 1842.

Owen names the Dinosauria

Between August 1841 and April 1842, Owen had a revelation. No longer were the giant saurians to be explained as some ancient and overgrown tribe of lizards. They clearly represented an entirely distinct group that had left no descendants. Owen's revelation came as he studied additional specimens of *Iguanodon* and *Megalosaurus*. He noted that the femur (thigh bone) looked more mammalian than reptilian in certain features. In particular, the head of the femur was curved inwards and had a ball-like structure that would fit into a socket made by the hip bones. These giant saurians apparently stood high on their legs, just as cows, humans and birds do; they did not sprawl, like living lizards and crocodiles.

Owen then saw a new specimen of *Iguanodon* that had come from the Isle of Wight, which preserved the hip region of the backbone. The sacral vertebrae, the elements of the backbone to which the hip bones are attached, were different from those of any living reptile. Normally, reptiles have two sacral vertebrae, and they are separate and distinctive bones. The new specimen showed that *Iguanodon* had five sacral vertebrae, and they were firmly fused together, and to the hip bones. Owen recalled that he had seen the same arrangement in a specimen of *Megalosaurus* in Oxford.

So *Megalosaurus*, *Iguanodon* and *Hylaeosaurus* differed from modern reptiles by being huge, by having a mammal-like femur and by their fused sacral vertebrae. Owen inserted a few brief paragraphs into the manuscript of his BA talk, and the deed was done. The giant land reptiles from the British Jurassic and Cretaceous were not huge lizards, but were representatives of a new group. He wrote:[9]

> The combination of such characters, some, as the sacral ones, altogether peculiar among Reptiles, others borrowed, as it were, from groups now distinct from each other, and all manifested by creatures far surpassing in size the largest of existing reptiles, will, it is presumed, be deemed sufficient ground for establishing a distinct tribe or suborder of Saurian Reptiles for which I propose the name of '*Dinosauria*'.

In a footnote, Owen explained his new term, based on the Greek words *deinos*, meaning 'fearfully great' and *sauros* meaning 'lizard', or in more modern terms, 'reptile'. So Owen was father of the dinosaur, and in naming the group he displayed his synthetic powers and his brilliance. Gideon Mantell, who had laboured for twenty years on the giant bones, had failed to see how his new fossils fitted into the broader picture of the history of life, and he was rapidly eclipsed by his younger, more urbane rival.

Owen's new understanding of the dinosaurs of the Jurassic and Cretaceous allowed him to determine that the bones shown to him by Murchison in 1845 were clearly not from a dinosaur. Indeed, he argued that they came from a reptile that was more ancient than a dinosaur, and this was a critical observation for understanding the place in the overall rock sequence of the new Russian rocks that Murchison had called the Permian.

Owen and Murchison's Russian reptiles

The Copper Sandstones of the Orenburg district in the Urals had, as we have seen, been a source of the bones of fossil reptiles[10] since 1770. But little serious notice was taken of the fossils until the 1830s. In 1838, S. S. Kutorga, a professor at the University of St Petersburg, gave the first scientific descriptions of the fossils, naming one *Brithopus*, on the basis of some fragments of the humerus (the upper arm bone), and the other *Syodon*, on the basis of a tusk fragment. Kutorga made the astonishing claim that these were early mammals – *Brithopus* a relative of the modern sloths and anteaters, and *Syodon* of the elephant. In retrospect, this seems bizarre, since we know now that the first sloths and elephants appeared only some 50 million years ago, not 260 million years ago in the Permian, before the age of the dinosaurs.

Kutorga's claim was not so remarkable in another way, however. In spotting some mammalian characters in the Copper Sandstone fossils, he had actually unwittingly hinted at what these extraordinary reptiles might be, something that Owen was not to realize for a long time. These were indeed members of the synapsid line of reptiles, and therefore nothing to do with frogs, dinosaurs, lizards or any of the other fossil groups that Owen was then studying. The Permo-Triassic synapsids are commonly known as mammal-like reptiles, to signify their evolutionary position on the way to mammals.

Independently, one of the directors of the copper mines in Ufa (now in Bashkortostan Province) and Orenburg, the splendidly named F. Wangenheim von Qualen, had amassed one of the largest collections of Copper Sandstone reptiles. As his name suggests, von Qualen was of German extraction, not so uncommon in Russia at that time, when many scientists and technical experts were imported from Prussia. Von Qualen published a series of reports in the Russian scientific journals on the bones he had found. He also speculated about the age of the reptile beds, and he correlated them with the Kupferschiefer ('copper shales'), a part of the Zechstein of Germany. As a mining engineer, von Qualen was familiar with the Kupferschiefer, a rock unit that had been mined for decades in Prussia. And his age correlation turned out to be inspired: the Zechstein is indeed now dated as Late Permian in age, as we shall see in Chapter 2.

Von Qualen's specimens were studied also by Kutorga, and by two other noted professors, German imports like himself: J. G. F. Fischer von Waldheim, from the Natural History Museum in Moscow, and E. I. von Eichwald, from the Academy of Science in St Petersburg. Waldheim named a new mammal-like reptile, *Rhopalodon*, and Eichwald named the reptile *Deuterosaurus* and the amphibian *Zygosaurus*.

Owen did not have a chance to see these Russian specimens at the time, though he knew of the publications, and in reporting on the bones brought back for him by Murchison, he identified them as belonging to Fischer's beast *Rhopalodon*. However, as we saw at the start of this chapter, Owen believed that this reptile was a crocodile of some kind. Whether it was a crocodile or a mammal-like reptile, Owen's studies confirmed for Murchison that the Russian sandstones of the south Urals were part of a new system of rocks which he called the Permian. The reptiles seemed to be most like those of the Triassic of England and Germany, and that was all he wanted to hear.

Revolution in palaeontology

In a few short decades, then, palaeontologists had resolved some major debates. Extinction was real, as Cuvier had convincingly shown in the 1790s, and fossils could be studied as the remains of plants and animals that had once lived. Owen and his contemporaries accepted that similarities among fossils indicated similarities in the ages of rocks in different parts of the world. Cuvier's new science of comparative anatomy gave Owen the tools to make these comparisons, and to begin to try to disentangle just what had been going on in the history of life through the Permian, Triassic, Jurassic and Cretaceous. In retrospect, some of his ideas may seem odd, but he was limited by the isolated fossil finds and the difficulties of travelling from place to place to make necessary comparisons.

So much for Owen's understanding of palaeontology and comparative anatomy. The other great revolutions in natural science in the early nineteenth century had turned geology on its head. These rapid advances allowed Murchison to identify rocks of a new geological system, the Permian, and that was a necessary precursor to our current understanding of the greatest mass extinction of all time.

2
MURCHISON NAMES THE PERMIAN

In December 1841, Roderick Murchison,[1] then President of the Geological Society of London, named the Permian System, based on rocks he had observed around the city of Perm in Russia. He did this in a short note of six pages in the *Philosophical Magazine and Journal of Science,* under the rather non-committal title 'First sketch of some of the principal results of a second geological survey of Russia. . . .'. It took the form of a letter written to M. Fischer von Waldheim, bearing the date 5 November.[2] This seems a modest way in which to make such a major announcement, and the publication of an apparently private letter might also appear odd. It also looks like it was done in haste, published exactly one month after it was sent in. What was going on?

After a lengthy tour of the Urals and southern European Russia, Murchison had spent a few days in Moscow, in October 1841, completing various pieces of business before his return home. It was here that he wrote the letter to Fischer von Waldheim, outlining his discoveries of the previous months. Johann Gotthelf Friedrich Fischer von Waldheim (1771–1853), a German palaeontologist, was Director of the Natural History Museum in Moscow, and, as we saw in the last chapter, he was also one of the Russian-Germans who had named reptiles from the Russian Copper Sandstones.

Murchison's express instructions were that his report, in the form of the letter to Waldheim, should be published at the same time in the English, German and Russian journals. The letter duly appeared in Russian in the *Gorny Zhurnal,* the house journal of the Mining Institute in St Petersburg in late 1841, and in English in the *Philosophical Journal* in December 1841 and, rather belatedly, in the *Edinburgh New Philosophical Journal* early in 1842. Today, such multiple publication of the same scientific paper would be severely frowned upon.

This was a major contribution to world geology. Naming a geological system is not an everyday occurrence, and by 1841 most of the geological timescale had been divided up and labelled by various geologists, including Murchison himself. So Murchison was well aware of the significance of his action. Why, then, did he choose to issue his article at high speed in a periodical that was not a major geological journal? Why, indeed, in such a brief form, merely a copy of a letter sent to a fellow scientist overseas? Was he forced into rapid publication?

The reformed fox-hunter

By 1841 Roderick Impey Murchison (1792–1871) was at the height of his powers (Fig. 3). He came from a moderately wealthy Scottish family, and had served in the army in the Peninsular War and in Ireland. When the fighting ended, he retired from the army, married an intelligent heiress, Charlotte Hugonin, in 1815, and settled down to life as a country gentleman. But the lack of activity bored him and he decided to take up scientific pursuits. Soon after his marriage, he sold the family seat at Tarradale, just outside Inverness, and moved to London, where he was able to mix with the leading scientists of the day (although they did not generally call themselves scientists – the word came into use by 1840 – but rather geologists, observers, savants or philosophers). He attended courses in chemistry at the Royal Institution, and, having decided that geology was the rising subject, joined the Geological Society of London in 1825. Geology appealed to his desire for physical activity, and it allowed him to pursue his passion for pheasant shooting and fox hunting. At this point, Murchison was comfortably well off, and he did not have to work to support his home and

3 *Portrait of Roderick Murchison around 1841, the year in which he named the Permian System.*

family, but he only became seriously wealthy after his mother-in-law died in 1838, and his wife Charlotte inherited her estate.

But Murchison was no idle dilettante. He engaged in lengthy field excursions to difficult parts of the world, and he worked with amazing speed. The scope of his interests was vast and, at any time, he was typically involved in two or three major fields of research. He was also a great controversialist. From the late 1840s, after he had finished his work in Russia, Murchison was President of the Royal Geographical Society for many years, and he sent out expeditions all round the globe. The explorers and cartographers named numerous mountains and lakes after him.

During the 1830s, Murchison concentrated on stratigraphy, the study of rock sequences and the timing of events in Earth history. This was an exciting time to be a geologist. Murchison and his colleagues were literally carving up geological time into its major divisions, and he in particular realized that this was a pursuit with global ramifications. In other words, close study of the rocks of one part of the world could provide evidence for a world standard of geological time. The standard geological timescale is such a basic part of geological understanding today that it is easy to forget that things were not so clear in the 1830s.

In 1839 Murchison had named the Silurian System,[3] a unit of geological time characterized by certain rocks in Wales (and named after the Silures, an ancient tribe of Wales). Also in 1839, Murchison, together with his sometime enemy and sometime friend Adam Sedgwick (1785–1873), Professor of Geology at the University of Cambridge, named the Devonian System,[4] which lay above the Silurian. This was characterized, as its name suggests, by particular rocks in Devon, southwest England.

Murchison's speed

In naming the Permian System in December 1841, Murchison perhaps felt he had to move fast. He gave a clue to his motivations in his Presidential Address to the Geological Society of London a few months later, in February 1842. In the published version,[5] he wrote:

> Whilst on the topic of Russia, I will now state, that if on account of the preparation of this discourse and other official duties I had not been greatly occupied, I might before now have presented to you some of the results of the second visit to

that country. . . . We offer . . . the chief results of our enquiries, and . . . place them on record as bearing date from September 1841.

Among the results I will now merely allude to the first announcement of some of them, in a letter of the above date, addressed to Dr Fischer de Waldheim in Moscow, in which . . . were . . . the classification of certain cupriferous deposits of sand, marl, limestone, &c, under the term of 'Permian system'. . . . I should not now occupy your time by alluding to it, had not the mention of the word already called forth from M. A. Erman the remark, that these deposits have been long known to other observers.

Murchison goes on, after some justification of the originality of his observations and of his new name, Permian, to say that 'I was, therefore, surprised to read the premature criticism of M. A. Erman'.

Georg Adolf Erman (1806–77), the same Monsieur (M.) Adolf Erman referred to by Murchison, was a German physicist and explorer who had travelled widely through Russia, Siberia and Kamchatka. Murchison clearly felt slighted by some of his remarks and sought to ensure priority in print for his term, Permian, in case some other geologist beat him to it. As in all things, in science priority is critical, and priority has to be established by the date of publication. A letter to a friend, or a speech at a scientific meeting are not adequate – the new idea or name has to appear in print. Two years previously, Sedgwick and Murchison had established the Devonian System in a short note in the same *Philosophical Journal*, again in haste, seemingly to ensure their priority over other potential competitors.

Conventions were certainly different then. Scientists such as Murchison exchanged formal letters describing their activities, and it was common for these letters to be presented to the scientific societies of other countries, such as the Geological Society of London. In this way, the gentlemen scholars who met to discuss the latest ideas in the meeting rooms of their societies in London, Paris, Berlin or Moscow could keep up to date with new ideas from overseas.

Murchison's speed of publication, and the desire for priority to be established, is clear, though whether he feared Erman, or some other scholar, cannot now be established. So while there is no doubt that Murchison rushed into print with a half-digested account of his new Permian System (Fig. 4), he did not act with unusual haste, judged by the *modi operandi* of his day. Such informal abstracts or outlines of new ideas

THE

LONDON, EDINBURGH AND DUBLIN

PHILOSOPHICAL MAGAZINE

AND

JOURNAL OF SCIENCE.

[THIRD SERIES.]

DECEMBER 1841.

LXII. *First Sketch of some of the principal Results of a Second Geological Survey of Russia. Communicated by* RODERICK IMPEY MURCHISON, *Esq., F.R.S., President of the Geological Society.*

To the Editor of the Philosophical Magazine.

DEAR SIR,

IT was my earnest wish to have complied earlier with your request when I left this country, to send you from the spot some account of my distant wanderings; but the desire to avoid communicating early conceptions which might be modified by subsequent observation, induced me to stay my pen until I could offer something worthy of a place in the Philosophical Magazine. The short sketch which follows was written at Moscow near the close of the journey, and is, with some very slight alterations, the translation of a letter addressed to M. Fischer de Waldheim, the venerable and respected President of the Society of Naturalists of that metropolis. Since then, besides the official report to the Minister of Finance, the Count de Cancrine, I have submitted to His Imperial Majesty, a tabular view of all the formations in Russia, accompanied by a general map and a section from the Sea of Azof to St. Petersburgh. These documents, which will be engraved in the course of the winter, are to be considered only as the prelude to a long memoir with full illustrations of the organic remains, mineral structure and physical features of the country, which will be laid before the Geological Society of London, as soon as, with the assistance of my fellow-labourers, I shall have prepared the materials for the public eye. In the mean time the friends of science must be happy to learn, that

Phil. Mag. S. 3. Vol. 19. No. 126. *Dec.* 1841. 2 E

4 *The title-page of Murchison's brief paper in which he established the Permian System.*

were the normal means of communication, often preceding the full-scale evidence, presented in the form of an illustrated memoir or monograph, by many months, or even years, if it appeared at all.

The Geological Society of London

The Geological Society of London[6] existed as a kind of gentleman's club, but one with a serious purpose. The meetings were convivial, and were held every two weeks during the winter season, November to June. From July to October, the fellows might be out of London, either engaged in field work or simply visiting their country estates.

The purpose of the society was genuinely to promote geological knowledge, and the fellows were expected to make original observations and bring accounts of their discoveries to the meetings. It was not merely a theoretical debating society, however. Indeed, the Geological Society of London, from its earliest days, established a format for their meetings that became the accepted norm elsewhere. Members were expected to give well-prepared papers and to submit themselves to incisive discussion and criticism, both at the meeting and afterwards. Nothing would be published by the Society that had not been vetted carefully: the Society was keen to uphold its reputation for high standards. This is very much the usual practice in all sciences today, but it was less so in the 1830s.

The discussion meetings were reported in brief form in the *Proceedings* of the Society, as well as in a number of gentleman's weekly magazines, and sometimes in journals overseas. When circumstances permitted, some of the papers were worked up into full-scale memoirs for publication in the *Transactions* of the society. But many authors chose not to prepare such full accounts, preferring to move on to new topics. Others, such as Murchison, however, were perhaps more serious, and they wanted to preserve a proper account of their work for posterity.

A memoir was a major undertaking, requiring the preparation of a full text presenting the evidence and arguing the case. A key problem then was the cost, both of printing and of illustrations. Geology is essentially a visual science. You cannot work as a geologist without maps, diagrams, cross sections and drawings of rocks and fossils. The geological author in the 1830s had to engage an artist to produce detailed drawings, and then an engraver to convert the drawings into printing plates (lithographs). The

Society bore some of the expense, which was covered by a high subscription rate for the fellows (three guineas, or £3.15 per year), but the authors had to contribute as well. Only a wealthy man, like Murchison, could seek to publish memoirs regularly.

Putting the rocks in order

So how had Murchison arrived at his new Permian System? Murchison's motivation in the 1830s and 1840s was to sort out the complexity of the geology of the world. A large task, and inconceivably grand for one person today. However, in his Presidential Address to the Geological Society of London in February 1842, as well as presenting his new Permian System, he enlarged a little on advances in geology during the previous year, touching on new discoveries from every continent. He surveyed progress up to 1841 in establishing an international scheme for the division of geological time:[7]

> I am encouraged to hope that the word 'Silurian', which has been warmly sanctioned by the classic authority of Von Buch, which E. de Beaumont and Dufrénoy have engraved upon their splendid map of France, and which our fellow-labourers in America have adopted, will not be obliterated to make way for other names which are not founded upon any *new* distinctions, stratigraphical or zoological. So long, gentlemen, as British geologists are appealed to as the men whose works in the field have established a classification, founded on the sequence of the strata and the imbedded contents, so long may we be sure that their insular names, humble though they may be, will, like those of our distinguished leader, William Smith, be honoured with a preference by foreign geologists.

An arrogant, insular view perhaps, but startlingly prescient. The geological timescale which Murchison and others had been drawing up in Britain did gain international acceptance. Everyone today uses Murchison's terms Silurian, Devonian and Permian to refer to rocks of a particular age, containing certain fossils, all round the world, not just in Britain, Germany, France and America, but from Argentina to Alaska, Australia to Azerbaijan.

Murchison clearly had a mature vision in 1842, but the early development of the geological timescale had been somewhat haphazard.[8] Miners and quarrymen had long given names to the rocks they worked for commercial advantage. Coal came from a sequence of rocks called the Coal

Measures by British miners, or 'carbonifère' ('coal-bearing') by French workers. We have already encountered another of these miners' terms, the Kupferschiefer of Germany, a term that is still used locally to refer to the lower part of the Zechstein succession of the Upper Permian. Terms such as these were certainly utilitarian, but they did not say anything about the history of the Earth, nor could they be applied on an international scale.

Mr Smith's practical stratigraphy

Murchison, in the quotation above, clearly identifies William Smith as 'our distinguished leader', and he is often now called the father of stratigraphy. Murchison spoke with added affection, since William Smith (1769–1839) had died only three years earlier, and he was right to honour him in this way. Smith was a working geologist, who earned his living surveying the routes for canals throughout England. He walked long distances, assessing the rocks that were to be cut through, estimating the costs of the workings and identifying potential problems for his paymasters. Murchison had first met Smith, then aged 57, in Yorkshire in 1826, when Smith showed the young tyro the rock successions along the coast.

As he traversed England, Smith had noticed certain repetitions in the rock types and in their contents. The rocks seemed to occur in recognizable sequences, and certain rocks contained predictable assemblages of fossils. From these observations he laid some of the key building blocks for geology as an organized science. Smith realized that fossils could be used to identify different ages of rocks, and that it would be possible to apply such schemes over long distances, perhaps even worldwide, to produce a standard view of the order of rocks, and thus the order of events in geological time. As Smith was aware, this would be an indispensable tool for the geological surveyor, who could enter a previously unexplored area, search for fossils, consult a standard compendium and then give a date for the rocks in front of him. In Smith's day, wealthy landowners were sinking boreholes on their land haphazardly in the search for coal. In their ignorance of the ages of the rocks, most of these boreholes were a complete waste of time since they penetrated rocks that were either too old or too young.

Unlike Murchison, Smith was not a keen writer; while he had just about formulated his model of the stratigraphy of England by 1799, he did not publish it until much later. In particular Smith had sorted out the

Jurassic and Cretaceous, and some of the divisions of these major systems, based on their rich fossil content. For example, in the Middle Jurassic around his home district of Bath and the Cotswolds, different rock units could be distinguished based on their particular assemblages of brachiopods ('lamp shells'), bivalves and ammonites. Modern geologists divide the Middle Jurassic into 11 or 12 major zones (Fig. 5), based primarily on ammonites, coiled molluscs that belong to a group now extinct, but related to modern squids, octopuses and the shelled *Nautilus*.

Smith could determine the lower part of the Inferior Oolite by the ammonite now called *Ludwigia murchisonae* (named, as it happens, after Murchison), and upper parts by *Stephanoceras humphriesianum* and

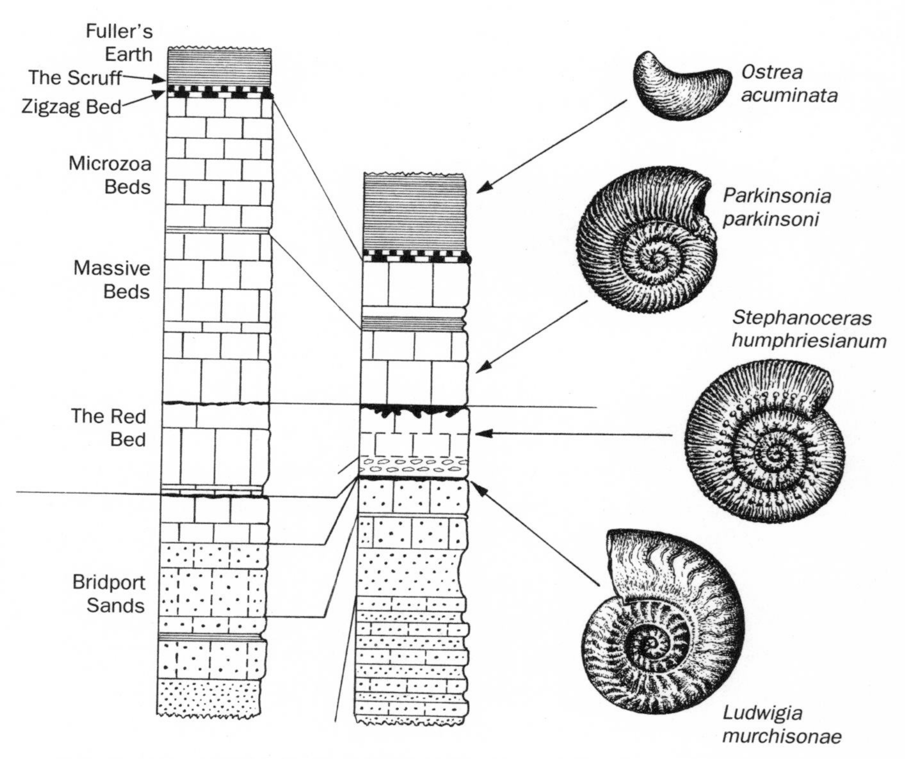

5 The basis of stratigraphy, with an example from the MIddle Jurassic of England, as established by William Smith. Two measured sections are shown, from Chideock (left) and Burton Bradstock (right) in Dorset. Lines linking the two sections indicate correlations. The main zones and some key fossils are shown.

Parkinsonia parkinsoni respectively. The overlying Lower Fuller's Earth of Dorset and Somerset contains very few ammonites, but is characterized instead by the small oyster *Ostrea acuminata*. And so it goes on, bed after bed. Smith could verify the sequence of rock types by observation in quarries and canal cuttings: everywhere the Lower Fuller's Earth lay on top of the Inferior Oolite, and never the other way round. And everywhere the upper parts of the Inferior Oolite contained the ammonite *Parkinsonia parkinsoni*, among other characteristic fossils.

William Smith had laid the groundwork, and during the heady days of the 1830s, his dream of a universal, worldwide scheme for the division of geological time was taking shape. Already by 1830, Murchison, among others, had extended William Smith's stratigraphic work from southern England to other regions – to Scotland, France and Germany, and even further afield. Geologists hardly dared hope that the system might work worldwide, but it seemed to. With reports from the Americas and further regions of Europe, Smith's Jurassic succession seemed to be repeated. The same was true of the other units. This opened up vast possibilities for the furtherance of geology as a science, and it pointed the way for scholars to adopt an international outlook.

The New Red Sandstone

Murchison spent most of the 1830s sorting out the Silurian and Devonian periods. In doing so, he and other field geologists had resolved one of two great red sandstone units. These seemed to sandwich the Carboniferous, the succession of thick marine limestones and coal-bearing continental sediments that had long been established. Below the Carboniferous across much of Britain lay the Old Red Sandstone, and above lay the New Red, both of them composed of thick sandstones that appeared to have been deposited by rivers and lakes in tropical conditions. Murchison argued strongly that the Old Red Sandstone was equivalent in age to the marine Devonian of the county of Devon, and history has proved him right.

But what of the New Red Sandstone? The Carboniferous, lying between the Old and New Red Sandstones, with its economic coal deposits, was easy to recognize throughout Europe, and Murchison was not interested in going over old ground. The New Red Sandstone in turn was overlain by the Jurassic, which William Smith and other geologists in

England, France and Germany, had long ago worked out. Again, this was old news, and Murchison was not interested. The New Red Sandstone was for Murchison the only substantial gap in the global stratigraphical scale that he and others were rapidly completing. To plug that gap would be to complete the great international programme of a single timescale.

The upper parts of the New Red Sandstone had been termed the Triassic ('three-part') in Germany, by Friedrich von Alberti (1795–1878), Inspector of Saltworks at Friedrichshall, southwest Germany. Alberti chose the name because the German Triassic was characterized by a three-part sequence: a basal unit of red and yellow sandstones and conglomerates called the Buntsandstein ('coloured sandstone'); a middle marine unit of limestones and mudstones called the Muschelkalk ('shelly-limestone'); and an upper unit of red and purple mudstones and sandstones called the Keuper (an older quarryman's term for the varicoloured mudstones).

The German Triassic overlies a sequence of marine limestones, mudstones and evaporites (salts) called the Zechstein (an older miners' term), which in turn succeeds a succession of red and yellow sandstones termed the Rotliegendes ('red beds'). The Rotliegendes and Zechstein were tentatively identified in England, but it was hard to mark their boundaries precisely. The tripartite German Triassic could not be recognized since the Muschelkalk appeared to be missing. In the end, the British geologists were forced to continue to use the rather feeble, informal term New Red Sandstone.

Despite this lack of clarity, another British geologist had already recognized that something major was happening in the middle of the New Red Sandstone. Between the undefined post-Carboniferous sandstones and limestones of the lower part and the Germanic Triassic of the upper, a huge change had taken place in the typical animals found in the sea. All the old types had gone and new ones had appeared. This geologist was John Phillips, as energetic as Murchison, but from a very different background.

John Phillips and the value of fossils

John Phillips (1800–74) was of a lower social class than Murchison or Sedgwick.[9] But, by his energy and good sense, he was able to achieve high regard in his lifetime among fellow geologists. Phillips was in fact the nephew of William Smith, and he learned a great deal of practical geology from his uncle as they walked around Somerset and Wiltshire. As a young man, and with no formal training, Phillips was offered a museum post in York in 1824, where he was invited to rearrange and display the fossil collections. It was here that Murchison met Phillips in 1826, while he was engaged in this task. Phillips carried out similar curatorial work elsewhere, and he was secretary of the British Association for the Advancement of Science from its inception in 1831. He rose from these somewhat menial positions to become Professor of Geology successively at King's College, London in 1834, Trinity College, Dublin in 1844, and finally at Oxford in 1856.

Phillips studied mainly palaeontology and geology, but he also made contributions to meteorology and astronomy. He wrote a number of books on the geology of Yorkshire, the Oxford region and southwest England, as well as a series of highly successful textbooks aimed at keen amateurs, of which there were thousands, and school and university students. Phillips was a recognized palaeontological authority, and he argued repeatedly for the importance of basing stratigraphy on fossils. In other words, he insisted that units of geological time should be founded on their fossil content, and hence made internationally applicable.

There was no point, Phillips argued, in establishing a geological system for a merely local manifestation of rocks, perhaps only to be seen in Devon or in north Wales. Geological systems should be defined, top and bottom, by unique fossils that could be recognized readily wherever geologists went in the world. In this view, Phillips was upholding his uncle's classic work. Phillips influenced Sedgwick and Murchison deeply, and these essentially field geologists were fully convinced of the need to submit the fossils they found to expert palaeontologists, and to rely on their judgments in correlating rocks from place to place.

The end-Permian mass extinction recognized?

Phillips produced a famous diagram (Fig. 6) which is often taken as the first recognition of the great end-Permian and end-Cretaceous mass extinctions. He shows the diversity of life by the curve swelling out to the right. But most important to our eyes are the two drops in diversity. Surely here Phillips meant to show the two greatest cataclysms in the history of life at the end of the Permian and Cretaceous periods?

6 A diagram showing John Phillips' conception of the Palaeozoic, Mesozoic and Cenozoic, as shown in his textbook of 1860. The wavy line broadly represents the diversity of life, and the boundaries between the Palaeozoic and the Mesozoic and the Mesozoic and the Cenozoic are clearly indicated as times of reduction in diversity and turnover of faunas.

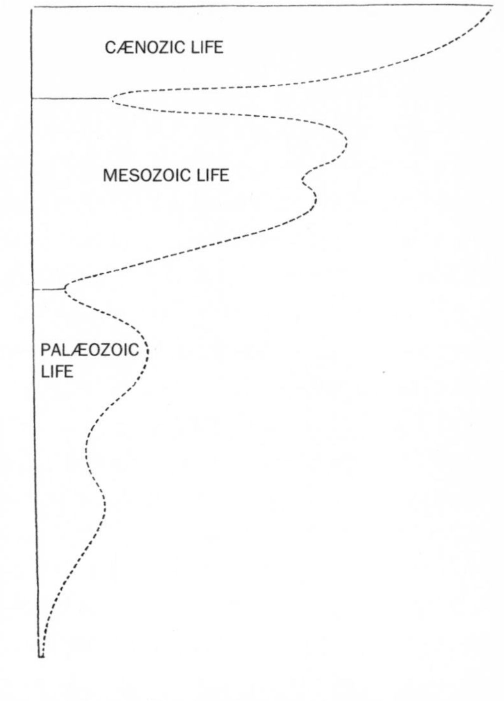

Not in the modern sense. Phillips intended the diagram – which appeared in one of his books in 1860 but surely represented his opinion in 1840 also – to show how geological time could be divided into three major units, or eras. The eras were more all-encompassing than the geological periods that Murchison and Sedgwick were establishing: the Silurian, Devonian and Permian were merely parts of the Palaeozoic Era, for example. Phillips' point was that the three eras could be recognized by quite different assemblages of fossils, and this would apply worldwide. The faunas and floras represented by 'Palaeozoic life' and 'Mesozoic life' were quite different, and linked by only the most tenuous transitional forms, but geologists and palaeontologists were not ready then for evolution and mass extinction as we understand them today.

Sedgwick had in fact named the Palaeozoic Era in 1838, and he intended this new unit to include all the pre-Carboniferous rocks that he and Murchison were then tussling over – in other words their Cambrian and Silurian. Phillips immediately grasped Sedgwick's new concept, and extended it, in articles he wrote in the *Penny Cyclopedia* in 1838 and 1840. Phillips was not a rich man, and he was happy to be paid for such hack writing, while the readers of the *Penny Cyclopedia* probably did not appreciate the originality of what they were reading.

For Phillips argued that the whole geological column could be divided into three great series, or eras, the first being Sedgwick's Palaeozoic. The second was the Mesozoic and the third the Kainozoic, both of which

Phillips named in one of his *Penny Cyclopedia* articles. These three terms are pure Greek, meaning respectively 'ancient life', 'middle life' and 'recent life', in reference to the relative antiquity of their fossils. The Kainozoic came to be more commonly spelled Cainozoic, sometimes even Caenozoic, but it is now generally rendered as Cenozoic. Each of the three divisions was then characterized by utterly different suites of fossils.

Phillips argued that Sedgwick's Palaeozoic could not be restricted simply to the Cambrian and Silurian, but that it had to be extended to the Carboniferous, since the fossils of the new Devonian system supplied a faunal link between the Silurian and the Carboniferous. Phillips went further, proposing that the Palaeozoic should extend into the New Red Sandstone.

This idea was revolutionary, and it placed a strong focus on the New Red Sandstone. If, as Phillips said, the division between the Palaeozoic and Mesozoic eras lay somewhere within this succession of red sandstones, where should it be placed precisely?

Murchison overseas

Despite Phillips' contributions, the New Red Sandstone still stood in everyone's way. Without a clear definition of the sequence of rocks between the Carboniferous and the Triassic, any internationally applicable system of time division would fail. Murchison realized that he could not sort out the New Red Sandstone in England, and he had already seen the Rotliegendes and Zechstein and the classic Triassic of Germany and central Europe. He would have to venture overseas again to find somewhere that displayed a complete sequence of rocks lying between the Carboniferous and the German-style Triassic.

Such an international endeavour suited Murchison well. He knew that the scientific leaders of the new Victorian age (Queen Victoria had acceded to the throne in 1837) would have to be world travellers, people with friends in all the major scientific capitals and who could speak several languages. Murchison was such a man. He spoke French fluently, happily giving talks to the scientific societies in Paris and publishing papers in French from time to time. He also knew enough German to get by. Despite his long visits to Russia, however, he did not speak Russian. But then the Russian savants did not either: they much preferred to speak and write in

French, or, failing that, German or English – anything other than the barbarous tongue of the serfs and peasants.

Murchison visits Russia

Murchison's first visit to Russia was not in fact to solve the New Red Sandstone problem. He was still reinforcing his Silurian and Devonian systems, and hoped to find new evidence to resolve the matter once and for all. Sedgwick and Murchison had geologized together in northern Germany and Belgium in the summer of 1839, and they had also visited Berlin and eastern parts of Germany. In the written account of their explorations, they repeatedly used their observations to confirm the validity of their recently established Cambrian, Silurian and Devonian systems.[10] Their rapid traverses over vast tracts of country tempted Murchison to head further east. He decided on an excursion across northern Russia in 1840, the planning for which occupied all his time through the preceding winter.

Murchison gave several brief accounts of this first trip to Russia. He spoke at the British Association meeting in Glasgow in September 1840, and then gave a joint presentation to the Geological Society of London with his French colleague, Edouard de Verneuil, in March 1841.[11] In their papers, Murchison and Verneuil described the extent of the information that had been available to them before their visit concerning what they might expect to find in northern Russia. In fact, there was essentially a single paper,[12] a short memoir in the *Transactions of the Geological Society of London*, published in 1822, with the remarkable title 'An outline of the geology of Russia' and a mere 39 pages long. The author was W. T. H. Fox-Strangways, the then British Ambassador in St Petersburg.

Strangways had written a highly competent account, including a map, but it was based on only limited observations that he had been able to make in the course of his official duties around St Petersburg, and he was not a professional geologist. Murchison and Verneuil had also seen the accounts of Russian fossils by J. G. F. Fischer von Waldheim, C. H. Pander and E. I. von Eichwald, but these memoirs said very little about the geology on the ground. On the whole, then, they had very little to go on when they entered Russia.

Murchison and Verneuil began their traverse at St Petersburg, at the head of the Baltic, where they confirmed that the rocks were Silurian in age, overlaid by Devonian. They then went on to describe the Carboniferous, New Red and Jurassic rocks that they identified in their northwards sweep to Archangel, and on their return southwards, across the Volga, and round through Nizhnii Novgorod and back to Moscow, a journey totalling some 6000 kilometres.

In a note added to their paper, dated 26 March 1841, Murchison gives a brief report of some new findings in the southern part of European Russia made by his Russian colleagues, Baron A. von Meyendorf and Count A. Keyserling. He also indicates that he and Verneuil planned to visit these tracts, as well as, significantly, the Ural Mountains and the region around Orenburg. When this had been accomplished, Murchison notes, 'all the chief facts will have been obtained for the construction of a general geological map of Russia in Europe'. Such splendid confidence! And he meant it. In 1840, with almost no previous work to build on, it was necessary to construct maps of huge territories on the basis of such rapid traverses. The detail could be added later.

Murchison's second visit to Russia

Murchison's second visit to Russia had two main purposes. He wanted to characterize his new Permian Period, and determine its exact relations to the Carboniferous below and Triassic above. He also wanted to cement his social relations with the Tsar and all his princes, and secure funding from them for his trip and for an ambitious publication he had in mind.

The published materials describing Murchison's second visit to Russia were rather more extensive than for the first.[13] As we have seen, he announced his new system, the Permian, in the note published in the *Philosophical Magazine* and elsewhere in late 1841. The first public presentation was given in his Presidential Address to the Geological Society of London on 18 February 1842, in which Murchison defended his new name 'Permian' as thoroughly original – and refuted Erman's suggestion otherwise.

The long-awaited full account of Murchison's and Verneuil's second trip across Russia was read at two consecutive meetings of the Geological Society, on 6 and 20 April 1842. In it, they reviewed the rock sequence in order, as before, from Silurian to Cretaceous, with some comments on the overlying Pleistocene. The mode of argument was also as before: Murchison recounted the rock types they had seen, and he used Verneuil's identifications of fossils to fix the ages by comparison with the sequences in England, France and Germany.

Murchison added some further explanation of the Permian System: he equated the Russian Permian with the Zechstein of Germany (already recognized as part of the New Red Sandstone) and with the Magnesian Limestone of England. These last two had long been regarded as equivalents. In Russia, Murchison noted the huge area covered by his new Permian, from Perm itself to Orenburg in the south, covering many hundreds of kilometres of the western side of the southern Urals. He had followed the sequences in the Urals up through undoubted Carboniferous rocks, as indicated by the presence of coal, but more importantly by the presence of specific fossils (marine shells and terrestrial plants) shared with the Carboniferous of western Europe. Above the Carboniferous he had seen Permian limestones, gypsum and salt beds, red and green sandstones, shales, conglomerates and copper-rich sandstones (the miner's so-called 'Kupferschiefer').

Bottom and top of the Permian: molluscs and reptiles

Murchison knew he had to do two things to identify the lower New Red Sandstone as an acceptable international stratigraphical system, his Permian. First, he had to define the base, and second he had to find characteristic fossils that placed the age of the rock units definitively between the Carboniferous and the Triassic. In stratigraphy, defining the base of a system is crucial. The top doesn't matter since it is defined by the overlying unit. In this case, the work had already been done, since the base of the Triassic had been fixed in Germany by Alberti when he named that system in 1834, as we saw above.

Murchison had studied the gypsum- and salt-bearing lower New Red units around Perm, and Keyserling had provided him with the relevant information on the fossil-bearing rocks around Arti. These were marine rocks, full of shells and other, identifiable fossils which allowed Verneuil to make direct comparisons with well-studied, definitive Carboniferous and Triassic rocks of western Europe. So Murchison had the Carboniferous/New Red transition documented south of Perm in marine rocks with fossils. He had seen the brachiopods and plants himself, and he now had the clincher, bones of fossil reptiles that Richard Owen had determined were truly intermediate in character between Carboniferous and Triassic forms. The shells, the plants and the bones all seemed to confirm his conclusion – the Russian Permian lay between the Carboniferous and the Triassic. He had plugged the stratigraphic gap.

This conclusion is borne out by the famous letter Murchison wrote to Fischer von Waldheim when he reached Moscow in October. In it he justified his new Permian system on the basis of the fossils, some of them Carboniferous in appearance (brachiopods), and others intermediate between Carboniferous and Triassic (plants, fishes, reptiles):[14]

> Of the fossils of this system, some undescribed species of *Producti* [brachiopods] might seem to connect the Permian with the carboniferous aera; and other shells, together with fishes and Saurians, link it on more closely to the period of the Zechstein, whilst its peculiar plants appear to constitute a Flora of a type intermediate between the epochs of the new red sandstone or 'trias' and the coal-measures.

These arguments were repeated in later papers, but Murchison had clearly formulated his interpretation while he was in the field near Orenburg. On 20 November 1841, less than three weeks after his return to England, and just over a week before the name 'Permian' was published in the *Philosophical Magazine*, Murchison wrote to Richard Owen:[15]

> I have been coming to you for some days because I wish you to see the bones of a saurian or two from the lowest deposits in which such animals are found in Russia, and which I believe to be on the level of our Magnesian Limestone.... I am personally much interested in the determination of these saurians as the yes or no to a certain query which may guide me very much in my *Imperial* classification.

The yes or no question was: are these reptiles intermediate between those of the Carboniferous and the Triassic? Owen's answer, as we've seen, was yes, and Murchison was vindicated, as he stressed in his President's Address to the Geological Society of London in February, 1842.

The Geology of Russia (1845)

After his first trip in 1840, Murchison planned to present the results of his researches in Russia simply as memoirs in the *Transactions* of the

Geological Society of London. However, as he saw the great support he was receiving in Russia, from the Tsar downwards, he conceived the idea at the beginning of his second trip of publishing a more substantial work, a large book that would stand beside *The Silurian System* as a worthy successor. Count I. F. Kankrin, the Russian Minister of Finance, had hinted both before and after the 1841 trip that the Russian government would underwrite the production of a map and report. They did not realize, perhaps, that their concept and Murchison's differed considerably in scope and magnitude.

Early in 1842, Murchison planned his great Russian book.[16] It would consist of 600 pages, divided into three sections, with plates of fossils and a large map. The first section was to be about the geology of Russia in Europe, the second on the Ural Mountains and the third on the fossils. Murchison wrote solidly through 1842, and delivered the first chapters for typesetting in November of that year. He continued sending chapters for typesetting during 1843, but the project was clearly getting out of control. He did further fieldwork: in Poland, Germany and Czechoslovakia in 1843, Denmark, Norway and Sweden in 1844, and these countries plus Germany in 1845. This new work forced him to reconsider his earlier chapters, and in the end he had to cancel the first 40 pages and replace them with nearer 80 pages of new text.

The fossils were another problem. Murchison was not a palaeontologist, and most of the fossil descriptions were to be done by Verneuil and others (although Murchison himself did describe the trilobites). He soon asked Verneuil to arrange for the fossils to be described in a separate volume which would be published, in French, in Paris, at the same time as his geological account. Production of two volumes, in two languages, by two authors, meant that there were all sorts of confusions and mistakes, but the work had to be pushed forward.

So in the end the planned 600-page volume had more than doubled in size. There were two volumes, totalling over 1300 pages, 60 plates and two large maps. Everything was finished in a rush. Murchison's volume 1 bears the date 'April 1845', but printing of revised pages and illustrations continued after that. Verneuil's volume 2 began production in 1844. In a great rush during June and July 1845, six copies of the complete work were put together, and Murchison carried them to St Petersburg on 29 July 1845. He presented the six sets to Tsar Nicholas I on 12 August 1845 who studied them with evident approval and intelligent appreciation.

The end-Permian mass extinction ignored

Thus the years 1840 and 1841 marked a critical time for the understanding of the end-Permian mass extinction. Phillips had established the exact extent of the geological series, the Palaeozoic, Mesozoic and Cenozoic, and he had made a firm link between these major time divisions and extinction. Murchison had named the Permian System, and he had shown the importance of Russia in understanding that span of time. However, neither Phillips, nor Murchison, nor any other geologist of the day, pointed to the huge mass extinction that had occurred at the end of Permian time.

This omission can be explained in two ways, one rather simple, the other perhaps more fundamental. It could simply be that the geologists and palaeontologists in 1840 did not have enough evidence to pin the extinction event down. After all, the Permian had only just been named, and Phillips had only just begun to document large-scale changes in the fossil record through time. More fundamental, though, had been a huge change in mindset that had come about during the 1830s. Whereas in 1830 it had been legitimate to talk about catastrophes in the geological past, by 1840 catastrophism had become a dirty word. This astonishing shift in geological philosophy had been engineered by one man, and his efforts set the tone of geological enquiry for 150 years.

3
THE DEATH OF CATASTROPHISM

I learned my geology at the University of Aberdeen in the 1970s. Our first-year lectures were delivered by a magnificent Scottish professor of the old school, who covered everything, from the formation of the Earth to the ice ages, from fossils to mineralogy. These were the days before colour slides were ubiquitous, and he laid out large drawings made on white boards around the room before his lectures began. Assiduous students arrived early to copy the complex diagrams into their notebooks.

I remember his introductory series of lectures, which covered the history of geology. He told us about Smith, Murchison, Sedgwick and Lyell, who had laid the foundations of the subject. These men had overcome every obstacle to reveal to us the truth of the history of the Earth. They allowed no mysticism, no wild notions, none of the strange ramblings that dominated the early nineteenth-century scene. These fine, upstanding geologists, fathers of the discipline, British through and through, had blazed a trail through the obscurantism and fog emanating from the geological societies of Germany and France. Indeed, not only were they British, but two of them, Murchison and Lyell, were Scots. Best of all, the leader of this group of reformers, Charles Lyell, had been born only a few miles south of Aberdeen. Scottish common sense had prevailed over foreign imagination.

Such was the story our professor told us – as it had been told to countless geologists over the years. I happily accepted it. On reflection, though, these simple stories are usually more complex. Can it really ever be the case that one group of philosophers or scientists is completely right, and the other completely deluded? Clearly not.

What had happened, it now seems, was that Lyell and his supporters rewrote the history of geology. In the best political tradition, geologists became divided sharply into two camps, and labels were applied. The good guys were the uniformitarians, and the baddies were the catastrophists. Murchison, Sedgwick, Phillips, and indeed most geologists in Britain, were actually catastrophists, but Lyell was able to wield such influence that they all rather cravenly accepted his labels and his strictures. Anything but be classified as a catastrophist!

And yet the catastrophists were right about extinctions. It is crucial to understand this dramatic switch in geological reasoning which happened around 1832. Virtually overnight, it seemed, Lyell engineered a new understanding in geology and, in doing so, he made it very hard for any rational person even to think about mass extinction. He was the victor, and his vision dominated geology for over 150 years.

Murchison and theories of the Earth

Roderick Murchison was a practical field geologist and he did not venture a great deal into theorizing about broad issues, about the history of the Earth, the progress of life or the meaning of time. But he lived through two major debates, both masterminded by Lyell and both of which made him rather uncomfortable. First, did time run in a single direction or not, and second, was the history of the Earth dominated by catastrophes or not?

Charles Lyell (1797–1878), suave and impressive, was five years Murchison's junior. He was the author of an influential textbook, published in three volumes between 1830 and 1833, entitled *Principles of Geology*,[1] which achieved iconic status and is still regarded by most people as one of the founding works of geology. And yet this epochal tome contains some unbelievable nonsense among its good material. For instance, Lyell wrote that if temperatures were to increase worldwide, ferns and other primitive plants might clothe the lands, and

> Then might those genera of animals return, of which the memorials are preserved in the ancient rocks of our continents. The huge iguanodon might reappear in the woods, and the ichthyosaur in the sea, while the pterodactyle might flit again through umbrageous groves of tree-ferns.

In his *Principles*, Lyell makes it clear that he sees groups of plants and animals as eternal. They come and go simply in response to physical conditions on the Earth. This seems incredible, but Lyell was not to be disputed in the 1830s and 1840s. Extraordinary as his comments might seem, most geologists accepted them. Perhaps in private they doubted his views, but they did not speak out in public.

The other debate that influenced Murchison was that between catastrophism and uniformitarianism. Lyell had pinned his colours to the uniformitarian mast in the subtitle of the *Principles*: 'being an attempt to explain the former changes of the Earth's surface, by reference to causes now in operation'. He argued strongly, and repeatedly, that every geological phenomenon could, and should, be explained by modern processes. This is the principle of uniformitarianism, 'the present is the key to the past', and it seems eminently sensible. Surely it is better for a geologist to refer to the rules of physics, and to the observations of modern geographers and biologists, than to imagine unknown processes?

French and other continental geologists, on the other hand, were at the same time postulating explosions, meteorite impacts, sudden extinctions and miraculous events in the geological past. This is catastrophism, and clearly it must be wild, non-scientific speculation. Uniformitarianism was the straightforward, common-sense, British viewpoint. Lyell had the right credentials to be considered sensible, twice over, as both a Scot and a barrister. To deny him, and to deny uniformitarianism, would be to label oneself a wild theorist.

Murchison could not risk such a derogatory label, although he had been a catastrophist himself in the 1820s. He, and most other geologists at the time, now strove to avoid the 'catastrophist' label by denying all reference to sudden events or major calamities in the past. They also accepted both Lyell's doubts about the extinction of major groups and the possibility of continuous recycling of groups on and off the scene, although many found this unpalatable. In following Lyell, Murchison ultimately had to deny the possibility of mass extinctions. He also had to reject the views of the highly respected Georges Cuvier, the greatest catastrophist of the day.

Cuvier and catastrophism

There was nothing new in catastrophism for Cuvier. Such notions had been around since the seventeenth century, when savants speculated about the origin of the Earth, and the nature of planets, comets and meteorites.[2]

Cuvier was later unfairly bracketed with those early speculators. Yet he was always thorough in marshalling his evidence before he drew any conclusions, and indeed he himself at first mocked the geologists of his day as wild speculators who based their ideas on minimal evidence.

Then in around 1800, Cuvier began to study the fossil mammals found in the Paris basin. Beautiful complete skeletons of small horse-like and dog-like animals had been excavated from the Eocene and Miocene limestones, and Cuvier could see that the skeletons were clearly those of extinct forms. He described the circumstances of their occurrence in a report in 1810:[3]

> The unknown bones are almost always covered by beds full of seashells. It is thus some marine inundation that annihilated the species; but the influence of this revolution, by its very nature, was not perhaps exercised on all marine animals.

Here he uses the word 'revolution' to mean some major physical calamity. He comes back to this question a few pages later:

> The primitive diminution of the waters, their repeated returns, the variations of the materials they deposited and which now form our strata; those of the organisms whose remains fill a part of these strata; the first origin of these same organisms: how are such problems to be resolved with the forces that we know now in nature? Our volcanic eruptions, our erosions, our currents, are pretty feeble agents for such grand effects.

Cuvier did not have strong evidence at this point, nor was he precise about the nature of the 'revolutions', but he was clearly positing processes, and scales of processes, operating in the past that do not operate today.

Après moi le déluge

Cuvier developed the theme of revolutions in much more detail in his more famous 'Discours préliminaire' to his great book *Recherches sur les ossemens fossiles*, published in several editions from 1812 onwards.[4] The preliminary discourse was a fuller manifesto than the 1810 report, outlining for the general public Cuvier's views on geology and evolution, and it was hugely influential, being translated and reissued many times, even long after Cuvier's death. In the fifth section of the preliminary discourse, Cuvier wrote:

> Thus the great catastrophes that produced revolutions in the basin of the seas were preceded, accompanied, and followed by changes in the nature of the liquid and in the materials that it held in solution. . . . During such changes in the general liquid, it was very difficult for the same animals to continue to live in it. And they did not do so. Their species and even their genera change with the beds.

Cuvier goes on to discuss the physical processes that act on the Earth today, and he makes a strong plea for consideration of more catastrophic processes than these in the past. The catastrophes seem to have affected land animals, and especially the mammals, much more than the lowly sea creatures:

> the nature of the revolutions that have altered the surface of the globe must have had a more thorough effect on terrestrial quadrupeds than on marine animals. Since these revolutions largely consisted of displacements of the seabed, and since the waters must have destroyed all the quadrupeds that they reached, . . . they could at least have annihilated the species peculiar to those continents, without having the same influence on marine animals.

I was intrigued to discover this passage, since it resonated strangely with some of Murchison's remarks in his Presidential Address to the Geological Society of London in 1842.[5]

Cuvier further clarified his views on the rapidity of the repeated inundations and retreats of the seas in later editions of the preliminary discourse:[6]

> the majority of the cataclysms that produced them were sudden. This is particularly easy to demonstrate for the last one. . . . It . . . left in northern countries the bodies of great quadrupeds, encased in ice and preserved with their skin, hair and flesh down to our own times. If they had not been frozen as soon as killed, putrefaction would have decomposed the carcasses. . . . This development was sudden, not gradual, and what is so clearly demonstrable for the last catastrophe is not less true of those which preceded it. . . . Life in those times was often disturbed by these frightful events. Numberless living things were victims of such

catastrophes: some, inhabitants of the dry land, were engulfed in deluges; others, living in the heart of the sea, were left stranded when the ocean floor was suddenly raised up again; and whole races were destroyed forever, leaving only a few relics which the naturalist can scarcely recognize.

Much of this is reasonable enough, and it hardly seems to justify Lyell's complete rejection of Cuvier's geological ideas as wildly dangerous.

It was clear to Cuvier, as to any geologist today, that the different layers of sediment are often separated by abrupt discontinuities. He had seen marine limestones in the Paris basin, full of seashells and fishes, which were capped by terrestrial mudstones and sandstones containing skeletons of land animals. Cuvier's only mistake was perhaps to try to visualize the rock sequences accumulating more rapidly than would be assumed by geologists today. This meant that he had to imagine that all sea-level changes, all shifts in the positions of coastlines, happened virtually instantaneously. In reality, it is more likely that such floodings and retreats of the sea took thousands or tens of thousands of years, as they have done more recently, in historical time.

Cuvier did not in fact claim that the Earth had been subject to impacts, vast disturbances or other unusual events. But he certainly claimed a role for events on a larger scale than are known today. His supporters, in France and in Britain, argued that the Earth's crust had been subjected to forces in the past that no longer obtained. Some noted the intense folding and the evidence of melting of rocks in the ancient mountain chains of France and Scotland, and in the Alps. Others assumed that the Earth had been hotter in the past, and more subject to violent volcanic eruptions. The last of the many extinctions to have affected the globe, indicated by the frozen mammoths and other large mammals that Cuvier had mentioned, was attributed to Noah's flood in the Bible. So the catastrophist doctrine could be tied to biblical orthodoxy, and for a while, in the 1820s and 1830s, diluvialists such as William Buckland, supporters of the Flood as the last of many catastrophes to affect the Earth, were in the ascendant.

Cuvier had influenced a generation of geologists, in France and Britain, and Murchison found it hard to adjust to abandoning this stance. But Lyell's critique in the 1830s was firm and persistent. He seized on Cuvier's *Discours* as his *bête noire*, and he did not give up. Cuvier had preached revolutions, evolutions and catastrophes. Lyell decided this was dangerous stuff, and he attacked it fiercely. Geology books today celebrate Lyell as the hero and Cuvier as the vanquished. And yet much of what Cuvier said was right and much of what Lyell said was wrong.

Lyell and Murchison go travelling

Charles Lyell (Fig. 7) was born at Kinnordy, Angus, to a landed family. He went to Oxford to study law, but practised as a lawyer unenthusiastically for only two years, before taking up his great love, geology, full time. Like Murchison, he had moved to London, and worked professionally as a geologist, but supported himself partly with his private wealth. This proved insufficient, however, and he relied also on income from his numerous textbooks on geology.

While at Oxford, he had attended geology lectures by the great, but eccentric, Professor William Buckland (p. 29). Buckland was one of Cuvier's leading supporters in England, and a strong proponent of the diluvialist viewpoint. Lyell was deeply influenced, and he switched to geology in the mid-1820s, about the same time as Murchison also moved to geology from his life of military manoeuvres and fox hunting.

Lyell visited France in 1823 and examined the rocks of the Paris basin, the very successions that Cuvier used as evidence for repeated revolutions, advances and retreats of the sea. He heard some dissenting views, but on the whole he was prepared to accept Cuvier's interpretations at that time. In 1828, Lyell and Murchison went together on a field trip to the Massif Central region of France, where they saw some rock sequences that apparently made Lyell begin to doubt Cuvier's revolutions. One succession of marls, over 220 metres (700 feet) thick, was composed of repeated thin layers, about 12 per centimetre (30 per inch) – a total of a quarter of a million individual layers. Lyell recognized the fossils in these marls from his studies of modern lakes in Scotland and he was able

7 *Charles Lyell in 1836, from a portrait by J. M. Wright.*

to conclude that the great thickness of ancient rocks had accumulated in a huge lake over a very long span of time. But surely each thin layer could not represent a revolution?

Lyell and Murchison continued their wanderings to southern Italy, and they saw active volcanoes around Naples and in Sicily. This impressed Lyell with the current forces of nature. He could work out the power of the volcanoes to raise the land up and to deposit vast amounts of lava. In places, he saw beds of rock containing marine shells that had been lifted high above the current sea level, but within historical time. The key to many of these observations was the fact that there were ancient Roman archaeological remains among the volcanoes and uplifted marine beds, proving that the upheavals and cataclysms had happened within the last 1500 to 2000 years. Lyell determined from this that modern agencies *were* adequate to explain all the phenomena observed by geologists, and that there was no need to call upon large-scale upheavals as Cuvier had done. When he returned to England in 1829, Lyell already had the plan for his book, his *Principles of Geology*, the first volume of which appeared in 1830.

Lyell's advocacy

The strategy of the *Principles* was eminently laudable, but Lyell introduced confusion where it did not exist. Lyell's main claim, throughout the three volumes of his book, was that every phenomenon in geology can be, and should be, interpreted in terms of processes that can be observed today – the principle usually termed 'uniformitarianism', though this was not his word. He put together all the evidence he had seen in his geologizing rambles round Britain, France and Italy, with reference to other geological work up to that time.

But Lyell's *Principles* is not a geology textbook, as is often thought: it is a legal argument, or a piece of advocacy. Indeed, critics at the time realized what was going on. Adam Sedgwick, in his Presidential Address to the Geological Society of London in 1831,[7] said that Lyell uses the 'language of an advocate'. Another critic, W. H. Fitton said[8]

> The book at first appeared to us to be the production of an advocate, deeply impressed with the dignity and truth of his cause. The tone was rather that of eloquent pleading than of strict philosophical enquiry.

Lyell, then, used his considerable skills as a writer and as a legal debater to present a powerful argument for a new way of doing geology, and the key was uniformitarianism. But he mixed up several different meanings of uniformity in his book.

The meanings of uniformity

Lyell used the term 'uniformity' in at least four ways in the *Principles*,[9] and in doing so he persuaded the geological world of some very odd conclusions. The first two meanings are certainly sensible and would be readily accepted by any intelligent field geologist then, and indeed now. But the other two meanings were something of a distortion of common sense. And, running through his argument, Lyell sowed the seeds of a false dichotomy between the heroes of uniformitarianism and the dupes of catastrophism.

Lyell's four usages of the word 'uniformity' are:

- *The uniformity of law:* the laws of nature are constant through time; it would be wrong to suggest that the orbits of the planets, gravity, mechanics, the interactions of chemical elements or any other such fundamental laws of nature have changed.

- *The uniformity of process:* the use of observations of modern phenomena to interpret the past, sometimes termed *actualism.* Geologists should not invent processes that cannot be seen today to explain ancient rock formations.

- *The uniformity of rates:* processes in the past must be interpreted to have occurred at the same rates as they do today, sometimes termed *gradualism.* So, geologists should not posit massive catastrophes, floods, volcanic eruptions or meteorite impacts on a larger scale than can be observed today.

- *The uniformity of state:* change on the Earth has happened in a cyclical way, and there is no evidence of linear, or directional, change, sometimes termed *nonprogressionism.* Lyell argued that change does not go anywhere, but that the Earth is in dynamic balance, with climates, volcanic eruptions, deposition of rocks and other phenomena proceeding at fairly constant rates somewhere, all the time.

The first two meanings of 'uniformity' are indeed sensible. Of course a true observational scientist should use the current laws of nature, and modern processes, in explaining past phenomena. This is a mixture of the actualistic methodological approach and the uniformitarian philosophy. A geologist clearly could not function by speculating rather than observing, or framing explanations in wild and imaginative theorizing.

But these same messages had also been the solid underpinning of everything Cuvier had written about geology from the start. Indeed, Cuvier's great initial distrust of geology and geologists stemmed precisely from his contempt for their wild theoretical models of the Earth, based on minimal evidence. This he repeated time and time again,[10] including in his 1810 and 1812 essays. Cuvier also stressed with great clarity that the geologist must base everything on modern processes and the modern laws of nature. So, Lyell and Cuvier were in agreement here: both were clear thinkers, and both relied on their own eyes. But still Lyell somehow succeeded in casting Cuvier, who was still alive in 1830, as the great speculator, the enemy of true science.

A claim too far

Lyell's third and fourth uses of the term uniformity – that rates of processes have not changed, and that the Earth cycles endlessly in a nonprogressive way – were statements of belief by Lyell, with no foundation. He simply made these as claims, and dared anyone to deny him. He presented no evidence, but argued them as inevitable extensions of his first two definitions. No change of rate (3) was clearly quite opposite to Cuvier's views. Cuvier had argued strongly that there *had* been major revolutions in the past, and Lyell denied this as wild speculation. But neither geologist had evidence, and we now know that Lyell was wrong. Cuvier, too, was wrong. Each man had pushed his claim too far.

There clearly have been episodes of vast vulcanism on the Earth, major movements of the continents, the uplift of mountain chains, huge meteorite impacts, great extensions of the polar ice caps, and there have been times of quiescence. It just happens that the present is one of those phases of modest activity. So Lyell was completely wrong in his claim for gradualism, or uniformity of rate. Cuvier too: he pointed to repeated revolutions marking every major change of sediment type, every postulated advance and retreat of the sea. This is also clearly wrong. But I don't think any modern geologist would claim that either Cuvier or Lyell was more or less wrong than the other, and yet Lyell is still regarded as the voice of reason, and Cuvier as the deluded, wild-eyed speculator.

Lyell employed the most amazing sleight of hand in trying to convince geologists of his notion of uniformity of state, or his grand cycles of history. A consequence of the stately steady-state Earth was, in his view, that nothing ever changes. Volcanoes erupt, glaciers advance, sea levels go up or down, but everything is in equilibrium. He viewed the Earth as a closed system with certain fixed laws, and that meant that there could be no progress or direction of change. And this he extended to life. Lyell could not in any way accept a time of origin for life as a whole, nor a decisive point of extinction for any group. So, the palaeontologist who tried to say that trilobites or fishes originated at a particular point in time, and changed in some definite way, and then became extinct, was denying Lyell's view.

Quite counter to Lyell's opinion, palaeontologists in the 1820s and 1830s were coming to realize that there was apparently some order to the fossils in the rocks, simple marine creatures first, then fishes, then animals on land, then reptiles, then mammals. Lyell would have none of such a progressionist view. As quoted earlier, he fully expected to find fossil mammals in the Silurian, or that dinosaurs might come back in the future. He continued to defend this view, against all odds, well into the 1850s.[11] To a certain extent, Lyell was supported in his anti-progressionism also by Richard Owen.

Professor Ichthyosaurus

Lyell's fellow geologists were always uncomfortable with grand cycles of earth history, but most of them felt somehow unable to speak their minds. And Lyell's arguments were so cleverly woven that it was hard to draw out only the nonprogressionist view of life or of the Earth, and deny them. To query the nonsense was to reject the whole edifice. Without care, such a stance might cast the critic into the crowd of speculators and creationists who had been so mercilessly dissected and rejected by Lyell. Murchison, Sedgwick and many other distinguished geologists were clearly unhappy about some of Lyell's claims, but they largely kept quiet about it in public.

One trenchant criticism, however, is encapsulated in a famous cartoon entitled 'Awful Changes' (Fig. 8). Professor Ichthyosaurus, wearing a patterned silk coat and a pair of pince-nez, is lecturing his class of marine reptiles, who sit and lie among the rocks. In front of the Professor is a human skull, about which he says, 'the skull before us belonged to some of the lower order of animals[,] the teeth are very insignificant[,] the power of the jaws trifling, and altogether it seems wonderful how the creature could have procured food.'

The cartoon was drawn by Henry De la Beche (1796–1855), another wealthy young man who, like Murchison and Lyell, had turned to geology

8 *'Awful Changes', a cartoon by Henry De la Beche depicting Professor Ichthyosaurus lecturing his class of marine reptiles on the remains of ancient humans.*

as the coming science in the 1820s. He was famous for his caricatures of contemporaries. He drew Professor Ichthyosaurus in 1831, and distributed it widely among his friends as a lithograph. For a long time this cartoon was seen as simply a gentle piece of fun, perhaps a spoof on William Buckland, who had indeed written extensively on ichthyosaurs.

Ironically, however, Professor Ichthyosaurus was directed not against Buckland the catastrophist, but against Lyell the uniformitarian. De la Beche's notebooks prove the point.[12] He was envisaging a time in the future, when ichthyosaurs might return to the Earth, very much Lyell's view, and themselves comment on the ancient remains of humans, in just the same way that geologists in the 1820s and 1830s wrote about the Jurassic ichthyosaurs. Murchison, doubtless, sympathized entirely with De la Beche's view.

Murchison's views

In 1830, when Lyell published his *Principles*, Murchison was a Cuvierian catastrophist and a progressionist. And yet his friend had just issued a polemic against both views. What was he to do? The critical response to Lyell was led by Sedgwick, who attacked Lyell's uniformity of rates, while others attacked Lyell's steady-state non-directionalism. But the opposition was not overwhelming. Even his critics praised his observational examples and his logical argument: Sedgwick accepted that it was wrong to try to link geology to the Bible through the Great Flood.

Lyell's book was generally very well received by the public, despite the private mutterings. Charles Darwin always pointed to the *Principles* as one of the key books that influenced his early work, even though Lyell's message about the history of life was so entirely at odds with Darwin's later views. Somehow, Lyell managed to pull off the amazing trick of pleasing nearly everybody, while at the same time proposing some incredible nonsense. Admittedly, it is easy to be wise now, and Lyell was certainly not the only geologist writing highly speculative and, to our eyes, absurd material at that time. But the irony is that his ideas were not rapidly rejected, but on the contrary continued to dominate geological thought for decades.

Murchison, perhaps surprisingly, seems to have stood somewhat in awe of the younger Lyell, both in 1830 and probably throughout his life. Murchison was later to become a grandee of geology, with a view on every subject, but in 1830 he was yet to prove himself. A young Scottish geologist, J. D. Forbes (1809–68), who later became Principal of St Andrews University but was then visiting London for the first time, meeting the great men of his science, wrote some perceptive comments in his notebook.[13] He stated that Murchison 'is a diligent observer . . . but does not appear possessed of much originality'. Lyell, on the other hand, was 'a remarkable man', but he was also 'sanguine in his opinions, hasty in his generalizations' and always insistent on finding a theory to explain every observation.

Murchison, unlike the other British catastrophists of the day, did not make any substantial public pronouncement against Lyell. And yet, while the second and third volumes of the *Principles* were in press, Murchison already had enough evidence from his studies of the Silurian rocks and fossils to demolish Lyell's gradualism and his nonprogressionism. Indeed, in private Murchison never wavered. In letters written in 1851, he commented to a correspondent that 'I have repeatedly shown in other works that operations of great violence, not of Lyell's quietude, have been repeated', and to Lyell he wrote in the same year denying the significance of some supposed Silurian vertebrate tracks from North America.[14]

Lyell's gradualism of processes, and especially his nonprogressionism, were anathema to stratigraphers such as Murchison and Sedgwick – such views went exactly counter to William Smith's earlier observation that successions of rocks were repeated in somewhat predictable ways, and that each rock formation had its own characteristic package of fossils that could be recognized, and used in establishing a relative date. Sedgwick and Murchison in the 1830s, as we have seen, showed quite clearly that even the most ancient rocks, the Cambrian, Silurian and Devonian were quite distinct units containing particular fossils. Lyell's nonprogressionism implied that the whole stratigraphic enterprise, as understood then, was wrong-headed. Murchison, however, preferred to mount his horse and go into the field, and set about his practical stratigraphic work, leaving the theorizing to others.

The death of mass extinction

Lyell's brilliant disposal of catastrophism in the 1830s made it impossible for anyone to study mass extinctions within the bounds of normal science. At the same time, his views undermined the work by Murchison, Phillips,

Sedgwick and others, who were creating the edifice of the standard, global geological timescale. To the stratigraphers it was clear that fossils changed through time, and that there were major gaps and horizons at which faunas and floras changed wholesale in geologically short intervals of time. But this had no real meaning in a Lyellian world of endless cycles of time.

Views about extinction were, however, very different, in 1840. Although, as we saw, the very possibility of extinction had been stoutly denied by many scientists until around 1800, that bridge *had* been crossed by 1840. But two others had not. It was not clear to Murchison, and to many of his contemporaries, that major, co-ordinated extinctions could occur. And, if such mass extinctions had happened in the past, geologists like Murchison and Phillips were unsure about how they could be linked to known modern extinctions. In light of this, and Lyell's baleful and persistent influence, how did geologists and palaeontologists explain the growing evidence that catastrophes *had* happened in the past?

4

THE CONCEPT THAT DARED NOT SPEAK ITS NAME

Today, the end-Permian mass extinction is generally accepted as having been the biggest crisis of all time. During this event, as we shall see, all forms of life – plants and animals, small and large, terrestrial and marine – were driven virtually to total annihilation. And yet, in 1987, in his influential *Vertebrate Palaeontology and Evolution*, the standard text in the field, the Canadian palaeontologist Bob Carroll wrote:[1]

> The most dramatic extinction in the marine environment occurred at the end of the Permian, wiping out 95 percent of the nonvertebrate species and more than half the families. Surprisingly, there was not a correspondingly large extinction of either terrestrial or aquatic vertebrates.

Carroll has since accepted the reality of the end-Permian mass extinction among vertebrates.[2] However, his statement reflects a long-standing and respectable position among geologists and palaeontologists – the cautious, or reluctant, acceptance of the reality of mass extinctions and catastrophes.

A key feature of the debate about mass extinctions, from 1840 to 1980, has been the imbalance in perceptions of the two sides. The proponents of catastrophe and sudden mass extinction were consistently regarded as lunatics. To link a mass extinction to cosmic rays, sunspots or meteorite impacts was to class yourself with the pseudoscientists and astrologers. The extinction-deniers were the level-headed, careful scientists. Far better to call for more evidence, to argue that an extinction happened gradually, perhaps over 5 or 10 million years, to seek explanation in slow-acting earth-bound processes, such as sea-level change or climatic deterioration, than to fly off wildly into the arms of the soothsayers, doom-mongers and apocalypse-merchants!

Why was mass extinction denied *a priori*? The explanation almost certainly lies deep within the Cuvier vs. Lyell debate of the 1830s. Even in the 1960s and 1970s the influence of Lyell's rewriting of history was still there. And it still is today. But, since 1980, it has at least become increasingly feasible to discuss the possibility of mass extinction and catastrophe without the need to be apologetic. In this chapter, we will trace the opinions about mass extinctions, both in general and concerning the end-Permian event, through the years from 1840 to 1980. Here and there, we can identify some remarkably prophetic publications, in which individuals dared to speak their minds, but at the time their papers were ridiculed or, worse, ignored. These were the twilight years for the catastrophists.

Victorian views: problems of dating

Palaeontologists and geologists after Murchison barely discussed mass extinction at all. They mostly accepted that fossil species had become extinct, but those extinctions were seen as sporadic events, occurring as a normal part of the 'life cycle' of a species. Without any concept of the duration of geological time, it was difficult for palaeontologists in the nineteenth century to speak with any conviction about the coincidence or rapidity of events.

If a nineteenth-century palaeontologist had found a layer of rocks in which 50 fossil species had seemingly disappeared, he would not have been in a position to declare confidently that he had found a mass extinction level. First, there was no way of estimating whether the rock layer represented a week or a million years of time. Second, he could not determine whether the contact with the rock unit lying above the rock layer in question indicated a smooth transition – a short span of time – or a gap of a week or 10 million years after deposition of the fossiliferous layer. Further, the disappearance of 50 species could indicate nothing: they might have survived into the time represented by the gap and lived on for centuries, dying out and being replaced in a normal, gradual, Lyellian, way. Equally, even if there were no gap in deposition, the nineteenth-century geologist would have been entirely unable to demonstrate that the 50 extinct species had not simply become extinct locally and survived happily elsewhere. Without refined dating methods, it is hard to demonstrate a global mass extinction.

The fundamentals of stratigraphy had been established by 1840, and the details were increasingly ironed out during the later nineteenth century. The Systems, Silurian, Devonian, Carboniferous etc, identified and named by Murchison and his colleagues in the 1830s, were subdivided into epochs, stages and zones, and these finer units were traced over large areas. In the case of widespread marine sediments, for example those of the Early and Late Jurassic, this exercise proved to be highly successful. Using ammonites, individual zones, and even narrow beds, could sometimes be tracked across Europe, from the windy cliffs of Dorset, in southern England, south to the Jura Mountains in southeastern France, and west across the wooded hills of southern Germany. However, until 1920, there was no method to establish time spans.

In their minds, most Victorian geologists had long abandoned the literal estimate, based on the Bible, of an Earth that was only 6000 years old. They had a concept of the vastness of geological time – the sheer thickness of the rocks proved that – and they were thinking of millions of years, but whether 5 million or 5000 million, no one could say. Most late Victorian geologists settled on a figure of about 100 million years. However, such efforts at dating were purely speculative, and most geologists thought it pointless to indulge in such wild discussions.

Precision in dating rocks came at last after the discovery of radioactivity in the late nineteenth century by Henri Becquerel in Paris. The first use of measurements of the timing of radioactive decay – radiometric dating – came in 1906, following fundamental work by Ernest Rutherford. Radiometric dating, from the start, provided exact age estimates in millions of years. Initial efforts set the broad scale – that the fossiliferous units of the Palaeozoic and Mesozoic were hundreds of millions of years old, and that the Earth was thousands of millions of years old.

Work since 1920 has increasingly refined those age estimates, and palaeontologists have been able to achieve some confidence in estimating the time represented by a single rock unit, or the probability of a gap. However, the Victorian problems are still live issues, since radiometric dating cannot be applied to most rocks, especially fossil-bearing sediments, and the precision is often poor. In recent years, most geologists have accepted a date of between 245 and 250 million years ago for the Permo-Triassic boundary. Recent, more precise dating methods, have yielded values around 252 million years ago, fantastically accurate in geological terms, but still not good enough to determine, for example, whether the crisis interval lasted for a few years or a million years.

Richard Owen and the extinction of his Dinosauria

But surely, despite these niggling problems, the Victorian palaeontologists must have been impressed by the extinction of the dinosaurs, just as everyone is today? Dinosaur extinction is such an icon that it cannot be ignored. Quibbling over precision of dating should hardly affect the magnitude of such an event.

Certainly, the great reptiles of the Mesozoic, the marine ichthyosaurs and plesiosaurs, the flying pterosaurs and the dinosaurs, were well known by 1840 from several dozen specimens. But, no one then seemed to have an inkling that their final disappearance might have been part of a mass extinction. This is really no surprise when the facts are considered. In 1840, only 20 or 30 species of these reptiles were known in total, from a range of rocks in Europe dating from the late Triassic to the Cretaceous. There was no need to postulate such a thing as a mass extinction. Each species of dinosaur or plesiosaur had existed for its span of time, whether 10,000 or 10 million years, and it had then succumbed to some local crisis or had been replaced by another species.

Richard Owen, who named the Dinosauria in 1842, had some words to say about the extinction of the giant Mesozoic saurians. Indeed, his comments have sometimes been taken rather out of context and misinterpreted as perhaps the first published hypothesis about dinosaurian extinction. Not so. Owen argued in 1842[3] that the Creator had chosen the Mesozoic era as suitable for dinosaurs because of its different atmospheric conditions. He believed that the air then was deficient in oxygen, and that this suited the dinosaurs. As reptiles, they had lower metabolic rates than the birds and mammals and could survive on less energy. Owen argued that oxygen levels rose during the Mesozoic and that the atmosphere became more 'invigorating'. The world then became uninhabitable for the huge saurians, and they died out, together with the giant marine reptiles and the pterosaurs.

This argument was essentially circular, since Owen's evidence for low oxygen levels in the Mesozoic was simply the presence of dinosaurs and other prehistoric reptiles and the virtual absence of mammals and birds.

He was explaining the existence of these early reptiles in Mesozoic rocks in terms of a precise matching by a benevolent Creator of living things and physical environments. This is pure Lyell. Owen was not primarily trying to explain why the dinosaurs died out when they did – for him, that would have been a preordained event in the Creator's plan.

Mr Darwin's view

Darwinism largely put an end to such odd mixtures of Christianity and Lyell, but Charles Darwin could not escape the problems of dating and the quality of the fossil record. Despite his brilliant insights in other evolutionary topics, he was not able to make the conceptual leap to deny Lyell and to accept the reality of mass extinctions.

Charles Darwin devoted two chapters of his *On the origin of species by means of natural selection* to palaeontology. He discusses the 'imperfection of the geological record' in chapter 9, and surveys the history of life in chapter 10. Here, Darwin addresses the issue of mass extinction, but tentatively:[4]

> the utter extinction of a group is generally, as we have seen, a slower process than its production. With respect to the apparently sudden extermination of whole families or orders, as of Trilobites at the close of the palaeozoic period and of Ammonites at the close of the secondary period, we must remember what has already been said on the probable wide intervals of time between our consecutive formations; and in these intervals there may have been much slow extermination.

In other words, apparent mass extinctions, such as the end-Permian event ('the close of the palaeozoic period') were probably illusory, the sum of many minor local extinctions partly hidden in a gap in the rock record. Darwin, of course, explained that species naturally became extinct from time to time as a result of competition with other species and their replacement by superior competitors.

Darwinian palaeontologists of the latter half of the nineteenth century followed the master in saying nothing about dinosaur extinction. Darwin's strongest supporter, Thomas Henry Huxley (1825–95), wrote a number of articles about dinosaurs, but never discussed their extinction. For example, in one of his first papers on the subject, published in 1870, 'On the classification of the Dinosauria', Huxley[5] described the 16 species of dinosaurs known up to that time from fossils of Triassic, Jurassic and Cretaceous age found in Europe, North America, Africa and Asia. Dinosaur extinction? Not a word.

Later Victorian accounts were no different. Othniel Charles Marsh (1831–99) was in a strong position to talk about the end-Cretaceous mass extinction of the dinosaurs, but he did not. Marsh was a leading North American vertebrate palaeontologist, famous for his involvement in the 'bone wars' of the 1870s to the 1890s, when he and his arch-rival, Edward Drinker Cope (1840–97), vied with each other to unearth and name as many dinosaurs as possible from the American Midwest. Marsh wrote a number of reviews of the diversity of the dinosaurs, covering the same ground as Huxley had, but incorporating also all the new North American finds. In 1882, Marsh listed 46 dinosaur genera, and in 1895 this had risen to 68.[6] In these papers, he showed how the different dinosaur dynasties had waxed and waned through the Mesozoic, but their final departure from the world was not discussed.

But what of the other mass extinctions that we now recognize? Virtually nothing was written about the end-Permian event and the others. But the most recent extinction event, when large hairy mammals, such as mammoths, mastodons, woolly rhinos and others died out did attract repeated attention in Victorian times.

Floods or glaciers?

The late Pleistocene extinctions, some 10,000 years ago, seem to coincide with the retreat of the last ice sheets from northern Europe and North America. It had been clear to early geologists and palaeontologists that strange, exotic beasts had once lived in these parts of the world. As we saw in Chapter 1, specimens of mastodons from North America, mammoths from Europe and other large hairy mammals were much discussed in the eighteenth century.

By 1850, many more specimens had been collected. William Buckland, for example, who described the first dinosaur, *Megalosaurus*, in 1824, was at the same time engaged in studies that he regarded as much more important. He had directed excavations at Kirkdale Cavern in Yorkshire, which had been discovered in 1821. Buckland found abundant bones of deer, hippos, rhinos and mammoths in the cave, together with hyaena bones and coprolites (fossilized faeces). A zookeeper had drawn Buckland's attention to the fact that the coprolites contained crushed shards of bone, and that they looked just like the excrement of the modern hyaenas that were in his care. Buckland formulated the view that the cave had been a hyaena den, and that those scavengers had dragged back carcasses, and parts of carcasses, to the cave. Here indeed was a scene that differed considerably from modern Yorkshire!

In his *Reliquiae Diluvianae*, published in 1822,[7] Buckland argued that climates had been warmer in Pleistocene times, hence explaining the exotic, rather African, fauna. More importantly, though, he accounted for the extinction (at least the local extinction) of these exotica by the universal Flood recorded in the Bible. The rising waters had trapped the hyaenas in their cave and had killed off the large, exotic mammals throughout England, and also in the rest of Europe.

Buckland lived to give up his flood-based viewpoint. Increasing evidence was found in the 1830s that northern Europe had been swept by vast ice sheets. The shape of the landscape, especially in Scotland, Scandinavia and around the Alps, showed that glaciers had gouged deep, smooth-sided valleys. In highland areas, great pavements of bare rock were to be seen, still bearing scratch marks produced by ancient glaciers. Erratic rocks lay everywhere – boulders that had been torn up by moving glaciers and dumped miles from their original source.

The glacial model was championed especially by Louis Agassiz (1807–73), the famous Swiss geologist and expert on fossil fishes. He argued that European climates had once been warm and equable – entirely suitable for large African and Asiatic mammals such as hippos and elephants. The cooling of the climate, and the march of the ice, had brought all this to an end:[8]

The appearance of this great cover of ice must have caused the extinction of all organic life on the surface of the globe. The territory of Europe, recently covered by tropical vegetation and occupied by herds of elephants, hippopotamuses and gigantic carnivores, found itself entombed under a vast mantle of ice that covered fields, lakes, seas and plateaus alike. To the movement of a powerful creation succeeded the silence of death. Springs dried up, streams ceased to flow, and the sun's rays, in rising over those frozen expanses (if they still reached there), were met only by the whistling of the northern winds and by the thunder of crevasses as they split the surface of this vast ocean of ice.

Pleistocene overkill

Geologists of the time generally accepted Agassiz's remarkable insight that Europe and North America had been covered by ice during the Pleistocene. But very few accepted his seemingly catastrophic viewpoint that large swathes of life were wiped out by the cold and the ice. Indeed, Charles Lyell, while admitting the truth of the ice age, based as it was on clear physical evidence scattered through the landscapes of Europe and North America, was extremely uncomfortable with the idea of a sudden intense freezing and global extinction. He noted, correctly, that the large mammals had lived on during the times of cold and that they had clearly been adapted to the conditions. Lyell's view was that the various large Pleistocene mammals had died out one by one, for a variety of reasons, but that there had not been any such thing as a single extinction event.

For the remainder of the nineteenth century most geologists preferred Lyell's viewpoint. They did not identify a single extinction event, but linked the extinctions to changing climates throughout the northern hemisphere. Extinctions had been gradual, and had been caused by particular changes in climates and local conditions – perhaps different causes for the extinction of each species.

A role for humans in these most recent extinctions had been discussed since the first discoveries of mastodons and mammoths. Could stone age peoples in different parts of the world have killed off the large mammals? After all, they would have made attractive propositions for dinner. Lyell rejected such a notion, since initially he was convinced that humans had not appeared until *after* all the late Pleistocene mammalian extinctions had taken place. By 1860, however, archaeological evidence proved that humans and mammoths, and other large Pleistocene mammals, had co-existed – bones of the giant mammals were found in close association with human bones and artifacts. Lyell was forced to change his mind, and he admitted a possible role for humans in the demise of the Pleistocene mammals. At the same time, a French investigator was coming up with convincing evidence for a different catastrophic explanation for the events in the late Pleistocene.

Boucher de Perthes: the last catastrophist?

Jacques Boucher de Perthes (1788–1868), a French civil servant, was instrumental in providing the evidence that early humans and the Pleistocene mammals of Europe had cohabited. He knew that in order to convince the doubters, such as Lyell, he had to work with scrupulous care. Boucher de Perthes excavated sites around Abbeville in the valley of the Somme, and he reported stone tools from layers *below* bones of woolly mammoth, woolly rhinoceros and hippopotamus. He argued that his excavations had uncovered remains from before the biblical deluge.[9] The Flood had then swept over the world, destroying the strange mammoths and rhinos in Europe, together with the primitive humans that hunted them. After the Flood, new, modern, animals filled up the territories of Europe. This idea of a catastrophic flood came directly from Georges Cuvier in the 1820s. In France and Germany such catastrophist viewpoints still held sway – Lyell had not been able to convince everyone.

British palaeontologists, however, were unable to accept a revival of catastrophism. Lyell himself argued that such a huge catastrophe could not be proposed since so many of the Pleistocene mammals still survived today – rats, mice, shrews, foxes, wolves. He noted that it was only the larger mammals that had been wiped out. In the 1860s, Richard Owen supported Lyell's viewpoint, and he went further in pointing to what he thought was convincing evidence for the overkill hypothesis. Based on his work in Australia, Owen argued that many of the large marsupials – giant kangaroos, huge wombat-like herbivores and others – had disappeared only when the first humans arrived in Australia. In New Zealand, where he studied the giant flightless birds, the moas, Owen found even more convincing evidence that the Maoris had killed off these birds in the past hundreds of years.

The great biogeographer and naturalist, Alfred Russel Wallace (1823–1913), was also convinced by the arguments of Lyell and Owen. He could see the great threats that human expansion posed for the wildlife of the tropical world in his day, and hence he found it plausible that earlier phases of human migration had killed off native plants and animals.

In late Victorian times, then, several influential commentators in England – Lyell, Owen, Wallace – were strong supporters of the overkill hypothesis. The latest of the documented extinction events, the death of large mammals on many continents in the late Pleistocene, had indeed been rapid. But Boucher de Perthes' idea that the rapidity implied a sudden global catastrophe was clearly unacceptable. Human activity provided a convincing explanation: the timings of human migrations seemed to tie in closely with the timings of the extinctions.

The overkill model also provided a clear way to maintain a gradualistic, uniformitarian view of the history of life. If the last of the great extinctions could be explained as a special case, a one-off caused by humans, then there was no need to invoke catastrophic explanations for earlier mass extinctions when humans were absent. But how did late Victorian and early twentieth-century scientists explain the extinction of the Permian reptiles, and of the dinosaurs, two of the other well-known great dyings of the past?

Directed evolution

In the late nineteenth and early twentieth centuries, many palaeontologists and biologists took up non-Darwinian viewpoints, with grand names such as 'orthogenesis' and 'finalism', both of which implied some kind of preordained plan to evolution.[10] These models assumed that evolution was directed in some way and that it proceeded according to regular patterns. The Permian synapsids, or the dinosaurs, could be viewed as primitive, lumbering beasts that had to give way to more advanced forms. Mass extinctions – the replacement of Permian plants and animals by those of the Triassic – just happened. It was part of the plan, and didn't really require an explanation.

This reversal in opinions about evolution may seem rather strange. Surely, after Darwin's pronouncements in the *Origin of Species* in 1859, and the debates that surrounded them, scientists had essentially accepted the correctness of his views? However, it seems that most biologists in the late nineteenth century, although they claimed to be Darwinians, weren't really. They accepted evolution – that organisms have changed through time, and that they are linked through lines of descent that form a huge branching tree – but not many could accept the seeming purposelessness of Darwin's evolution.

In general, though, the extinction of the dinosaurs remained a non-question at this time. Standard textbooks of general and vertebrate palaeontology of the latter half of the nineteenth century and the first decades of the twentieth barely mention the subject at all. And if they do, the explanation is brief. The dinosaurs simply came and went in their due time. As an example of this view, Arthur Smith Woodward (1864–1944), chief of palaeontology at the Natural History Museum in London, and later to be famous for having been duped into describing the famous forgery, Piltdown man, as a new species of hominid, wrote in his 1898 textbook[11] that 'toward the close of the Mesozoic period . . . the Dinosaurs gradually became extinct'. And later he states that the dinosaurs of the Cretaceous 'became more specialized and almost fantastic just before they disappear'.

This low-key approach to the extinction of the dinosaurs continued remarkably late in the twentieth century. I scanned a dozen standard textbooks, some of them published in England, others in the United States and in Germany, dating from 1902 to 1968, and there is barely a mention of the great end-Cretaceous mass extinction event. This absence of debate in the textbooks probably reflected the general opinion of vertebrate palaeontologists. None the less, during the first half of the twentieth century, discussion about the demise of the dinosaurs did continue, and writers concentrated on the 'excess spinescence' (growth of spines) in the ancient reptiles.

Racial senility

When I was young I remember being puzzled by newspaper headlines that said such things as, 'British trades unions are dinosaurs', 'Party leader is *Brontosaurus*, doomed to extinction'. As a mad dinosaur fanatic at the age of eight, I knew that dinosaurs had been vigorous and successful for over 150 million years, so why did adults consistently use dinosaurs as a metaphor for redundancy and inefficiency?

The popular metaphor stemmed from scientific views of some seventy or more years earlier. The orthogenesis and finalism of the turn of the twentieth century led to views of *racial senility* – the belief that certain long-lived groups of animals became old and their store of evolutionary novelty dried up. There was a parallel here between the life span of an individual plant or animal and that of an evolutionary stock. Youth and early racial vigour were equated, and seen to be just as inevitable as the old age and death of an individual and the racial senescence and eventual extinction of a major group. According to this view, the dinosaurs had been around for a long time, and they simply ran out of adaptability. The remarkable horns, frills and spines of some late Cretaceous dinosaurs were occasionally cited as evidence for this racial senility.

A typical early account was given by Arthur Smith Woodward in 1909 in his address to the British Association for the Advancement of Science.[12] He pointed to the great spinescence, excess growth and loss of teeth of the later dinosaurs as evidence. At the same time, the American palaeontologist Frederick Loomis expressed similar views when he wrote about the bony plates along the back of *Stegosaurus*:[13] 'with such an excessive load of bony weight entailing a drain on vitality, it is little wonder that the family was short-lived.'

These arguments are easy to refute. For example, *Stegosaurus* lived during the Late Jurassic, some 90 million years before the extinction of the dinosaurs. Likewise, there is no evidence that the last dinosaurs, from the Late Cretaceous, were any more spinose, crested or toothless than their forebears. None the less, expressions of pure racial senescence are to be found in the writings of many distinguished geologists and palaeontologists through the 1920s and 1930s, even though their views were challenged at the time.

In the end, such ideas of orthogenesis and the racial senility of the dinosaurs were effectively demolished by the advent of the modern synthesis or neo-Darwinian model of evolution in the 1930s and 1940s. This was a revolution brought about by the arguments of a whole host of brilliant young evolutionists – Theodosius Dobzhansky, Ernst Mayr, George Gaylord Simpson, Julian Huxley – who saw that orthogenesis, and related ideas, were pure mysticism. They pulled in evidence from the new laboratory science of genetics, and combined it with the kind of field observations that Darwin had made.

The modern synthesis position, which is the view held today, represented a return to pure Darwinism, with the addition of new sciences that Darwin could never even have dreamed of. There was no longer any place for preordained patterns in this view of evolution. Of course, although the scientists rejected racial senility long ago, it is hard for many people to shed such views completely: surely we have to believe in *progress* of some kind? Surely the dinosaurs were simply doomed to extinction by their large size?

Baron Franz Nopcsa: spy and theorizer

During the 1920s and 1930s, a number of authors eschewed racial senility, and concentrated instead on biotic and physical factors that might have caused the extinction of the dinosaurs. Baron Franz Nopcsa (pronounced 'NOP-sha'; Fig. 9) was one of the first. Indeed, Nopcsa was an original in many ways. His full name was Baron Franz (or Ferenc) Nopcsa von Felsö-Szilvás (1877–1933), hinting at his aristocratic origins. He was in fact the last in a long line of noblemen with estates in Transylvania, lands that are now shared between Romania and Hungary.

In 1895, Nopcsa's sister, Ilona, found some huge bones at Haţeg (pronounced hat-SEG) on the family estates. She showed them to Franz, who took them to Vienna, capital of the Austro-Hungarian empire, for identification. No one could help young Franz, and in the end he decided to learn about palaeontology and study them himself. What Ilona Nopcsa had found were some of the last dinosaurs to survive in Europe, from rocks dating from the very end of the Cretaceous.

Thus began Nopcsa's distinguished career as a maverick, but brilliant, palaeontologist. He moved freely throughout Europe, using his urbanity and his astonishing command of languages to attach himself to scientific societies in all countries: his publications appeared in perfect German, English, French or Hungarian, according to his interests at that moment. At the same time, Nopcsa lived the life of a freebooter, offering his services when the First World War broke out to the Austro-Hungarians: he suggested that he could secure the friendship of Albania by becoming its king. This proposal was rejected, but Nopcsa acted as a secret agent for the empire.

Nopcsa was really the first to explore reasons for the extinction of the dinosaurs in a serious way. He suggested, for example,[14] that the great amount of cartilage that he believed was necessary

9 *The brilliant Transylvanian palaeontologist, Baron Franz von Nopcsa, an early dinosaurian palaeobiologist.*

for growth to huge size 'perhaps ... was one of the causes for the rapid extinction of the Sauropoda'. Later, Nopcsa summarized a number of views on the extinction of the dinosaurs: their 'low power of resistance', their huge size, a shortage of food, or 'a reduction in their sexual functions'. He focused particularly on the supposed 'increase in function of the hypophysis' – this is the pituitary gland, located in the head, which controls growth. Nopcsa believed that some malfunction of the pituitary caused the very large body size of the dinosaurs. Secretions from the pituitary caused giantism, partly by the production of large masses of cartilage as precursors of bone, and partly by a form of acromegaly, or pathological excess thickening and overgrowth of limb bones and facial bones. Nopcsa wrote that 'the increase in weight of the limbs in the dinosaurs recalls the eunuch condition'. Quite where Nopcsa got his information about eunuchs is unclear.

After the war, Nopcsa lost huge amounts of money, mainly because various governments confiscated his estates in the ensuing chaotic conditions. He was put in charge of the Hungarian Geological Survey in 1925, and this saved his finances for a while. But, in 1929, he left in a rage, and embarked on a 5600-kilometre tour through Italy and southern Europe. He travelled in a motorcycle combination with his faithful secretary and lover, an Albanian called Bajazid, at his side. The two men returned to Vienna, but, beset by financial problems and poor health, Nopcsa finally shot Bajazid, and then himself, in 1933.

The German school of Paläobiologie

Most other palaeontologists of the early twentieth century were neither as exotic, nor as imaginative, as Franz Nopcsa. Mass extinctions, and the extinction of the dinosaurs in particular, were not discussed by many people at all. Of those who did consider the question, most preferred to focus on climatic changes, rather than internal, biological causes as Nopcsa had done.

As an example, the noted American expert on fossil mammals, William Diller Matthew, presented evidence in 1921 for a model of dinosaurian extinction that involved gradual topographic change and progressive replacement by mammals.[15] His study of the late Cretaceous and the succeeding Palaeocene in North America suggested that there was extensive mountain building and continental uplift. The dinosaurs, which were adapted to lowland and marsh situations, were displaced, and the placental mammals, which were adapted to upland zones, moved in.

Other suggestions were that climatic cooling was the cause, or that disease levels had risen markedly in the late Cretaceous dinosaurs, or that early mammals ate all the dinosaur eggs, or that volcanic eruptions were responsible.[16] In a sense, the debate was heating up (though not very much, as we shall see). But at least some geologists and palaeontologists were identifying the fact that something unusual had happened, and that it deserved an explanation. Indeed, some, but not all, of these authors, recalled that it wasn't just the dinosaurs that died out 66 million years ago. Clearly, any satisfactory explanation had to take account of all the other victims on land and in the sea.

In 1929, a remarkable, but largely forgotten, paper appeared in a German scientific journal called *Palaeobiologica*. This journal was set up in 1928 to mark a new wave of thinking in German palaeontology. The discipline was clearly named palaeobiology – not palaeontology – to set it apart from traditional approaches. Palaeobiologists were not simply interested in fossils for the purpose of dating rocks; they wanted to treat them as once-living organisms. This school of thought hoped to take over the world of palaeontology with its new breed of young scientists who used biological and biomechanical approaches in studying the life of the past.

The paper, by Alexander Audova, an Estonian palaeobiologist, presented a 61-page review of the whole question of the extinction of the dinosaurs.[17] Audova rejected racial senility and simple natural selection as explanations, and focused on environmental change. His favoured view, after surveying geological evidence on palaeotemperatures and physiological evidence on the thermoregulation of modern reptiles, was that temperatures had declined gradually worldwide, and that this acted directly on the dinosaurs and other Mesozoic reptiles by preventing proper embryonic development.

One hundred theories for the death of the dinosaurs

In the time from 1920 to 1990, at least one hundred theories for the extinction of the dinosaurs were proposed, although clearly at the low rate of only one or two per year. No stone or fossil was left unturned in the search for ideas, and these ranged from the purely biological, to interactions among species, to changes in environments, to extraterrestrial causes. It's impossible to catalogue in detail all the suggestions here. In 1964, the American dinosaur expert G. L. Jepsen was able to list 40. In 1990, I was able to identify over 100 separate theories.[18]

I did not include casual remarks made by palaeontologists at scientific meetings, nor did I include any of the constant flood of speculative pieces in the newspapers, with the exception of one from 2000, which suggested that the dinosaurs had gassed themselves out of existence. A French palaeontologist had calculated that a cow produces enough gas each week to fill a barrage balloon. A dinosaur weighing fifty times as much as a cow would produce fifty times as much gas; billions of gallons of methane would therefore enter the atmosphere each year through the digestive systems of the dinosaurs, replacing the oxygen in the atmosphere, asphyxiating the dinosaurs. While this theory was never recognized in a scientific journal, some years later, David Wilkinson from the University of Liverpool,[19] and colleagues, presented a paper in the highly reputable *Current Biology* proposing that dinosaurian methane contributed to warming Mesozoic temperatures.

My 1990 list does not include hearsay and student japes, only theories that were seriously proposed through the normal channels of scientific publication – which means that the papers should have been checked by at least two or three experts before proceeding to publication.

One hundred theories for the extinction of the dinosaurs, presented from 1842 to 1990.

Biotic causes (26): Medical problems: slipped vertebral discs; malfunction or imbalance of hormone systems; overactivity of pituitary gland and excessive (acromegalous) growth of bones and cartilage; limb bones too heavy; pathological thinning of egg shells; diminution of sexual activity; cataract blindness; disease (caries, arthritis, fractures, and infections); epidemics; parasites; AIDS caused by increasing promiscuity; change in the ratio of DNA to cell nucleus. Mental disorders: dwindling brain and consequent stupidity; absence of consciousness, and absence of the ability to modify behaviour; development of psychotic suicidal factors; *Palaeoweltschmerz*. Genetic disorders: excessive mutation rate induced by high levels of cosmic rays; abortion of embryos by cosmic rays.

Racial senility (6): evolutionary drift into senescent overspecialization, as evinced in gigantism, spinescence or excess armour; racial old age (Will Cuppy:[20] 'the Age of Reptiles ended because it had gone on long enough and it was all a mistake in the first place'); increasing levels of hormone imbalance leading to ever-increasing growth of unnecessary horns and frills; head too heavy to lift.

Biotic interactions (6): competition with mammals; competition with caterpillars which ate all the plants; overkill capacity by predators (the carnosaurs ate themselves out of existence); egg-eating by mammals; consumption of all plants by giant dinosaurs; methane poisoning from dinosaur flatulence.

Floral changes (11): spread of angiosperms and reduction in availability of gymnosperms and ferns, which led to a reduction of fern oils in dinosaur diets, and to lingering death by terminal constipation; loss of marsh vegetation; increase in forestation, leading to a loss of habitat; reduction in availability of plant food as a whole; presence of poisonous tannins and alkaloids in the angiosperms; presence of other poisons in plants; lack of calcium and other necessary minerals in plants; rise of angiosperms, and of their pollen, led to extinction of dinosaurs by terminal hay fever.

Climatic change (12): climate became too hot (high temperature inhibited spermatogenesis, unbalanced the male:female ratio of hatchlings, killed off juveniles, or led to overheating in summer); climate became too cold (too cold for embryonic development, dinosaurs too large to hibernate, froze to death in winter); climate became too dry; climate became too wet; reduction in climatic equability and increase in seasonality.

Atmospheric change (7): changes in the pressure or composition of the atmosphere; high levels of atmospheric oxygen leading to fires; low levels of carbon dioxide removed the 'breathing stimulus'; high levels of carbon dioxide in the atmosphere and asphyxiation of dinosaur embryos in the eggs; extensive vulcanism which produced volcanic dust; selenium, or other toxic substances, which caused thinning of dinosaur egg shells.

Oceanic and topographic change (12): sea levels rose; sea levels fell; floods; mountain building; drainage of swamp and lake habitats; stagnant oceans caused by high levels of carbon dioxide; bottom-water anoxia; spillover of Arctic water (fresh) from its formerly enclosed condition into the oceans, which led to reduced temperatures worldwide, reduced precipitation, and a 10-year drought; reduced topographic relief, and reduction in terrestrial habitats; break-up of supercontinents.

Other terrestrial catastrophes (5): sudden vulcanism; fluctuation of gravitational constants; shift of the earth's rotational pole; extraction of the moon from the Pacific Ocean; poisoning by uranium sucked up from the soil.

Extraterrestrial explanations (15): entropy (increasing chaos in the Universe and hence loss of large organized life forms); sunspots; cosmic and ultraviolet radiation; destruction of the ozone layer by solar flares and entry of ultraviolet radiation; ionizing radiation; electromagnetic radiation and cosmic rays from the explosion of a nearby supernova; interstellar dust cloud; flash heating of atmosphere by entry of meteorite; oscillations about the galactic plane; impact of an asteroid; impact of a comet; comet showers.

Problems with the 'dilettante' approach

Why focus here on the extinction of the dinosaurs? After all, the topic of this book is the end-Permian event, some 185 million years earlier. However, very little was ever said, or written, about the end-Permian mass extinction until recently. Despite the fact that the end-Permian event was so much vaster than the end-Cretaceous crisis, all attention focused on the demise of the dinosaurs. And this was most unfortunate, since it lent an air of amateurism to the whole topic of mass extinctions.

According to the varied list of reasons for dinosaurian extinction (see Box), anything is possible. Climates became too wet, too dry, too hot, too cold: you can take your pick. Evidently, something is wrong here, and this feeling of a subject that was running out of control convinced many serious-minded geologists and palaeontologists in the 1960s and 1970s that they should keep clear of anything to do with mass extinctions, not just the extinction of the dinosaurs. I have labelled this phase of speculation the 'dilettante' approach, meaning that it was beset by dabblers, by people who thought about the topic for a while, published their pet idea, and then moved on to another subject.

Certain of the suggestions in the list are indeed perfectly reasonable ideas on the basis of present knowledge, but the obviously ludicrous nature of many had important consequences. While serious geologists were quietly appalled by the random speculation, others, often outside the direct fields of expertise required, thought that mass extinctions, and particularly the extinction of the dinosaurs, were fun, speculative topics in which anyone could join.

Many of the ideas listed were presented by non-palaeontologists, and certainly most of the authors had little first-hand knowledge of the late Cretaceous fossil record of dinosaurs – hence the 'dilettante' sobriquet. A large number of the theories, all of which were published in standard scientific journals by scientists who were no doubt expert in their own fields, show a remarkable relaxation of scientific standards. It was as if, at the mere mention of 'dinosaur extinction', scientists breathed a sigh of relief and felt freed from the straitjacket of normal scientific hypothesis-testing.

It is important to remember that the extinction of the dinosaurs 66 million years ago was part of a larger mass extinction, during which numerous other marine and terrestrial groups died out. In the sea, the marine reptiles, particularly the plesiosaurs and mosasaurs, disappeared, as did some major mollusc groups, the free-swimming ammonites and belemnites, and the bottom-living rudists (large, reef-building molluscs that were fixed to the sea floor). Even more striking was the loss of huge diversities of microscopic plankton, in particular the foraminiferans, tiny shelled protozoans. On land, of course, the dinosaurs died out, but so too did the flying pterosaurs, as well as some groups of birds and mammals. The whole mass extinction is generally termed the KPg event (K for *kreta*, Greek for 'chalk', a common rock in the Cretaceous, and Pg for Paleogene, the subsequent geological period). The KPg event used to be called the KT event, where T stood for Tertiary, but the Paleogene (the first portion of the Tertiary) is now the accepted time division. The oft-quoted age of 65 myr. has similarly been revised to 66 myr. according to the latest dating evidence.

I believe four main arguments support the view that the study of the KPg event had in fact run out of control by the 1960s and 1970s.

- Many of the authors demonstrated an ignorance of basic palaeontological data. For example, the hypotheses were often restricted to explaining why the dinosaurs alone died out, and no mention was made of the marine plankton and other animals that also disappeared. The question of the survivors of the KPg event was often not tackled: some scenarios were so extreme or catastrophic that it is hard to understand how the land plants, insects, frogs, lizards, snakes, crocodiles, turtles and so on were not detectably affected. In other cases, the timing of evolutionary events is wrong: for example, the flowering plants appeared 40–50 million years before the KPg event, the mammals 150 million years before. Neither group could have caused the demise of the dinosaurs unless some other major evolutionary innovation in one or the other is proposed.

- A number of the theories apparently ignored basic biological principles. Could caterpillars really compete with herbivorous dinosaurs and eat all the plants? Could dinosaurs really have been like automata, and unable to modify their behaviour? Is it possible to model a terrestrial biosphere in which a single factor – epidemics, parasites, glandular malfunction, competition, predation – would lead to a complete ecological breakdown?

- The mode of argumentation in many papers was by strong advocacy – the technique of course which Charles Lyell had used to such effect in the 1830s when he removed catastrophism from the field of reasonable scientific discourse. An argument by advocacy runs like this : 'If it is assumed that dinosaurs were endothermic/that UV radiation was increasing during the Cretaceous/that caterpillars competed for food with plant-eating dinosaurs, then it follows that. . . . If it is further assumed that climates were becoming warmer, or colder, or drier, or wetter, then it follows that. . . .' It is rare to find careful weighing of evidence both for and against particular hypotheses.

- There is also the assumption by some authors that the whole subject is really just a parlour game, and not terribly serious. If a dinosaur palaeontologist were to write an account of his or her theory of the origin of the universe or of a cure for cancer or of why caterpillars turn into butterflies, he or she would probably fail to get into print in a reputable scientific journal. However, most of the dilettante theories of the extinction of the dinosaurs were published in very reputable journals: *Science*, *Nature*, *American Naturalist*, *Journal of Paleontology*, *Evolution*, and so on. How did these speculators get away with it?

Otto H. Schindewolf: crazy theorist or visionary?

In the mid-twentieth century, at a time when most English-speaking palaeontologists steered well clear of any talk of mass extinction, a powerful thinker in Germany challenged their timidity head-on. Otto H. Schindewolf (1896–1971; Fig. 10) had a long career by any standards, publishing his first paper in 1916, aged twenty, and continuing to publish his ideas on palaeontology and stratigraphy until 1970, a span of 54 years of professional activity. After the Second World War, he became professor of palaeontology at the University of Tübingen, and he dominated his staff and students in the classic German way: what he said was law, and could not be debated or discussed. And yet, what he was saying then was completely at odds with what was being said in the English-speaking world.

While the leading American and British scientists were establishing the modern synthesis, linking Darwin's classic writings with the new discoveries in genetics, ecology and palaeontology, Schindewolf remained all his life an anti-Darwinian. He had been influenced as a young man by older German ideas that species, and larger groups, such as the dinosaurs, the ammonites, the trilobites, passed through a 'lifespan' akin to the lifespan of an individual. The group began with an initial phase of explosive evolution (= youth), followed by a period of stability (= middle age)

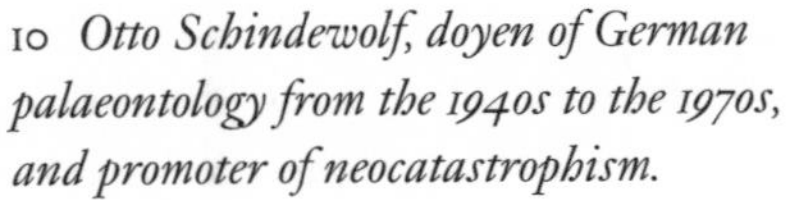

10 *Otto Schindewolf, doyen of German palaeontology from the 1940s to the 1970s, and promoter of neocatastrophism.*

and ended with degeneration and extinction (= old age and death). This concept received the grand title of typostrophism.

Whatever it was called, typostrophism was simply a version of the early twentieth-century idea of racial senility that had been roundly rejected by the neo-Darwinians. Such a pre-programmed history could have no place in the world of Darwinian evolution, since there is no genetic mechanism to record and promote a ready-made history for a group.

There have been many different views of Schindewolf's role. It has been suggested that he virtually single-handedly held back the development of modern evolutionary theory in German palaeobiology, and even after his death it was considered sacrilegious to criticize his ideas. However, despite the serious anachronism of typostrophism, Schindewolf was also something of a lone voice speaking up for mass extinctions at a time when there was a deathly hush on the subject in the English-speaking world.

In his most influential text, *Grundfragen der Paläontologie*, published in 1950 and essentially the Bible for German palaeontologists for decades, Schindewolf gave his view of the end-Permian events:[21]

> The Permian system constitutes the end of the Palaeozoic, the age of ancient animals, and in fact, there is at that point a break of major importance in faunal evolution. In the Permian we find the last of the *trilobites*, which were so thoroughly characteristic of the Palaeozoic. Large groups of *hydrozoans*, *brachiopods*, *crinoids* and *bryozoans* of the old stamp die out. . . . Several of these ancient groups [of amphibians and reptiles] died out in the Permian and were replaced in the Triassic system by numerous new forms. In short, we encounter almost everywhere a radical contrast between the old and the new.

In his book, Schindewolf did not discuss mass extinctions in any further detail – but that was soon to come.

In the 1950s, Schindewolf developed his idea that extinctions, as well as the other phases of the typostrophic cycle, were not affected by physical processes such as climatic change, sea-level change, vulcanism or the like. Instead, extinction was part of the evolutionary cycle, and its causes were innate to the organisms. But, to explain *mass* extinctions, when many different typostrophic cycles seemed to come to an end simultaneously, Schindewolf argued that cosmic radiation following the explosion of supernovas was the cause. The sudden bursts of cosmic radiation, he argued, led to an increase in mutation rate within many groups of organisms. This forced them into the declining phase of the typostrophic cycle, when the runaway mutations caused overspecializations, deleterious organs and finally extinction.

To supplement the development of his ideas about mass extinctions Schindewolf embarked on a series of studies of the Permo-Triassic boundary. This programme of work was, in retrospect, highly innovative. He and his students tracked down high-quality rock sections around the world that traversed the boundary. He concentrated in particular on the Salt Range successions in Pakistan, and documented in detail how the faunas changed. Schindewolf began to pull together the evidence from different sections, to produce global-scale documentation of the magnitude of the end-Permian mass extinction. But he did not convince many geologists outside Germany.

Neokatastrophismus?

In a provocatively titled publication in 1963 – 'Neokatastrophismus?' – Schindewolf claimed a place for catastrophism in geology. In this paper, as in others at the time, Schindewolf argued with his critics, some of whom were rightly sceptical about his theory of cosmic rays, and others of whom even denied that there had been an extinction at the end of the Permian.

This denial of the end-Permian event was particularly prevalent among vertebrate palaeontologists. Two distinguished experts on fossil amphibians and reptiles, the American Charles L. Camp and the senior British palaeontologist D. M. S. Watson, both argued in the 1950s that the apparent changeover among dominant vertebrates at the Permo-Triassic boundary was probably more to do with the imperfections of the fossil record than anything else. Camp suggested that the Permian amphibians and reptiles that apparently disappeared might just have lived in habitats that were not preserved in the Triassic. Watson, similarly, suggested that, for vertebrates at least, the apparent end-Permian mass extinction might be little more than a gap in the record.

Schindewolf was incensed by what he regarded as such woolly thinking.[22] He argued strongly that there were no gaps – the rock successions in the Salt Range, and in South Africa and Russia, where the amphibians and reptiles have been most discussed, were continuous, with no obvious breaks. There truly had been a mass dying.

In the introduction to his paper, Schindewolf noted, perhaps a little coyly, that he had been termed 'the most important and most consequential spokesman for the idea of neocatastrophism in current palaeontology'. Indeed, his espousal of mass extinction, and of cosmic rays from exploding supernovas, was reminiscent of the dreaded catastrophism that Lyell had so successfully disposed of in the 1830s (although only in England – not in Germany). Schindewolf compared his standpoint with that of Cuvier, and he defended the term neocatastrophism, as opposed to simply catastrophism, in that Cuvier and his supporters had operated at an early stage in the development of geology as a science, and that their views had of course taken no account of evolution.

But Schindewolf went further. Yes, Darwin had written about the extinction of species and genera, and such extinctions were clearly a normal part of Darwinian evolution. As new species arise, older ones die out for a variety of reasons. But, Schindewolf noted, over 100 years had passed since Darwin wrote his classic work in 1859, and geology and biology had moved on – we should no longer feel bound to accept what Darwin had said on these subjects.

This kind of anti-Darwinian remark did not promote the acceptance of Schindewolf's ideas by palaeontologists in the English-speaking tradition, nor by German biologists. After all, the discovery of the structure of DNA in 1953, and the rapidly developing science of molecular biology, were confirming Darwinism and the modern synthesis time and time again. Schindewolf made himself something of an outcast, except within his own milieu. In doing so, he seemed to confirm to others that catastrophism, whether it was restyled as neocatastrophism or not, was still wild and dangerous speculation. Lyell had surely been correct. But, in insisting on the reality, and the great magnitude, of the end-Permian event, it turns out that Schindewolf was right, and his critics were wrong.

The end of catastrophism . . . or not?

Charles Lyell's rejection of catastrophism in the 1830s, and Darwin's espousal of gradualism in evolution in 1859, seemed self-evident by 1900. True, catastrophism reared its ugly head from time to time, whether in discussions of the extinction of the dinosaurs, or in Otto Schindewolf's writings, but it was easy to ridicule such ideas. And ridiculed they were through much of the twentieth century. I remember discussing a paper written in 1956 by the American palaeontologist, M. W. de Laubenfels, in which he suggested that the dinosaurs had been wiped out by a giant meteorite. One of my senior colleagues, a palaeontologist, said he had written to de Laubenfels at the time to ask him why the turtles and crocodiles had not also been wiped out by the great atmospheric disturbances. Had the meteorites, he asked, simply bounced off the turtles' backs? My colleague received no answer to his enquiry.

Nationality no doubt played a part. For a time, most of the neo-catastrophists were, like Schindewolf, German. The American and British commentators simply chose to ignore their work: many indeed could not read German, and the majority were probably largely unaware of the German literature since it appeared often in regional German specialist journals that were not taken by many libraries outside Germany. It was easier, too, to think of the catastrophists as crazy if they remained anonymous, and in a different country. But everything changed on 6 June 1980. De Laubenfels was dead by then, but he had been right. The Earth had been hit by a giant asteroid 66 million years ago, and that impact did kill off the dinosaurs. This marked the beginning of a new era of serious research into mass extinctions.

5
IMPACT!

The modern era of mass extinction studies began in 1980, with the publication of the proposal that the dinosaurs had been wiped out by the impact of a huge meteorite, or asteroid, on the Earth. The paper finally re-established catastrophism at the core of geology. This was one of the most daring papers ever published, it was wide open to refutation, it had immensely high heuristic value, it raised a storm of protest, and it was one of the most influential publications in earth sciences in the twentieth century.

The paper[1] was titled 'Extraterrestrial cause for the Cretaceous-Tertiary extinction', and it appeared in the leading American weekly journal *Science*, in the issue dated 6 June 1980. The first author of the paper was Luis W. Alvarez, who had won the Nobel prize for physics in 1968 for his work in identifying subatomic particles. The other authors were his son Walter, a professor of geology, and their colleagues Frank Asaro and Helen V. Michel, all at the University of California at Berkeley.

The paper was daring because the authors had very little evidence to back up their very large claims. This in turn meant that it was wide open to refutation: science does not work by *proving* cases – that is for lawyers. Scientific theories are the best explanations for a series of observations, but a counter-observation might *disprove* the theory at any time. So, Luis Alvarez and his colleagues had really stuck their necks out, making large claims and predictions that could so easily have been demolished. The paper was heuristic, meaning that it opened up a whole array of new problems and predictions, and essentially launched a new branch of earth science studies, the true 'neocatastrophism', as opposed to Otto Schindewolf's earlier brand of the 1960s. The word heuristic is derived from the Greek *heuriskein*, 'to find', as in *eureka*, 'I have found it'.

The proposal that an impact had killed the dinosaurs offended many, palaeontologists in particular, since it came from a physicist, and an explosion of controversy followed in the 1980s.[2] The paper was hugely influential since it presented a hypothesis that could have been refuted readily – but it was not, and indeed new evidence arrived all the time that bolstered it.

Turning the hypothesis on its head

The whole hypothesis presented by the Berkeley team turned on a reversal of their starting idea. Walter Alvarez and his colleagues were seeking an independent way to calculate rates of sedimentation in ancient rock sequences. Geologists can easily measure the thicknesses of beds of sandstone, mudstone or limestone, but the thickness is not really proportional to time: a thin band of mudstone might represent hundreds of years of slow deposition of fine particles in the deep ocean, while a massive bed of sandstone, 100 metres thick, might have been dumped from a catastrophic slumping event in a matter of minutes or hours. How could a chronometer be devised that told time accurately?

Luis and Walter Alvarez reasoned that iridium might offer a solution. Iridium is a rare platinum-group metal that occurs in only minute quantities on the Earth's surface – in fact, in parts per billion. When the Earth formed, iridium was present, but it then segregated into the core. Iridium is rare now on the surface of the Earth since it arrives essentially from extraterrestrial sources, in the fine rain of small meteorites, tektites and cosmic dust that settles over the surface in a slow rain. If the rate of arrival of iridium was known – say one microgram per square kilometre per hundred years – then the amounts in different thicknesses of sediment could be measured and the length of time each bed of limestone, mudstone or whatever had taken to be deposited could be calculated.

The problem, or one of many problems, was how to measure the quantities of such a rare element. Up till then there were no analytical machines available that could even begin to detect such tiny quantities. It was here that Luis Alvarez's knowledge of experimental physics came into play. He and his colleagues were able to build a neutron activation machine that could achieve the necessary levels of precision.

The geologists decided to test their new chronometer on some rock sections that Walter Alvarez had been studying near Gubbio in north Italy.

Here, in the valleys to the east of the medieval walled town, were great sequences, over 400 metres thick, of thinly bedded mudstones and limestones that happened to span the Cretaceous-Tertiary boundary. They seemed to represent millions of years of slow accumulation of sediment in a moderately deep tropical sea. The geologists sampled bed-by-bed, and brought the rock chippings back to California.

After painstaking treatment in the laboratory in Berkeley, the team plotted the values of iridium in the sediment samples. They expected to find variations through time – low values in rapidly deposited beds, and higher values in slowly deposited units – which would allow them to calculate the relationships between time and thickness for the first time. Normal concentrations of iridium were very low, averaging 0.3 parts per billion. But, at the KPg boundary, they found a surprising result. Through a thickness of 10 millimetres, the iridium values shot up to 9 parts per billion, a vast increase of 30 times the 'normal' level (Fig. 11). How should this be interpreted?

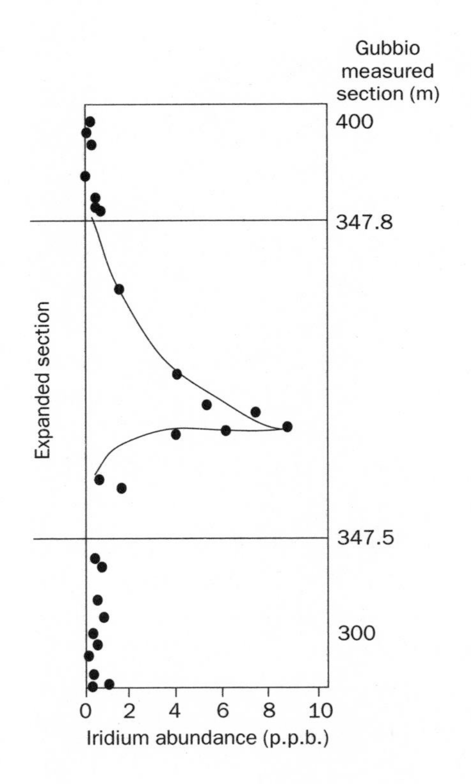

11 *The iridium spike, as recorded by Alvarez and colleagues in 1980 in the Gubbio section, northern Italy. In the Cretaceous-Tertiary boundary interval, a thin clay band contains much higher values of iridium than the units above and below (but note the variable vertical scale, in which the KPg boundary unit is much exaggerated). The discovery of this iridium enhancement led to the proposal of a massive KPg impact.*

Following the protocol of the research programme, Alvarez and his team should have reasoned that a 30-fold increase in iridium concentrations meant that the thin clay band at the KPg boundary had simply taken 30 times as long to be deposited as the clays and limestones above and below. Such a conclusion would have been within the bounds of expectation, since it is clear that sedimentation rates can vary enormously in the sea. However, they turned the theory on its head, and made their daring prediction.

The iridium spike

Iridium levels that shot up to 30 times their normal level could mean that the rock unit had taken 30 times as long as normal to be deposited. Or, Alvarez thought, perhaps it meant that the arrival of extraterrestrial material had suddenly increased. In other words, the Earth had been hit by an asteroid – a huge meteorite.

The team cross-checked their results at another KPg section, the Stevns Klint cliff section in Denmark. If they found a similar iridium enhancement in the boundary clay they were vindicated. If not, then the iridium spike at Gubbio was merely a local phenomenon, and probably indeed caused by unusually low sedimentation rates at that location. The Danish values came in: background iridium abundances of 0.26 parts per billion, pretty much the same as at Gubbio, and then a spike in the boundary clay of 42 parts per billion. So, at Stevns Klint the spike represented an enhancement of 160 times background levels, even more dramatic than at Gubbio. That was enough – the team rushed their results into print.

They calculated the size of the impacting object by various means, based on the predicted volume of material dumped on the Earth, the size of the explosion required to send sufficient material into the atmosphere so that it encircled the Earth, and the known relationship between impacting objects and the energy they transmit. The theory was based on observations of the effects of huge volcanoes, such as Krakatoa, a volcano located in the Sunda Strait between Java and Sumatra, which had erupted in 1883.

The Krakatoa eruption shot an estimated 18 cubic kilometres of molten rocks and ash into the air, of which about 4 cubic kilometres reached the stratosphere, the higher parts of the atmosphere, and remained there for two years or more. This fine dust encircled the globe and the effects of the eruption could be detected as far away as Europe. Writers and artists in Europe recorded unusually brilliant sunsets for weeks after the eruption. So, reasoned the Alvarez team, a major impact would produce just the same effects, only more so.

The equation

So far, we have managed to avoid any mathematics. Now is the time to introduce the only equation in the book. The equation is so daring, and really so simple, that it is a good idea to work through it. We are talking here of an enormously bold prediction from really quite minimal evidence.

The Alvarez team took their two thin boundary clays, from Italy and Denmark, and proposed to use them to support the notion that a giant rock, 10 kilometres in diameter, had hit the Earth 66 million years ago, punching a vast hole in the atmosphere, slamming into the Earth's crust, instantly vaporizing, excavating a huge crater 100 to 150 kilometres across, and flinging millions of tonnes of rocks and dust into the atmosphere. The dust encircled the globe, blacking out the sun for a year or more, thus preventing normal photosynthesis in the plants, and hence cutting the base from food chains in the sea and on land. Mass extinction followed.

The calculation was based on the occurrence of the 1-centimetre thick boundary layer in Italy and Denmark. The Alvarez team argued that this clay was no ordinary marine clay, derived from normal sources. It was the ash or dust of the impact, material that had been lofted into the stratosphere and deposited over a few years, carrying with it the iridium that marked its extraterrestrial source. Their formula was:

$$M = \frac{sA}{0.22f}$$

where M is the mass of the asteroid, s is the surface density of iridium just after the time of the impact, A is the surface area of the Earth, f is the fractional abundance of iridium in meteorites, and 0.22 is the proportion of material from Krakatoa that entered the stratosphere (in other words 4 divided by 18 cubic kilometres). The surface density of iridium at the KPg boundary was estimated as 8×10^{-9} grams per square centimetre, based on the local values at Gubbio and Stevns Klint. Measurements of modern meteorites gave a value for f of 0.5×10^{-6}.

Running all these values in the formula gave an asteroid weighing 34 billion tonnes. The diameter of the asteroid was at least 7 kilometres. Other calculations led to similar results, and the Alvarez team fixed on the suggestion that the impacting asteroid had been 10 kilometres in diameter. This dimension led to further simple calculations – but there is no need here for any more equations.

The relationship between impacting objects and crater size is known. Observations of recent craters, and experiments with massive cannon that fire bullets and large steel balls into clay boards show that the crater is always 10 to 40 times the diameter of the impacting object, 10 times for large impacts, 40 times for smaller. So, a 10-kilometre asteroid excavates a crater 100 to 150 kilometres across. Equally, the speeds of asteroids and the energy they transmit are also known. The KPg asteroid probably entered the atmosphere at a speed of 25 kilometres per second, and its energy was equivalent to 100 million megatonnes of TNT, perhaps 30 times the explosive power of all nuclear warheads currently in arsenals around the world today.

How did the asteroid kill all the dinosaurs? Obviously it would have killed all life in the immediate vicinity of the impact, and probably for 1000 kilometres all around, because of the blast and burning that followed the impact. The Alvarez group, following earlier studies of the wider effects of volcanic eruptions, reasoned that a large enough impact would fill the whole upper atmosphere with dust, so blacking out the sun for more than a year. Absence of sunlight would prevent photosynthesis in green plants both on land and in the sea (the microscopic phytoplankton). If plants were unable to grow, plant-eating animals would die off and, in turn, so too would the flesh-eaters. Absence of sunlight would also lead to freezing. Either way, much of life on land (Fig. 12) and in the sea would die out.

The scale of the postulated event was huge, matched by the audacity of the Alvarez group. Here, using the simplest of equations, they had reconstructed one of the most dramatic and devastating events in the history of the Earth. And based on what? A couple of clay layers in Italy and Denmark. No wonder, in 1980, their paper was met with intense outrage. Now, 20 years later, we know that they were largely right. Catastrophism is re-established at the core of earth sciences. Cuvier and Buckland must be smiling in their graves, Murchison would probably be greatly relieved, while Lyell, his friend, would no doubt be squirming and seeking to find how he could accommodate himself to the new evidence.

12 *The impact and its consequences?* Triceratops *and* Tyrannosaurus rex, *two of the last dinosaurs, face-to-face with their fate.*

Dangerous catastrophists

Recall that Otto Schindewolf, the 'neokatastrophist' of the 1950s and 1960s, had been shunned except by people associated with his power base in Germany. Other catastrophists had met the same fate. For example, Ken Hsü, professor of geology at the Swiss Federal Institute of Geology and a convinced catastrophist, records[3] how, when he arrived in the United States in 1948 to begin graduate studies, he was warned away from such dangerous doctrines. His supervisor, Edmund Spieker, told him to steer clear of the work of T. C. Chamberlin, who, at the beginning of the twentieth century, had proposed that sudden events – catastrophes – provided ideal marker horizons over wide areas that could be used in correlating rocks. 'He has done more harm to geology than any person' with such ideas, declared Spieker.

At the same time, in the United States, geologists were debating the origin of Meteor Crater in Arizona. To us it looks like a crater, with its deep scooped centre, its raised rim and its circular shape. Amazingly, influential geologists constantly opposed such a view. G. K. Gilbert, for example, the Chief Geologist of the United States Geological Survey around 1900, consistently argued that this crater, and others, had been produced by volcanic processes. Such a suggestion seems to go against all common sense, but such was the fear of catastrophism that most people denied that the Earth had ever been hit by a large meteorite.

The evidence for impacts had been so obvious that it could be denied only by a supreme effort of will. Meteor Crater looks so like an impact crater that many geologists really had to twist and turn before they could accept G. K. Gilbert's pronouncements. Opinions shifted relatively rapidly in the 1960s. Information was leaking out about the great Tunguska impact in 1908, when a large meteorite had exploded over a remote part of Siberia, flattening trees for miles around. But the real turning point came with Gene Shoemaker's studies of the Ries crater in Germany.

The Ries crater

Here was a structure in southern Germany, surrounding the medieval market city of Nördlingen, that looked very like a crater. Nördlingen stands off-centre in a vast circular structure, measuring 22–23 kilometres across (Fig. 13). The bowl of the crater is shallow and filled with lake

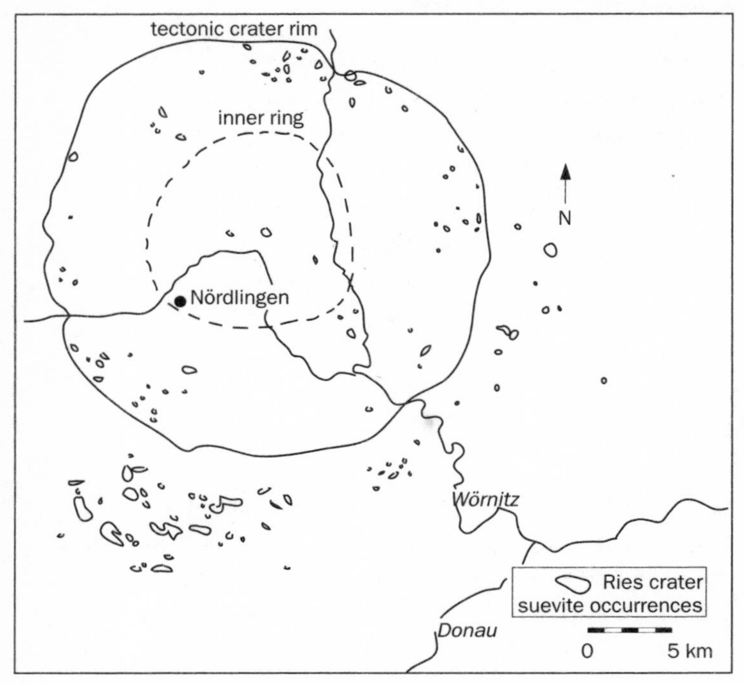

13 *Map of the Ries crater in west Bavaria, south Germany. The inner and outer rings of the crater are clearly visible, and the ejecta (suevite), material thrown up by the back-blast following the impact, can be mapped out (irregular zones inside and outside the crater). Of course, much of the ejecta has been eroded away or covered up by later sediments.*

sediments that have turned into rich soils for farming. As you drive away from Nördlingen in any direction, you come to a sharp incline. The road zig-zags its way up and then down on to the surrounding plain which lies several tens of metres higher than the lands round Nördlingen. This is the crater rim. Giant rocks, some as large as a house, stand at crazy angles both inside and outside this zone.

Again, as with Meteor Crater in Arizona, German geologists had tussled with many explanations for the Ries structure. Most argued for a volcanic origin – that the structure had resulted from some kind of volcano that had erupted and then collapsed back on itself, leaving no more than a shallow bowl-like structure. The volcanic cone had sunk back into the Earth, or perhaps the volcano had been explosive and had never built a cone. Interestingly, all traces of volcanic ash and lava had also somehow disappeared. In 1911, Artillery Major W. Kranz carried out experiments with gunpowder buried in mud. He believed he could show that an underground explosion could have produced the Ries structure. Some doubters suggested the possibility of an impact, but their voices were ignored.

In 1960, Gene Shoemaker (1928–1997), the eminent astrogeologist who worked for the United States Geological Survey, visited Nördlingen. He later became a household name with the discovery with his wife of the asteroid Shoemaker-Levy that crashed into Jupiter in 1994. He had read all he could about the Ries structure, and was convinced that it was a crater. It was ideal for his studies since it was not particularly old – the sediments that filled the structure contained abundant fossils of molluscs, fishes and other animals and plants that gave a Miocene age of some 14.7 million years ago. Shoemaker arrived at the suevite quarry at Otting, on the eastern crater rim, and set to work. He selected the suevite quarry as most likely to convince the doubters. Suevite is a melt breccia made up from a jumble of angular shards of sedimentary rocks – limestones, mudstones and sandstones – set in a glassy matrix, which formed at temperatures as high as 600°C. The term suevite came from the Latin *Suevia*, the ancient Roman name for the region of Germany called Swabia in English and Schwaben in German.

Shoemaker, and his colleague E. C. T. Chao from Washington, had already explored the mineralogy of rocks from the bottom of Meteor Crater in 1959. There they had discovered an unusual form of quartz called coesite, which formed only under very high pressures of up to 300 kilobars. Quartz is the commonest mineral in rocks on Earth. Only the high energy of a meteorite hitting the Earth's crust was enough to modify normal quartz to coesite, so coesite was a clear marker of impact.

Shoemaker arrived at the Otting suevite quarry on 27 July 1960 late in the afternoon and, as he records:[4]

> Quickly I took three suevite samples as the darkness fell, and then we camped in a wood nearby. On the next day we drove to Nördlingen, and I sent the samples to Chao in Washington. Within a few days they reached him and quickly he had found coesite by X-ray tests.

Shoemaker and Chao wrote up their results and published their paper, entitled 'New evidence for the impact origin of the Ries Basin, Bavaria, Germany', in the *Journal of Geophysical Research* in 1961. Their evidence was so conclusive that there was no serious opposition. Geologists had been living a lie for so long – forbidding themselves to admit the obvious – that it was impossible to deny. If it looks like an impact crater and smells like an impact crater, it is an impact crater.

Megablocks, suevite and melt bombs

The paper by Shoemaker and Chao was just the beginning. Once geologists had accepted the obvious, they flocked to the Ries crater to find out everything they could about how large impacts work. Unlike older craters, many of the consequences of the impact could still be traced in the geology of the area around Nördlingen and could thus be identified.

The geometry of the crater and of the fallout rocks reveal what happened. The true crater edge is clearly marked, and is 22–23 kilometres in diameter (see Fig. 13). But there is a smaller inner ring, some 11–12 kilometres across. This resulted from bounce-back of the crater floor immediately after the meteorite had punched deep into the crust. The ejecta – the rocks thrown out after the impact – extend up to 70 kilometres from the middle of the crater. As would be expected, the ejecta layer becomes thinner the further one goes from the crater, and the mean size of the rocks also diminishes – huge house-sized boulders in and around the crater, and smaller gravel-sized material 40 kilometres away. Some finer-grained mineral materials called moldavite have been identified 600 kilometres to the east in the Czech Republic.

Six unique rock types in and around the crater tell the story of the events. First are the *megablocks*, huge irregular boulders, up to 25 metres across, composed of the underlying sediments, which lie at all angles just inside the crater rim. These were clearly torn out of the centre of the crater, thrown into the air and then fell back into the crater.

Second, the *bunte breccia* ('coloured breccia') are multicoloured blocks composed of fragments of the rocks from beneath the crater – red and yellow Triassic sandstones and mudstones, grey and black Jurassic mudstones and limestones – in irregular hand-sized chunks set randomly in fine-grained rock dust. The bunte breccia is found around the crater edge and in the surrounding area, and it was clearly made from material that was gouged out by the impact, thrown up as loose blocks, and then fused into an irregular mixed breccia as it landed.

Third, the *crystalline breccias* consist of fragments of deeper rocks, granites and other igneous rocks that lie some 400 metres below the surface. These were formed as the meteorite drilled deep below the sedimentary rocks and threw the basement crystalline rocks up into the area. Because of their original depth, and the force of their removal from such low levels, these rocks show evidence of metamorphism, that is, melting and distortion produced by high temperatures and pressures. The crystalline breccias fell back into the crater, and some are found outside it.

Fourth, the *suevite*, the rock type named first from the Ries crater, is a breccia composed largely of basement rocks, granites and gneisses, with rare pieces of sedimentary rock. The rocks were heated and up to 70% of any sample has been turned into glass. The separate fragments have been fused, but there are many holes, indicating that the suevite contained a great deal of gas during its formation. The suevite is found in great abundance, both inside and outside the crater; it had evidently been thrown into the air during the impact and had then fallen back. The good citizens of Nördlingen and surrounding towns had used the suevite as a major building stone since the Middle Ages – its warmth and easy workability made it ideal – though little did they know of the origin of the unusual stone!

Fifth, the *impact melt* materials are natural glasses. These were formed from the local rock that had been melted by the impact and had then fused as a uniform glass. The molten material is frequently shaped into 'bombs', disc-shaped glass bodies that had flown through the air, solidifying as they cooled. The external shape of the bombs – something like a frisbee – shows how they solidified in mid-flight, and flow patterns inside the bombs confirm how the molten glass flowed in the seconds before it solidified.

The sixth, and final, impact rock consists of the *distant ejecta*, material that was flung out after the impact and travelled some distance from the crater. Some blocks of the underlying sediment have been located as much as 70 kilometres away from the impact site, south of the Danube. More controversial are bentonite clays which are found close to the sediment blocks: these may be the product of fine-grained ash-like material that was blasted into the atmosphere and which may have blanketed a wide area around the crater. Finally, the moldavite tektites, small meteorite-like melt fragments, of the Czech Republic, mentioned earlier, may also have been thrown several hundred kilometres by the force of the blast.

Geologists have now spent thousands of hours anatomizing the Ries crater and its ejecta. Their maps, boreholes, measurements and geochemical studies of the extraordinary rocks in and around the crater have produced one of the most detailed stories of an event that took no more than a few seconds. What happens during an impact?

Anatomy of an impact

The Ries asteroid struck 14.7 million years ago, and was 500 to 700 metres across – more than half a kilometre. It punched through the outer atmosphere and hit the Earth's crust at a speed of 20 to 60 kilometres per second. The energy released was 100 megatonnes, equivalent to the explosive power of 250,000 Hiroshima bombs. One hundred and fifty cubic kilometres of rock were thrown out of the crater and the resultant blast killed every living thing for 500 kilometres around. Dust clouds travelled round the globe, and would certainly have caused startling sunsets worldwide, as the eruption of Krakatoa did in 1883. The effect of the Ries impact was considerably greater, however, and the dust may have been sufficient to black out the sun for a few days at least.

The story of the Ries impact can be dissected into six phases. The asteroid hurtles down from the sky, passing through the atmosphere in one to two seconds (phase 1). It penetrates through the Earth's crust to a depth of about 1 kilometre, cutting through 700 metres of Triassic and Jurassic sediments and 300 metres of underlying crystalline basement rocks beneath (phase 2). At this early phase, small particles of molten material, the moldavite tektites, are thrown out, and a huge shock wave is generated downwards and sideways.

At the instant of impact, the pressure is about 5 megabar, equivalent to 5 million times normal atmospheric pressure, and the temperature rises to 20,000°C. In phase 3, the meteorite and the surrounding rocks to a depth of 1 kilometre are compressed to less than a quarter of their original volume in under one-fifth of a second, and they vaporize explosively. The shock wave front is generated in all directions, travelling at a speed of 20 to 30 kilometres per second, but it fizzles out after a few kilometres. Below the deepest point of the crater, different zones of elevated pressure and temperature can be identified, the pressure and temperature diminishing away from the crater.

Then the ejection phase (4) begins, just two seconds after impact. As is often said, to every force there is an equal and opposite reaction. The crater floor bounces back, flinging up huge amounts of rocks, melt products, ash and gases from the vaporized meteorite and surrounding rocks. The mass of rocks and gas is shot upwards and outwards, and the ejecta front forms a conical shape. The circular cone of hot ejecta races sideways at the speed of an express train, stripping the soil and rocks as it goes, and killing everything. At first, the largest megablocks and bunte breccia rocks fall back in and around the crater, piling up to thicknesses of over 100 metres around the crater rim. The thickness of the ejecta blanket, and the average size of the rocks in it, decrease away from the crater, since the energy of the ejecta front also diminishes as it races outwards.

The large, essentially unmelted, blocks of sediment are followed rapidly by the suevite and melt products, including glass bombs, which are also dumped in and around the crater for a distance of up to 100 kilometres. Finer-grained ash travels further, perhaps up to 500 kilometres. After the ejecta blanket has passed, the landscape is devastated – irregular lumps of rock scattered at random are cloaked in a fine coating of white ash and burning stumps of trees penetrate the ash here and there.

The entire process is over in 10 minutes (phase 5). The meteorite has gone, the double crater rim has formed – the inner rim created by the bounce-back process, some 11 kilometres wide, and the outer rim, formed by sinking around the inner ring, about 23 kilometres across – and the rocks and ash have mainly fallen from the sky. The finest dust probably remains in the stratosphere for much longer, affecting climates for some years after the impact.

Over the past 14.7 million years (phase 6), the crater has been much modified. It filled with a lake in the years following the impact, and thick layers of lake muds and limestones built up. These sediments contain fossils of green algae, reeds, freshwater snails, fishes, and even tortoises, snakes, birds, hedgehogs, hamsters, bats, martens, muntjac deer and other animals that lived around the lake.

The Ries crater is still one of the best documented of all craters. But, after the epochal paper by Shoemaker and Chao in 1961, geologists began to spot craters all over the place. In fact, crater hunting became something of a minor industry, although only for a limited fraternity. At first, it was easy. Crater fanatics identified all the obvious modest-sized and relatively recent craters they could see. They then scanned aerial and satellite photographs for circular structures that were too big, or too old and too eroded, to be seen on the ground. Many more were located in this way, some of them as much as 100 kilometres across. But this new wave of craterology did not make the reception of the Alvarez paper in 1980 any easier. Most geologists were still uncomfortable with catastrophism, and could accept only a very limited number of craters and impacts on the Earth.

Obdurate opposition?

In retrospect, it would be easy to characterize the opponents of the KPg impact model presented by Luis Alvarez and his colleagues in 1980 as luddites or reactionaries. The impact proposal was in line with all the new research on the Ries crater, and on all the other craters that had been identified by 1980. At the time, admittedly, there was no large crater of the correct age that could be pointed to as the smoking gun that killed the dinosaurs. But that wasn't a strong criticism: in the course of 66 million years, even a crater 100 kilometres across could have been covered over by subsequent sedimentation, or it could lie under the oceans. The opponents of impact, however, had another line of argument.

The standard view in 1980 has been termed the gradualist ecological succession model. It was promoted especially by geologists and palaeontologists such as Bill Clemens, Leigh Van Valen, Robert Sloan and others, who had studied the last dinosaurs and the early mammals of the same period.[5] This model proposes that the dinosaurs, and other fossil groups that became extinct, were in decline long before the KPg boundary, perhaps for as much as 5 million years. The concept of ecological succession proposes that the typical Late Cretaceous floras and faunas gave way to new floras and faunas over a long span of time – so slow that an observer would not have seen what was happening. But, after 5 million years, the dinosaur-dominated communities had given way to mammal-dominated communities.

This gradual ecosystem evolution model is largely based on the progressive appearance of a mammal community (the *Protungulatum* Community) of distinctive Tertiary aspect in the last 300,000 years of the Cretaceous in Montana. As the mammals increased in abundance, the dinosaurs apparently declined, until they disappeared altogether. This gradual replacement is explained in terms of diffuse competition between dinosaurs and mammals set against a major change in habitats. The lush subtropical dinosaur habitats were apparently giving way to cooler temperate forests which favoured the mammals.

There is some doubt now about the correctness of the dating of the different communities involved in this particular example from Montana: perhaps some of the mammal fossils occur in much younger river channels that cut down into Late Cretaceous sediments. This uncertainty about dating, however, does not remove all evidence for gradual extinction.

The gradualist model has been extended to cover all aspects of the KPg events. In the seas, planktonic extinctions took 10,000 years, and various groups were already declining well before the boundary. A variety of sea-bottom dwellers and filter-feeders died out, but sea-bottom predators and detritus-feeders were little affected. Extinction patterns of many marine groups show gradual declines throughout the Late Cretaceous. The gradualists explain the long-term patterns of extinctions by pointing to cooling climates and major changes in sea level at the end of the Cretaceous.

Gradualists also argue that the fact that many groups did not go extinct at the KPg boundary is hard to understand in the face of some of the devastating catastrophist scenarios. On land, placental mammals, lizards, snakes, crocodiles, tortoises, frogs and other freshwater organisms show little sign of extinction, and the plant record reveals only modest and gradual changes. So, argue the critics of a simple impact model for extinction, how did all these groups that lived side-by-side with the dinosaurs survive if the Earth had been devastated by global blackout, freezing, catastrophic tidal waves, fires, mass poisoning and other disasters?

Also, and perhaps rather flippantly, the critics asked how the dinosaurs, ammonites and planktonic beasts knew that the meteorite was about to strike? After all, these creatures had all begun to decline tens of thousands, or even millions, of years before the impact.

Ugly disputes: physics versus palaeontology

Was there something more to the furious debate that followed the Alvarez paper in 1980? Perhaps the protagonists were not simply debating scientific evidence. The two main models for dinosaurian extinction are based on rather different kinds of data: essentially palaeontological and stratigraphic for the gradualist models, and mainly geochemical and astrophysical for the catastrophic models. This meant that it was hard for the proponents of one view to assess the evidence that supposedly favours the other view. There is, however, apparently a more fundamental source of potential conflict between certain biologists and physicists, or 'soft' scientists and 'hard' scientists respectively, as they are often termed.[6]

The initial publication by Alvarez and colleagues was greeted sceptically by many palaeontologists and geologists with long-term expertise on aspects of the KPg boundary. No doubt they resented the intrusion into their subject by a group of physicists, and Luis Alvarez's lengthy catalogue[7] of his team's credentials (a physicist, two nuclear chemists and a geologist) may not seem so unusual in view of this resentment: 'suddenly I realized that we combined in one group a wide range of scientific capabilities, and that we could use these to shed some light on what was really one of the greatest mysteries in science – the sudden extinction of the dinosaurs.'

The crux of the dispute was outlined by Robert Jastrow, an astronomer and science writer, in a report in the popular magazine *Science Digest*:[8]

> Professor Alvarez was pulling rank on the palaeontologists. Physicists sometimes do that; they feel they have a monopoly on clear thinking. There is a power in their use of math and the precision of their measurements that transcends the power of the softer sciences.

The very titles of the Alvarez papers could be seen to exemplify this: the 1980 paper is titled 'Extraterrestrial cause for the Cretaceous-Tertiary extinction – Experimental results and theoretical implications', while an overview from 1983 is 'Experimental evidence that an asteroid impact led to the extinction of many species 65 million years ago'. Leigh Van Valen, a gradualist critic commented[9] that 'to call [the Alvarez] evidence "experimental" is misleading propaganda; it refers merely to the fact that some observations were made in the laboratory rather than in the field, not to an active experimental test.'

Luis Alvarez is surprisingly revealing throughout his 1983 paper, and he is dismissive of his critics:[10]

> I think the first two points – that the asteroid hit, and that the impact triggered the extinction of much of the life in the sea – are no longer debatable points. Nearly everybody now believes them. But there are always dissenters. I understand that there is even one famous American geologist who does not yet believe in plate tectonics. . . . People have telephoned with facts and figures to throw the theory into disarray, and written articles with the same intent, but in every case the theory has withstood these challenges.

He later outlines the advantages of physics in comparison with palaeontology:[11] 'The field of data analysis is one in which I have had a lot of experience' and 'In physics, we do not treat seriously theories with such low *a priori* probabilities'. He further writes, 'That is something that made me very proud to be a physicist, because a physicist can react instantaneously when you give him some evidence that destroys a theory that he previously had believed. . . . But that is not true in all branches of science, as I am finding out.' Public utterances from Luis Alvarez about his 'opponents' were frequently more critical than these examples, to the point of being libellous, as reported at the time by the journalist Malcolm Browne in the *New York Times*.[12]

On the other hand, much of the distrust of the physicists by certain palaeontologists has surely been unfounded, as David Raup, a palaeontologist and a catastrophist, notes. He quotes at length a statement by Robert Bakker, a dinosaur palaeontologist, first published in the *New York Times*:[13]

The arrogance of those people is simply unbelievable. They know next to nothing about how real animals evolve, live and become extinct. But despite their ignorance, the geochemists feel that all you have to do is crank up some fancy machine and you've revolutionized science. The real reasons for the dinosaur extinctions have to do with temperature and sea level changes, the spread of diseases by migration and other complex events. But the catastrophe people don't seem to think such things matter. In effect, they're saying this: 'We high-tech people have all the answers, and you paleontologists are just primitive rock hounds'.

Here is a mixture of righteous indignation and real concern. Part of the dispute revolved around different ways of working, plus there was a real expression of the supposed pecking order in science, in which physics is seen as better, more reliable, than palaeontology. But much of the clash between the palaeontologists and the Alvarez camp in the early 1980s could clearly be traced back to Lyell, and the long-standing distrust of catastrophism of any stripe.

Styles of argumentation

Elisabeth Clemens has analysed the nature of the debate about 'asteroids and dinosaurs', and she argues[14] that there are many non-scientific undercurrents, such as styles of argumentation and the role of professional and popular publication. She wanted to investigate just why the impact theory gained such rapid acceptance by diverse groups of scientists and by the public, but was initially rejected by most geologists and palaeontologists who were actually close to the question. Why the mismatch?

It is important to note first that the broadly based research enterprise that has developed around the question of the extinction of the dinosaurs – geologists, palaeontologists, chemists, physicists, astronomers – is not a single community of scholars, all of whom have had the same training. It is a body consisting of several factions, each going in different directions, and with very little communication between them. Clemens suggests that the Alvarez theory gained rapid notice and acceptance in many quarters because catastrophism in geology was becoming intellectually fashionable. We have seen how Shoemaker's work opened the door in 1961 for the acceptance of the possibility of meteorite impacts on the Earth, and how he, and a growing army of geologists, had developed a sophisticated new branch of impact geology by 1980. But there was more.

There had already been precursors of Alvarez. As we saw earlier, de Laubenfels had seriously suggested in 1956 that an impact had wiped out the dinosaurs. He was a respected palaeontologist, but still the idea was fairly comprehensively ignored. However, a flurry of extraterrestrial models had been published in the 1970s and an influential current of opinion[15] had advanced the case for a supernova 65 million years ago. The argument was that an exploding star had blasted the Earth with cosmic radiation. Although this idea did not gain many adherents, it rested on much of the same evidence used by the Alvarez group (but not the iridium spike). Perhaps the supernova theory of the 1970s paved the way for Alvarez and impact.

Clemens argues that it was the mode of presentation of the Alvarez hypothesis that won it such wide attention and acceptance: 'In a sense, the problem of the K-T boundary was framed so as to be amenable to the methods of particle physics'. The bulk of the long 1980 paper (14 pages in all) is confined to the geological and physical evidence for an impact, and the physical results of the impact. The discussion of its biological results occupies only half a page. The paper is restricted then to a rather simple astrophysical hypothesis which could be tested in many ways, while the more complex aspects of stratigraphic imprecision and complexity of the evolution of biological communities are largely omitted. These issues had to be taken on board later, however.

In subsequent publications in 1984, the Alvarez team[16] allowed from 10,000 to 100,000 years for the overall length of time involved in the extinctions, and they noted that 'the paleontological record thus bears witness to terminal-Cretaceous extinctions on two time scales: a slow decline unrelated to the impact and a sharp truncation synchronous with and probably caused by the impact.' However, by 1984, the simplicity of the 'instant-extinction' model of 1980 had ensured its general acceptance by many scientists. The later modifications are rather *ad hoc* qualifiers that tend to protect the impact theory from refutation by stratigraphic or palaeontological evidence.

The style of the arguments on both sides of the debate in the professional scientific literature may have been important. But the role of the media, responding to intense public interest in the debate, may also have fed back into the science.

The role of the professional and popular press

According to Elisabeth Clemens, the nature of the professional and popular press has largely shaped the development of models of dinosaurian extinction since 1980. She points out that the 1980 *Science* article was twice as long as such articles usually are, and was published in a prominent position, at the start of the issue. Each week, when the leading science journals *Science* and *Nature* appear, the publishers choose one or two articles for press coverage, and those articles are then reported worldwide, eclipsing the other articles, some 20 or 30 each week, which then do not attract such press attention. This is no reflection on the quality of the other papers, just a fact of life.

The intense press coverage inevitably has a feedback effect, creating a flurry of excitement around one or two publications. This in turn inevitably affects scientists – who are merely human, and read the blogs and newspapers like any other citizen. This particular article by the Alvarez team gained a very wide readership, especially in the United States, while other articles that presented similar theories at the same time[17] were much less widely read.

Following the success of the Alvarez paper, it has been alleged that pro-impact papers were much favoured by the editorial board of *Science*, and the argument spilled over into the commentary and review sections of leading journals and into the newspapers.[18] Clemens suggests that the very format of publication has had a restrictive effect, since most of the debate has been carried on so far in the pages of *Science* and *Nature*, both of which normally publish only very short papers of four or five pages in length, and both of which require papers to be readily understandable to a wide audience. It is easier to present a simple, clear view, such as the impact, she argues, than to debate the imprecision of methods for dating rocks, or the complexity of biological communities.

Iridium, shocked quartz, glass beads and the fern spike

After 1980, and the immense commotion surrounding the Alvarez paper, geologists and palaeontologists pursued many lines of investigation into the KPg event. It became ever clearer that, although there may have been sociological, methodological and presentational criticisms of the strategy and style of the Alvarez team, they had been essentially right. It would have been so easy for the impact theory to collapse at any point, but it did not. In fact, new evidence supported it. Most impressively, kinds of evidence that had not been predicted by Luis Alvarez and his colleagues lent independent corroboration. What had been one of the most daring hypotheses in the earth sciences, founded on extremely slender evidence, was vindicated.

First, during the 1980s, geologists sampled KPg boundaries all over the world – they found the clay layer nearly everywhere they looked, and it was enriched in iridium. The mere fact of the occurrence of the clay is telling; the iridium enrichment is even more impressive. The clay and the iridium were found in all environmental settings, from rocks that had been deposited deep in the oceans, in shallow seas, on land, in lakes and rivers – everywhere. And it was a worldwide phenomenon – from Canada to New Zealand, from Russia to the South Atlantic. The clay layer proved that some agency had sent a plume of dust into the stratosphere, and that the dust or ash had fallen out uniformly, cloaking the whole Earth in a white blanket at least a few millimetres thick. In addition, that ash had brought with it iridium, a marker of impact. This was what the Alvarez team had anticipated. But some discoveries were unexpected.

In many KPg sections, as a result of intensely careful search, geologists found shocked quartz and glass beads. As we have seen, quartz is the most common mineral on Earth, and it normally occurs as largish grains, sometimes with irregularities and inclusions of other minerals, but never with any regular internal structure. At the KPg boundary, in the clays, geologists found quartz grains bearing crisscrossing lines, indicating that high pressures had been applied. Some grains had four or five sets of regular parallel lines running across: the more sets of lines, the higher the pressure. Shocked quartz can be produced during high-pressure volcanic eruptions, but volcanic shocked quartz usually has only a few sets of lines. So the KPg shocked quartz was independent confirmation of the high pressures of an impact.

In the Caribbean region and in the southern United States, geologists also found tiny glass beads, less than 1 millimetre across. These looked just like the results of melting during a volcanic eruption, but the KPg beads could not have been produced by a volcano. When geochemists analysed the glass, they found that it did not have the chemical composition of a lava, but it matched the composition of limestone and salt beds. How could a volcano produce molten limestone? These glass beads must indicate that the meteorite hit limestones and salts, melted the underlying rocks, and flung glass beads back out of the crater.

The fern spike was another unexpected finding. Palaeontologists found an abrupt shift in pollen ratios at some KPg boundaries, showing a sudden loss of the pollen of flowering plants and their replacement by ferns, and then a progressive return to normal floras. This fern spike is interpreted as indicating the aftermath of a catastrophic ash fall: ferns recover first and colonize the new surface, followed eventually by the flowering plants, such as grasses, bushes and trees, after soils begin to develop. This was illustrated after the eruption of Mount St Helens in the western United States in 1980. The blast of the eruption flattened the trees and killed off the grass and flowers for miles around. The landscape was shrouded in ash. Within a year, however, the first plant life had returned: ferns had somehow survived beneath the covering, and they began to uncurl through the ash. It took a few years for the other plants to return.

The Chicxulub crater

There was no need to find the crater, but a crater would surely help to clinch the KPg impact model. After several false leads, the crater that killed the dinosaurs was identified in 1991 on the Yucatán peninsula in southern Mexico, centred around the village of Chicxulub (Fig. 14). The Chicxulub crater was buried beneath Tertiary sediments and could not be seen at the surface, but boreholes and geophysical evidence suggested that it was 150 kilometres across – just the size Luis Alvarez had predicted.

The crater was located in a piece of brilliant detective work by a young Canadian graduate student, Alan Hildebrand.[19] He, along with others at the time, had realized that the crater must lie somewhere in the Caribbean area. There were two main lines of evidence: tumbled rocks produced by ancient tsunamis, and indicators of proximality in the boundary beds.

First, along the shores of eastern Mexico, and in Texas and other southern states of America, geologists had identified unusual, chaotic beds

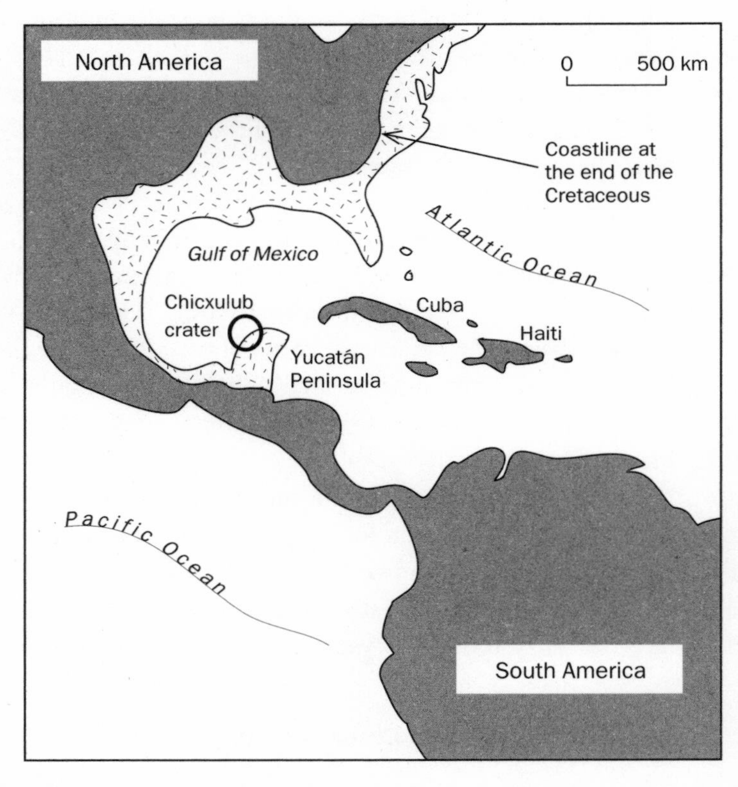

14 *The KPg impact site identified: the location of the Chicxulub crater on the Yucatán Peninsula, Mexico, partially underwater, with the coastline as it was at the end of the Cretaceous also shown.*

at the KPg boundary. Mixed with boundary clays and the iridium were tumbled blocks of the local limestones. The blocks had been torn up and dumped by some dramatic physical event, and it had been suggested that the crisis might have been a huge tidal wave, a tsunami. The sedimentological evidence has been disputed for many of these chaotic beds, but perhaps an impact somewhere out in the proto-Caribbean sea had sent out a vast tsunami that pounded the shores?

Second, Hildebrand and others had noticed that the KPg boundary layer became thicker towards the centre of the Caribbean. So, whereas the boundary bed was about 1 centimetre thick in most parts of the world, in Texas and Mexico it was a metre or more thick in places. Geologists are alert to indicators of proximality: the closer you approach a volcano, the thicker the lava beds; the closer you are to the head of a submarine sediment slide, the thicker the layers of sediment; not only in terms of thickness, but also in grain size. In the thicker boundary layers around the Caribbean, geologists found accumulations of glass beads, and these were not found further afield.

Hildebrand found some borehole records that had been recorded by the Mexican oil company Pemex in the 1960s. The oil geologists had identified a large circular structure deep beneath Chicxulub, which they thought might be an oil trap. They drilled on either side of the structure, and right through the middle of it (Fig. 15). To the sides, the boreholes penetrated 3 kilometres of early Tertiary sediments and underlying Cretaceous limestones. But in the middle of the structure, the drill first encountered *late* Tertiary sediments, then early Tertiary, and then *no Cretaceous*. Instead of the expected Cretaceous limestones, they hit a breccia and then a strange melt rock. From the descriptions, Hildebrand guessed that the melt rock was suevite. He had his evidence to identify the impact site.

The story seemed simple. The meteorite had struck in the proto-Caribbean, evaporating the water and punching kilometres down into the Cretaceous limestones and salt beds. Great boulders and debris were hurled in the air and fell around the crater. Melt products – glass bombs and beads – flew through the air and spread over an area at least 1000 kilometres across. The finer material entered the stratosphere, and transported shocked quartz, ash and iridium around the entire globe.

After the impact, sediments were deposited over the crater and the surrounding landscape. By mid-Tertiary times, some 40 million years

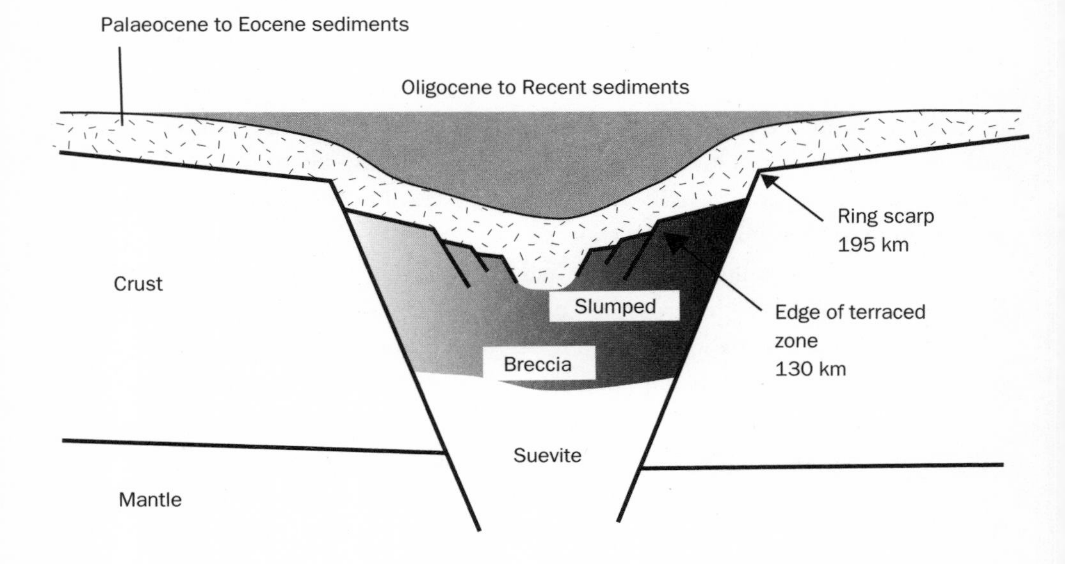

15 *Cross-section of the Chicxulub crater, showing the shape of the crater from geophysical surveys. The suevite (melt rock) and breccia below the crater have been recovered from boreholes. Note how the crater filled with sediment after the impact.*

after the impact, there was still a noticeable depression lying over the crater. Late Tertiary sediments filled it up, but sediments of that age were stripped off the surrounding landscape by normal erosion processes. Today, there is no physical sign of the crater since the ground is entirely level. In any case, half the crater lies offshore, beneath the modern Caribbean; the only way to study it is by geophysics.

Geophysical traverses of the Chicxulub crater show that it has a triple-ring structure. The inner ring, produced by the springback of the Earth's crust within seconds of the impact, is 80 kilometres across. A zone, from 100 to 130 kilometres across, represents the original crater edge, and a zone of collapse into the crater itself. Finally, at a diameter of 195 kilometres, is an outer ring which probably marks the extent of the force zone of the impact. It seems that the vast energy of the impact created a crater 130 kilometres across, but also pulled down a circle of crustal rocks 195 kilometres across.

Where are we today?

The KPg event is not completely resolved. One thing is clear, however: there was an impact centred on Chicxulub in Mexico 66 million years ago (radiometric dating confirmed the age of the melt rocks). This impact had all the global effects predicted from close study of the Ries crater, and modelled by Luis Alvarez and his colleagues in 1980. The impact seems to have sent out vast tsunamis that had devastating effects in the proto-Caribbean, and ash and iridium circled the globe. The ash certainly blocked out the sun, leading to global darkness and freezing. Evidence for freezing comes from studies of plant stems and leaves preserved at the moment of impact: the cell walls have been burst by growing ice needles in the sap. But did this cause the extinctions?

Some of the criticisms offered by palaeontologists in 1980 are still valid today. Many groups of organisms were indeed in decline before the impact, and these declines may relate to deteriorations in climate or changes in sea level. It is important also to recall that many plants and animals were seemingly unaffected by the impact, so the killing model has to take account of that. Ever more detailed studies of fossil occurrences up to the KPg boundary may shed further light on what was going on.

In addition, there were also major volcanic eruptions occurring at the same time. The Deccan Traps in India cover a huge area in the north of the country, and they erupted over about a million years of time spanning the KPg boundary. These were basalt eruptions, that produced huge volumes of lava in a number of bursts of activity. Associated with the erupting lava would have been vast volumes of carbon dioxide, sulphur dioxide and other gases. At one time, a group of geologists suggested that there had not been an impact at all, and that the Deccan volcanics could explain it all – the global ash layer, the iridium and the rest.

This position is clearly untenable now, since volcanic eruptions cannot explain the shocked quartz or the glass beads with geochemical compositions of limestone. But there is no doubt that the Deccan eruptions happened, and that they began at a time of climatic deterioration and decline of many fossil groups. Maybe – and this is not yet resolved – the eruptions did set the extinctions rolling, and the asteroid impact finished them off?

What a changed scientific world in the course of 35 years! In 1980, despite the work of the craterologists and the suggestion of a supernova explosion 66 million years ago, most earth scientists were still firmly in Charles Lyell's camp. When I learnt my geology in the 1970s, my professors did not even mention impacts, craters or mass extinctions.

Now, *my* students hear about catastrophes, asteroids, giant eruptions, death and destruction every week in their lectures. But were all mass extinctions like the KPg event? How does it fit into the grand scheme of the history of life, and how does it compare with the even bigger end-Permian mass extinction?

6

DIVERSITY, EXTINCTION AND MASS EXTINCTION

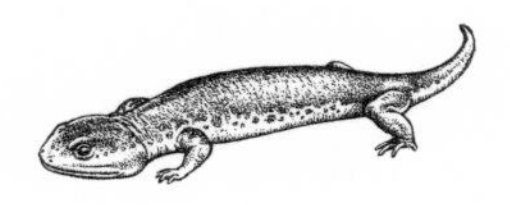

Charles Lyell's famous dictum 'the present is the key to the past' is drummed into all trainee geologists during their first lecture. This injunction makes sense. If a geologist means to try to understand events and processes that happened in the past, it is self-evident that he or she should compare them with what we know of the world today. The best way to understand the productions of ancient volcanoes is to examine as many modern volcanoes as possible. But there are some events and processes that are so unusual, or so huge, that they cannot be studied in the modern world.

The grand pattern of the evolution of life over millions of years is too big and too long-term for any biologist to understand simply by looking at living plants and animals. Mass extinctions, great impacts and world-scale volcanic eruptions happen only rarely – perhaps on timescales of tens or hundreds of millions of years – and again studies of the puny events that we witness from time to time can only give hints of what the larger catastrophes were like. Indeed, for studies of modern biodiversity and extinction, the past can be an essential guide to the present.

Life today is hugely diverse, and that diversity has been achieved in fits and starts. There have been times of rapid rises in diversity, for example when animals with skeletons invaded the oceans, when life moved on to land, when the first coral reefs and the first forests developed, and so on. But there have also been many times of extinction, times when global diversity fell back. Georges Cuvier actually got it right when he wrote in 1812:[1]

Thus life on earth has often been disturbed by terrible events: calamities which initially perhaps shook the entire crust of the earth to a great depth, but which have since become steadily less deep and less general. Living organisms without number have been the victims of these catastrophes.

As we have seen, it was remarks like these that so infuriated Charles Lyell. But Cuvier, and the stratigraphers like Murchison, knew that there were marked disjunctions in the record of the rocks, and in the record of life – times when all the typical fossils of one time unit seemed to disappear, to be replaced by another assemblage. As Cuvier surmised, many of these do indeed indicate times of extinction. But we now know that there were extinctions and extinctions.

The big five

Five extinction events during the past 500 million years stand out from the rest. These particularly large extinctions, commonly called mass extinctions, have been termed the 'big five'. The end-Permian mass extinction was the biggest of all, and it has been called the 'mother of all mass extinctions' by Doug Erwin,[2] who has written a great deal about it. The four smaller mass extinctions include the KPg event, and much studied as it is, the KPg event was by no means the largest. The other three mass extinctions happened in the Late Ordovician, the Late Devonian and at the end of the Triassic, some 444, 374 and 201 million years ago respectively. In addition there were dozens of smaller events, such as the loss of large mammals at the end of the ice age, 10,000 years ago.

In order to understand mass extinctions, it is important to comprehend the concepts of time and diversity. These are the two axes of the graph: geological time, measured in hundreds of millions of years, is the *x* axis, and diversity, measured in numbers of species or families, is the *y* axis. Extinction happens all the time, and there is an expected, or background rate. Then, from time to time, something extraordinary happens, and there is a higher-than-normal rate of extinction. This might then produce an extinction event or, if it is big enough, a mass extinction.

Some mass extinctions have been examined in great detail, but there are many questions still to be resolved. For example, the definition of a mass extinction is not entirely clear: just how big and how fast does the event have to be to qualify? How ecologically diverse should the victims be? There are major questions about the quality of the fossil record, and just how well it represents what really went on: how close can you get to an event that happened millions of years ago? Can geologists see what happened day by day, or should they step back and determine what went on over longer periods, and in a more fuzzy, but more accurate, way? Are there any common factors to mass extinctions? For example, are some kinds of organisms more or less likely to be victims? All of these issues are important in trying to understand mass extinctions in general, and as a grounding for a close examination of the end-Permian event. They also have important implications for understanding the modern biodiversity crisis.

Diversity and biodiversity

Biodiversity is one of the fastest-growing words in the English language, indeed in any language. The word was introduced in 1988 attached to a report[3] on the growing ecological crises on Earth, and since then it has entered the public consciousness. 'Biodiversity' can be read nearly every day in the newspaper, and it is used in thousands of scientific papers and reports every year. 'Biodiversity' means no more than the older, simpler term 'diversity' as applied to life.

Simpler in form, but not simpler in meaning. There are a hundred and one meanings of the term 'diversity'. Diversity can be assessed as the number of species in a single area, such as a forest or a lake, or it can be a larger-scale term, such as the number of species in a country, a continent, even the world. Diversity can refer to just one group, such as the diversity of birds or the diversity of grasses in a field or a county, or it can refer to all groups, from microbes to mammals. Diversity can be simply a count of species, or a higher taxonomic grouping such as a genus or a family. In some ecological contexts, diversity also includes an assessment of abundance, the number of individuals of a species in an area – an attempt to record the richness of life. In other cases, diversity includes a measure of the ecological range of organisms observed, whether the fauna includes large and small animals, for example, or a more limited scope. Diversity is increasingly given a genetic or molecular connotation these days too. Biologists try to assess the amount of unique genetic material in a species compared to others as a means of determining where to expend most conservation effort.

Palaeontologists generally use the term diversity in the simplest way, to mean a count of numbers of species, sometimes termed species richness. Species can be assessed in a local sample of fossils that are well preserved. However, for global studies, it is normal to climb up the taxonomic hierarchy to genera or families. These are larger groups, each containing several, or numerous, species, and they are simpler to compare from continent to continent. Their record is also more complete than that of species. Here are two issues, then, the taxonomic hierarchy, and the issue of quality of the fossil record. We will look at important issues of quality later in this chapter.

Naming the beasts

From the earliest times, thinking people have wondered at the diversity of life, and they have named the plants and animals they see around them. Such an instinct dates from the earliest times, as is shown by the writers of Genesis (chapter 2, verse 19):

> And out the ground the Lord God formed every beast of the field, and every fowl of the air; and brought them unto Adam to see what he would call them; and whatsoever Adam called every living creature, that was the name thereof. And Adam gave names to all the cattle, and to the fowl of the air, and to every beast of the field.

Naming and classifying are basic human activities. Roman citizens of two millennia ago could name and identify hundreds of plants and animals. Uneducated medieval peasants in Europe could do the same. In their tropical forests, natives of the Amazon and of New Guinea identify the plants and animals around them with great precision. Indeed, when biologists from the West visit these remote regions, they find that their formalized schemes of classification, based on decades of collecting and scientific study in the laboratories of Europe and North America, are no better than those of the locals.

Biologists around the world have agreed to a formal system of naming and recording new species, and the formal starting dates for published names are 1753 for plants and 1758 for animals.[4] Those dates mark the publication of two fundamental books, the attempts by the Swedish biologist Carl von Linné (Carolus Linnaeus, in the Latin form, the language he, and scholars from all countries, still used then to communicate much of their work) to instil order into the classification of plants and animals – his *Species plantarum* (1753) and *Systema naturae per regna tria naturae, secundum classes, ordines, genera, species, cum characteribus, differentiis, synonymis, locis* (1758). In these, Linnaeus presented outline descriptions of all the plants and animals he knew, an incredible achievement for the time. However, the key contribution he made was to establish a standard way of naming organisms.

Linnaeus characterized each species in his classifications by a two-part Latin epithet, such as *Homo sapiens* for humans and *Canis canis* for the domestic dog. The first term, written with an upper case initial letter, is the genus name, the second the species. Each genus contains at least one species, usually several, sometimes many. The species epithet is indicated with a lower case initial letter. The whole name is always given in italics.

The system is international and is recognized by scientists worldwide, whatever their native language or alphabet. You see the Latin names of species, written in the roman alphabet, in the middle of scientific texts in Chinese or Russian. This makes for perfect ease of communication between nationalities, at least in terms of the formal names of organisms. Priority is important, and once a species has been named, no one else can give it a different name, although a species may be moved to a different genus as a result of further study. The full form of a species epithet includes the name of the author and date of first publication, for example, *Homo sapiens* Linnaeus, 1758.

From one species to many

New species evolve constantly, and life has diversified astonishingly since its origin some 3500 million years ago. In that vast and unimaginable amount of time, diversity, by definition, has gone from one species to many millions, perhaps 20–100 million today. One species originally? Charles Darwin was one of the first to dare to speculate, in 1859,[5] that all life had evolved from a single species, or at most only a small number. Where that species came from, he did not care to commit himself on paper, and we need not enter into the various current hypotheses about the origin of life. Nevertheless, one species it was. What is the evidence?

All living species, from the simplest single-celled virus or slimemould to human beings and oak trees, share the molecule DNA, which codes genetic information, together with all the complexities of the protein manufacture system in the cells. In addition, all living things today share two further characters: similar cell membranes with mechanisms to control the passage of chemical ions and molecules in and out of the cells; and an energy transfer system involving the molecule adenosine triphosphate (ATP) by which energy is stored and released from a phosphate bond. These complex features of all life probably arose once only, and hence they point to a single ultimate ancestor of all life.

What about the fossils? Of course, DNA, membranes and ATP molecules cannot be detected in the rock record. But all fossils known to date belong to living groups. Even the bizarre fossil groups which are entirely extinct, such as dinosaurs, hard-shelled fishes, trilobites, rhyniophyte plants and the weird beasts of the ancient Precambrian and Cambrian sea floors, clearly fit lower down within the evolutionary tree of modern forms. That is not to say that life did not originate more than once in the primaeval ooze 3500 million years ago, but any other experiments in life have gone undetected, or they led to nothing.

Expansion

How did life diversify after the origin of the first species? Some key steps can be identified (Fig. 16). Many of the events happened in the long span of the Precambrian, but these are the world's most ancient rocks and so the fossil record is patchy, and radiometric dating is difficult. Fossils of tiny microbes are known from some amazing, but very rare, deposits, where they were instantly entombed in glassy cherts. But such chance discoveries, while hugely important in documenting what went on, can give no true picture of diversity worldwide.

It is only by the end of the Precambrian, during the Vendian period, that fossils become more abundant, and rather larger. Then, from the beginning of the Cambrian period, some 540 million years ago, fossils are found all over the place. Numerous groups of animals in the sea seemed to acquire mineralized skeletons of one sort or another – shells or carapaces made from calcite, shells or bones made from phosphate, or hard protein-rich cuticles – at about the same time. Again, this is not the place to consider why that might have happened, merely to record that it did.

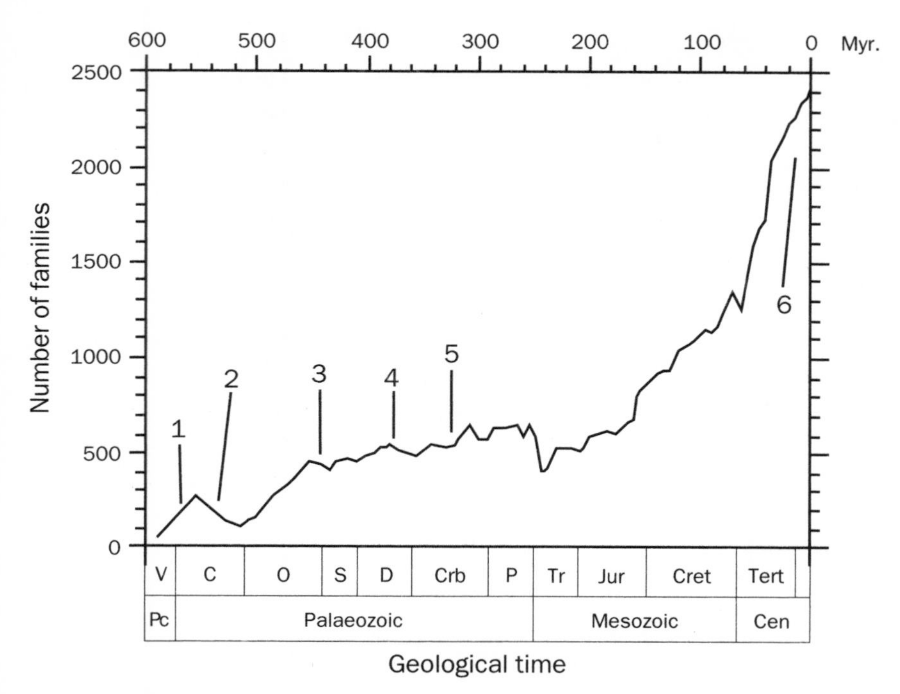

16 *The expansion of life, showing six major innovations. Before the Vendian (V) there had been over three billion years of the history of life, which saw the origin of life, of complex cells and of multicellular life. In the past 600 million years, one can note the Cambrian (C) explosion (1); the diversification of reefs (2); the move of green plants on to land in the Ordovician (O) and Silurian (S) (3); the origin of land vertebrates in the Devonian (D) (4); the spread of forests in the Carboniferous (Crb) (5); and eventually, the origin of humans in the late Cenozoic (Cen) (6). Other abbreviations: P, Permian; Tr, Triassic; J, Jurassic; Cret, Cretaceous; Tert, Tertiary.*

Through the last 540 million years, termed the Phanerozoic ('abundant life'), life in the sea expanded in diversity in pulses. At first, marine organisms swam or crawled on, or near, the sea bed. Then, free-swimming forms became established, which fed on the rich plankton at the sea's surface. Larger predators evolved to feed on the smaller animals. Next, some groups began to burrow into the sea floor, seeking organic matter. Others built up huge skeletal structures called reefs. Through time, burrowers burrowed deeper, and reefs grew taller. Fishes evolved, then great marine reptiles to feed on them, then sharks, whales and seals, even sea birds, came on the scene.

Life expanded on to land rather later. Perhaps some simple algae spread a green film over the watersides in Precambrian time. But life did not move on to land in a serious way until the Ordovician and Silurian, some 450 to 400 million years ago. First, small plants grew up around the sides of lakes and rivers. They were accompanied by ancestral spiders, mites, millipedes and scorpions. Then came larger plants, insects and amphibians, the first land vertebrates. Ecosystems on land became more complex, with the evolution of trees and forests. Insects diversified hugely, and so did land plants and land vertebrates. Evolutionary radiations of conifers and seed ferns, dinosaurs and pterosaurs, were followed by flowering plants and deciduous trees, mammals and birds.

These patterns of expansion in the sea and on land have been documented many times by palaeontologists.[6] The simplest way to show the patterns is by a plot of the numbers of families through time. Time here is taken as the last 600 million years (myr.) or so, the Vendian plus Phanerozoic span, when fossils have been abundant. Such graphs show (Fig. 17) how diversification was episodic. There was no uniform expansion in a steady way through time. Bursts of diversification followed major innovations, such as skeletons or the ability to fly, or the evolution of a major new ecosystem, such as reefs or forests. Diversification has also suffered numerous setbacks, some small, some larger. The larger drops in diversity, of course, mark the mass extinction events.

Extinctions and mass extinctions

The fossil record becomes hazier the further back in time one goes. There were almost certainly numerous major setbacks in the history of life in the Precambrian, but the precision of dating, and the abundance of fossils, are too inadequate to be sure. The major extinction events (Fig. 17c), including the big five, will be reviewed here in time order, starting with the oldest.[7]

17 *Diversification of life in the sea (a), on land (b) and in total (c) through the past 600 million years, the Vendian and the Phanerozoic. Extinction events, major, intermediate and minor are indicated in (c). Abbreviations: C, Carboniferous; Cen, Cenozoic; Cm, Cambrian; D, Devonian; J, Jurassic; K, Cretaceous; O, Ordovician; P, Permian; Pc, Precambrian; S, Silurian; T, Tertiary; Tr, Triassic; V, Vendian. As John Keats wrote: 'life is but a day; A fragile dew-drop on its perilous way'.*

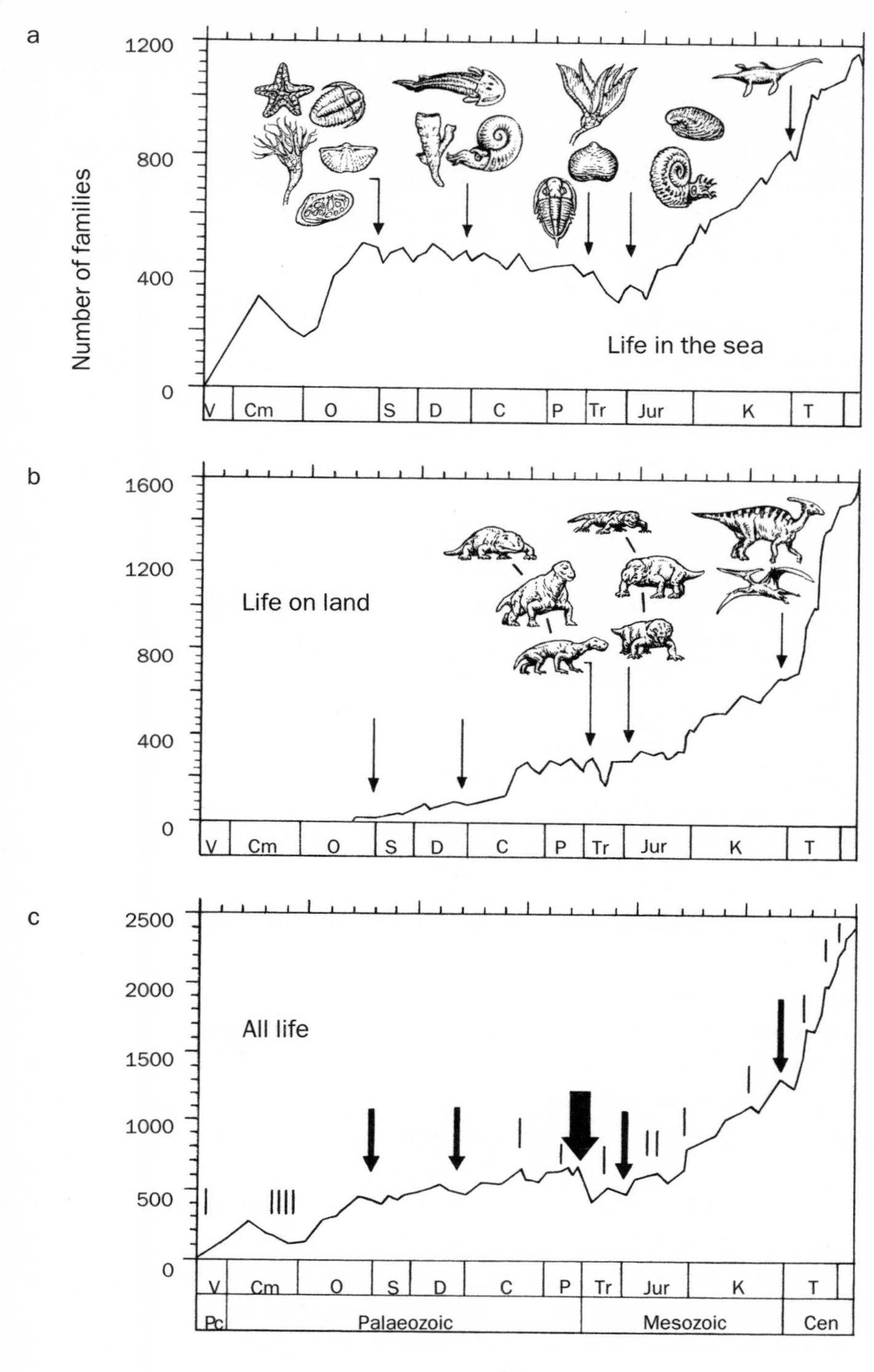
a
Number of families
1200
800
400
0
Life in the sea
V
Cm
O
S
D
C
P
Tr
Jur
K
T
b
1600
1200
800
400
0
Life on land
V
Cm
O
S
D
C
P
Tr
Jur
K
T
c
2500
2000
1500
1000
500
0
All life
V
Cm
O
S
D
C
P
Tr
Jur
K
T
Pc
Palaeozoic
Mesozoic
Cen
Time

The *Late Precambrian* event is ill-defined in terms of timing, but such an event clearly occurred about 560–550 myr. ago, when early kinds of animals disappeared. These early animals, sometimes termed a 'failed experiment', were found first in the Ediacara Hills of Australia, but have since been identified from around the world. The Ediacara beasts include worm-like animals, probable jellyfishes, strange frond-like forms and others – the first animal fossils that can be seen with the naked eye. But the animals of the Ediacara type disappeared, and the way was cleared for the dramatic radiation of shelly animals at the beginning of the Cambrian period (and the beginning of the Phanerozoic Eon).

A series of extinction events occurred during the *Late Cambrian*, perhaps as many as five, in the interval from 497 to 485 myr. ago. There were major changes in the marine faunas in North America and other parts of the world, with repeated extinctions of trilobites. Trilobites were complex animals that looked superficially like woodlice, with multiple limbs and an external hard skeleton. Their distant living relatives include crabs, insects and spiders. Following the series of Late Cambrian extinctions, animals in the sea became much more diverse, and groups such as articulate brachiopods, corals, fishes, gastropods and cephalopods diversified dramatically. The gastropods, including modern snails, slugs and whelks, and the cephalopods, including modern squid and octopus, are both subgroups of molluscs, the most diverse group of shelled animals.

In the *Late Ordovician*, about 444 myr. ago, further substantial turnovers occurred among marine faunas. This extinction event is the first of the 'big five' mass extinctions. All reef-building animals, as well as many families of brachiopods, echinoderms (sea urchins, sea lilies and starfishes), ostracods (microscopic crustaceans, distantly related to crabs and shrimps) and trilobites died out. These extinctions are associated with evidence for major climatic changes. Tropical-type reefs and their rich faunas lived around the shores of North America and other land masses that then lay around the Equator. Southern continents had, however, drifted over the south pole, and a vast phase of glaciation began. The ice spread north in all directions, cooling the southern oceans, locking water into the ice and lowering sea levels globally. Polar faunas moved towards the tropics, and warm-water faunas died out as the whole tropical belt disappeared.

The second of the big five mass extinctions occurred during the *Late Devonian*, and this appears to have been a succession of extinction pulses lasting from about 374 to 360 myr. ago. The abundant free-swimming cephalopods were decimated, as were the extraordinary armoured fishes of the Devonian. Substantial losses occurred also among corals, brachiopods, crinoids (sea lilies), stromatoporoids (colonial sponge-like animal colonies), ostracods and trilobites. Causes could be a major cooling phase associated with anoxia (loss of oxygen) on the sea bed, or massive impacts of extraterrestrial objects. The Late Devonian extinctions have figured, on and off, in the heated debates about whether meteorite impacts cause mass extinctions or not.

The largest of all extinction events, the *end-Permian* or *Permo-Triassic (PTr)* event, 252 myr. ago, is the third of the big five. The dramatic changeover in faunas and floras at this time has long been recognized, and was used to mark the boundary between the Palaeozoic and Mesozoic eras. During the end-Permian event, most of the dominant Palaeozoic groups in the sea disappeared, or were much reduced: corals, articulate brachiopods, bryozoans (or 'moss animals', generally small colonial animals), stalked echinoderms, trilobites and ammonoids (cephalopod molluscs with coiled shells). There were also dramatic changes on land, with widespread extinctions among plants, insects, amphibians and reptiles, which led in all cases to dramatic long-term changes in the dominant replacing forms. The causes will be explored later.

The *Late Triassic* events include the fourth of the big five mass extinctions. A marine mass extinction event at the Triassic-Jurassic boundary, 201 myr. ago, has long been recognized by the loss of most ammonoids, many families of brachiopods, bivalves, gastropods and marine reptiles, as well as by the final demise of the conodonts, a shadowy group of primitive fishes that lived throughout the Palaeozoic, but which are known almost exclusively only from their hardened tooth plates. An earlier event, near the beginning of the Late Triassic, 230 myr. ago, also had effects in the sea, with major turnovers among reef faunas, ammonoids and echinoderms, but it was particularly important on land. There were large-scale changeovers in floras, and many amphibian and reptile groups disappeared, to be followed by the dramatic rise of the dinosaurs and pterosaurs, the flying reptiles. At this time, many modern groups arrived on the scene, such as turtles, crocodilians, lizard ancestors and mammals. The cause of these events may have been climatic changes associated with continental drift. At this time, the supercontinent Pangaea was beginning to break up, with the unzipping of the Central Atlantic between North America and Africa. Although meteorite impact has again been pointed to as a possible suspect, the evidence is weak.

Extinctions during the *Jurassic* and *Cretaceous* periods were minor in extent. The Early Jurassic and end-Jurassic events involved losses of bivalves (two-shelled molluscs, like clams, mussels and oysters), gastropods, brachiopods and free-swimming ammonites as a result of major phases of anoxia. Free-swimming animals were unaffected, and the events are undetectable on land – they may be partly artificial results of incomplete data recording. Events have been postulated also in the mid-Jurassic and in the Early Cretaceous, but they are hard to determine. The *Cenomanian-Turonian* extinction event, 94 myr. ago, associated with extinctions of some floating planktonic organisms, as well as the bony fishes and ichthyosaurs (dolphin-like marine reptiles) that fed on them, may turn out to be an artifact of a major gap in the rock record.

The *Cretaceous-Paleogene (KPg)* mass extinction, 66 myr. ago, is by far the best-known of the big five events. As well as the dinosaurs, the flying pterosaurs, the marine plesiosaurs and mosasaurs, the ammonites and belemnites (both common cephalopod groups in the Mesozoic), various major reef-building groups of bivalves and most planktonic foraminifera (microscopic abundant shelly plankton) disappeared. The postulated causes, as we saw in the previous chapter, range from long-term climatic change to instantaneous wipeout following a major extraterrestrial impact.

Extinctions since the KPg event have been more modest in scope. The *Eocene-Oligocene* events, 34 myr. ago, are marked by extinctions among plankton and open-water bony fishes in the sea, and by a major turnover among mammals in Europe and North America. Later *Tertiary* events are less well-defined, which is surprising in a way since the rock and fossil record generally improves towards the present. There was a dramatic extinction among mammals in North America in the mid-Oligocene, and minor losses of plankton in the mid-Miocene, but neither event was large. Planktonic extinctions occurred during the Pliocene, and these may be linked to disappearances of bivalves and gastropods in tropical seas.

The latest extinction event, at the end of the *Pleistocene*, while dramatic in human terms, barely qualifies for inclusion. As the great ice sheets withdrew from Europe and North America, large mammals such as mammoths, mastodons, woolly rhinos and giant ground sloths died out.

Some of the extinctions were related to major climatic changes, and others may have been exacerbated by human hunting activity. The loss of large mammal species was, however, minor in global terms, amounting to a total loss of less than 1% of species.

There have been many extinction events, then, in the history of the Earth, the big five mass extinctions, and many smaller events. Do the mass extinctions share any characteristics?

Background extinction and mass extinction

Extinction is normal. Species do not last forever. Indeed, the average life span of a species is perhaps 5 myr., with a range from 100,000 years to 15 myr., depending on what you are, whether a microbe or a flowering plant. Species come and go and, even though the overall diversity of life still seems to be increasing, there is a steady rate of background, or normal, extinction. The background rate of extinction may be only 10 to 20% of species per million years – 10 or 20 species out of every 100 disappear every million years, which translates to one or two species per 100 every 100,000 years, or 0.01–0.02 per 100 per 1000 years, or 0.00001–0.00002 per 100 per year. Tiny rates when measured in human terms, but more noticeable on geological timescales of millions of years.

Background extinction, extinction events, and mass extinctions. There have clearly been times when extinction rates have gone up, times that stand out as extinction events of some kind. Extinction events are usually restricted in some way, perhaps the loss of all the large, cold-adapted mammals at the end of the Pleistocene, or the loss of life on some Pacific islands as a result of a huge volcanic eruption or other catastrophe. A mass extinction is the largest kind of extinction event. It is important to try to define the meanings of these terms, not least because many biologists have claimed that we are living through the sixth, human-induced, mass extinction today.

The big five mass extinctions share many features in common, but differ in others. Three things happened in all the mass extinctions of the past:

- many species became extinct, generally more than 40 to 50%
- the extinct forms span a broad range of ecologies, and they typically

include marine and non-marine forms, plants and animals, microscopic and large forms

- the extinctions all happened within a short time, and hence relate to a single cause, or cluster of interlinked causes.

These points all seem clear, but palaeontologists have struggled to be more precise. Just how many species should disappear, and how fast, for an event to stand apart from background extinctions as a mass extinction? Attempts have been made to find a more quantitative definition of which extinctions are truly mass extinctions, and which are more localized or ecologically restricted events, but none of these efforts has really been satisfactory.

A statistical test?

In 1982, the palaeontologists David Raup and Jack Sepkoski of the University of Chicago[8] claimed that they had found such a test for mass extinctions, and their idea was simple. They argued that if times of mass extinctions were associated with exceptionally high rates of extinction, then these should stand out clearly from normal background extinction rates. Raup and Sepkoski calculated a mean rate of disappearance of marine animal families, per million years, for each geological stage (average duration, 5–6 myr.). So, over a time span of the last 600 myr., they assessed about 100 separate extinction rates.

Raup and Sepkoski believed that they could apply a simple statistical technique to these measurements: regression analysis. This is a grand term for fitting a line to a set of points on a graph. The ideal is that a straight line can be fitted, and that most of the points will lie close to it. There was absolutely no reason why Raup and Sepkoski could have fitted a line to their measurements of extinction rates through time, but they succeeded. Most of their 100 extinction data points fell very close to a straight line, and that straight line was declining through time (Fig. 18). In other words, they argued, the likelihood of extinction was diminishing, which they attributed to 'improvements' in the ability of organisms to avoid extinction. What of the points that did not fall close to the regression line? These turned out to correspond to the big five mass extinctions: Late Ordovician, Late Devonian, end-Permian (PTr), Late Triassic and Cretaceous-Paleogene (KPg).

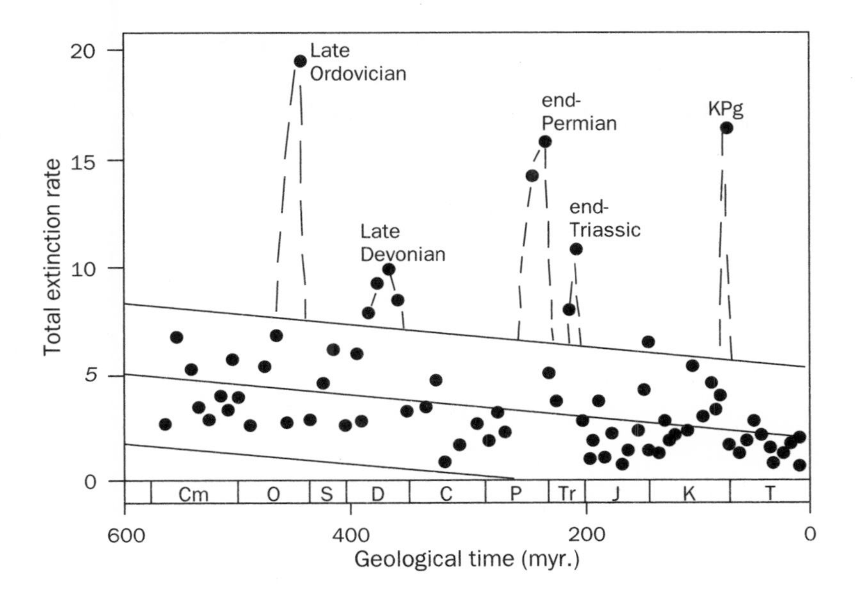

18 *Total extinction rate, measured as extinctions per million years, for marine invertebrate families. Most points fall on either side of a regression line that declines through the last 600 million years of the good-quality fossil record. Five sets of statistical outliers, lying above the 95% confidence envelope (enclosed by dashed lines), correspond to the five major mass extinctions.*

Statistically, the case seemed clear. In formal terms, the regression line fitted the data with 95% confidence, meaning that at least 95% of the points lay very close to the regression line. The remaining 5% of points, five in this case since the total number was 100, are termed statistical outliers. They lie away from the general trend, and must be explained in some other way. Perhaps they were wrongly measured or, the interpretation generally accepted, perhaps they were truly different from the other points. If so, that would mean that mass extinctions had been identified in a quantitative way, and they had been set apart. The implications would be profound if the observations were correct.

The five mass extinctions had been picked out as something different from normal extinction, certainly much bigger, but perhaps more than that. Raup and Sepkoski hinted clearly that mass extinctions were not only quantitatively different from normal extinction events (i.e. bigger), but perhaps they were also qualitatively different. A whole different class of phenomena. If they were right, this would suggest that the normal rules of evolution broke down during mass extinctions, and that unique kinds of causes should be sought. Perhaps, indeed, all mass extinctions might themselves share characteristics with each other, and not with other smaller extinction events. This important idea was actively discussed, but ultimately rejected, by most palaeontologists in the 1980s and 1990s.

The implications of Raup and Sepkoski's 1982 paper were profound, but these authors had to admit, when challenged by a statistician, that the method was flawed. The technique of regression analysis could not be applied to their data.[9] The idea was strong, but the statistics were weak. None the less, their studies led Raup and Sepkoski to think more about the fact that there had been repeated mass extinctions through time: what if they followed a regular pattern?

Periodicity, and the next asteroid to strike the Earth

Could all mass extinctions be explained by a single cause? Some geologists had already suggested that the history of life, including phases of diversification and phases of extinction, might be controlled by changes either in temperature (usually cooling) or in sea level. The idea was simply that the physical environment on the Earth was sometimes attractive for life, and at other times it was not. Geologists in particular found such ideas easy to accept, that life was merely a part of the huge complex of the Earth's systems, an intimate weaving of the geosphere and the biosphere. Biologists tend to find such ideas of external controls on large-scale evolution harder to accept, since they are used to seeing changes brought about by interactions among species, for example competition or predation. The contrast is between external control, the geologists' view, and internal control, the biologists' view.

The revolution in KPg studies brought about by the Alvarez paper of 1980 clearly set palaeontologists and geologists thinking. What if the history of the Earth, and in particular the history of life, were controlled in certain ways by repeated impacts of asteroids or comets?

The search for a common cause gained great credence with the discovery in 1984 of an apparently regular spacing between mass extinctions during the last 250 myr. Raup and Sepkoski found a regular period of 26 myr. separating peaks of elevated extinction rates for the record of marine

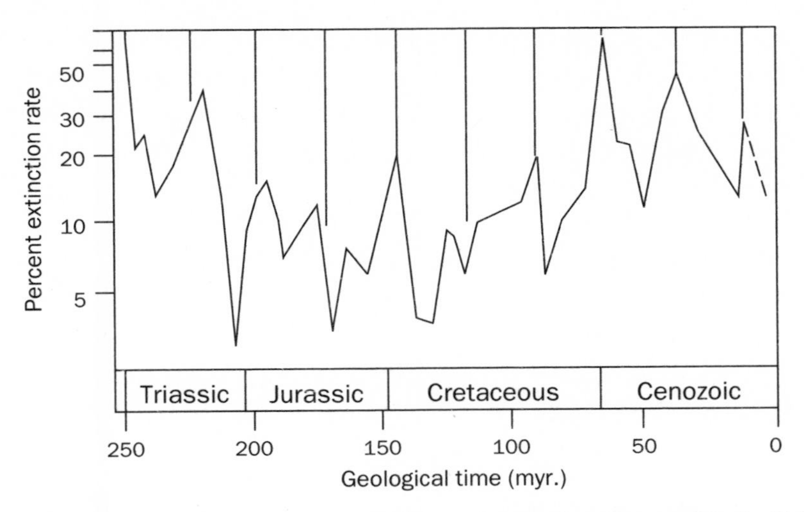

19 *Periodic mass extinctions? This is the famous diagram from Raup and Sepkoski (1984) that launched a thousand astronomical studies. If there is a regular 26-million-year periodicity, an astronomical cause seems essential, whether it be tilting galaxies, a twin star Nemesis or Planet X.*

animals (Fig. 19). Initial responses to this proposal were polarized. Many enthusiastic geologists and astronomers accepted the idea, and its clear implication that a regular periodicity in mass extinctions implied a regular astronomically controlled causative mechanism. It is worth pausing here for a moment.

If Raup and Sepkoski were correct, their simple palaeontological study would imply a whole new galaxy-scaled system of processes that controlled the entire history of the Earth and of life. Also, certain people thought, if we know when the last crisis happened, we can therefore work out when the next monster asteroid will hit. The Alvarez proposal of 1980, that the Earth had been hit by an asteroid 65 million years ago, was nothing when compared to the grandeur of the new proposal by Raup and Sepkoski.

Numerous astronomical models to explain the periodicity of mass extinctions were proposed. All these models involved a regularly repeating cycle which disturbed the Oört comet cloud and sent showers of comets hurtling through the solar system every 26 myr. The Oört comet cloud is a shell of comets at the fringes of our solar system, lying beyond

Neptune and Pluto. The disturbing factor might have been the eccentric orbit of a sister star of the Sun, dubbed Nemesis (but not yet seen), tilting of the galactic plane or the effects of a mysterious Planet X which lay beyond Pluto on the edges of the Solar System.

The astronomers enthusiastically polished their telescope lenses and pointed them skywards in search of Nemesis, Planet X or tilting galaxy edges. Geological supporters of periodicity went out to find evidence for massive impacts at mass extinction boundaries to match the physical evidence that had already been established for the KPg event. But the critics of periodicity argued that each mass extinction was a one-off, and that there was no linking principle. The 26 myr. cycle discovered by Raup and Sepkoski was, they argued, a statistical artifact or the result of limited data analysis.

So what is the current view of periodicity? I think that most palaeontologists and geologists have just quietly let it drop. Close analysis of the fossil data has failed to confirm periodicity. Indeed, scrutiny of some of the extinction peaks in Fig. 19, such as the three in the Jurassic, has suggested that these are largely artifacts of the data collecting. Also, the searches for Nemesis and Planet X have not been successful; nor has the search for indicators of impact at the time of the other identified mass extinctions. Iridium, shocked quartz or craters have been found for only two or three of the ten postulated mass extinction peaks that are elements of the periodic cycle, but this evidence is feeble in the extreme when compared to the manifold lines of evidence for impact at the KPg boundary.

Periodicity of mass extinctions was an intriguing idea, but it has been almost conclusively rejected now. But that does not mean that one should not take an overview of all the extinction events of the past in search of common factors.

Scaling and taxonomic targets

Extinction events of the past vary in magnitude, and they may usefully be sorted into major, intermediate and minor events, based on their magnitudes (Fig. 17c). The end-Permian mass extinction is in a class on its own, since it is known that 60–65% of families disappeared at that time, and this scales up to a loss of 80–95% of species. The four intermediate mass extinctions are associated with losses of 20–30% of families, and perhaps 50% of species. The minor extinction events experienced perhaps 10% family loss and 20–30% species loss, but these cannot be called mass extinctions.

What do these species loss figures mean? Are we really sure? Well, in assessing mass extinctions, palaeontologists generally rely on larger subsets of the classification of life. It would be marvellous if the fossil record were complete, and all species that ever existed were laid out to be seen and recorded. There would then never be any doubt about what happened, and when, and even why. However, most organisms that ever lived do not become fossils. Indeed, perhaps even most species that ever existed do not become fossils, and this thought certainly gives palaeontologists sleepless nights.

Fortunately, there is an important scaling issue that saves the palaeontologist's sleep: the bigger the taxonomic target, the more chance you have of hitting it. As you climb the taxonomic hierarchy, the chances of preservation increase. So, the chance of finding any particular fossil species might be one in a hundred, or 1%. But if there are ten species in a genus, then the chance of finding that genus increases to 10%. If there are ten genera in a family, the next traditional rank in the taxonomic hierarchy, then the chance of finding at least one representative of that family is 100%. So, palaeontologists make their initial global studies at the level of the family. Once that survey is complete, they may attempt to drop to the genus level, or even to the species level, perhaps for local studies.

It is possible to use this scaling logic in reverse to interpret the family-level data to the species level. In every case, it is clear that a higher proportion of species than families are wiped out in any extinction event, simply because families contain many species, all of which must die for the family to be deemed extinct. Hence, the loss of a family implies the loss of all its constituent species, but many families will survive even if most of their contained species disappear. So, a loss of 60% of families in the end-Permian event must mean a loss of a higher proportion of the species, and that has been variously estimated as a loss at the level of 80–95%.

Seeing what you want to see

The first defining character of mass extinctions is that many species should have disappeared. In broad terms, the scale of extinctions ranges from losses of 20–60% of families, and estimated losses of 50–95% of species overall. In detail, however, these rates are not uniform across all groups: extinction rates vary from 100% of species within clades that disappear entirely to 0% of species in other clades. The term 'clade' has recently come into use more and more, and it is a useful expression. It refers to a group of any size that had a single ancestor, and includes all of its descendants. So, *Homo sapiens* is a clade, as is the Family Canidae (dogs and relatives), the Class Aves (birds), the Kingdom Animalia, and so on.

Good-quality fossil records indicate a variety of patterns of extinction. Detailed collecting of planktonic microfossils based on centimetre-by-centimetre sampling up to, and across, crucial mass extinction boundaries offers the best evidence of the patterns of mass extinctions. In detail, some of the patterns reveal rather sudden extinctions (Fig. 20). Study of the rock succession across many KPg boundaries suggests that the sequences are as complete as could ever be achieved, and that the total time intervals involved are in the range of 0.5–1 myr. The precision of dating may be as

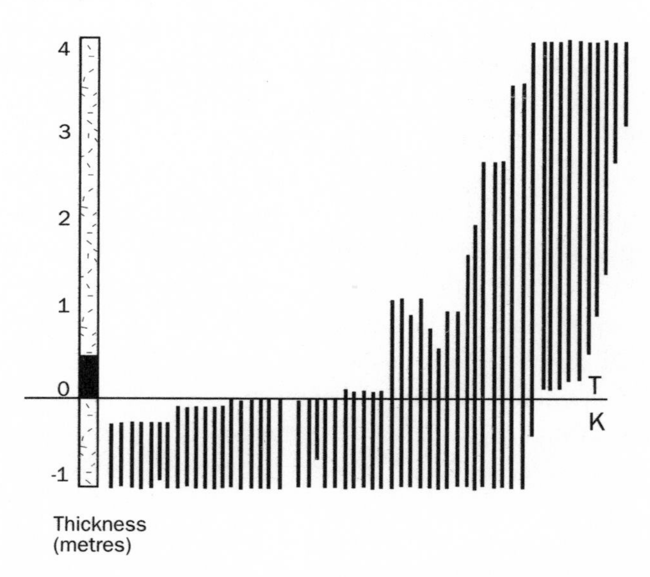

20 *A catastrophic pattern of extinction. The recorded time spans of different species of foraminifera in the classic KPg section at El Kef, Tunisia, based on centimetre-by-centimetre sampling through 5 metres (about 1.5 million years) of sediments, show a species loss of 65% at the KPg boundary.*

fine as 50,000 years in these cases. This is amazing accuracy for a geologist, but pretty hopeless for a biologist.

Does this superb field example (Fig. 20) show a catastrophic or gradual pattern of extinction? The problem concerns definitions: how sudden is sudden? Palaeontologists with different biases could argue that Fig. 20 shows a catastrophic or a gradual dying. A gradualist would argue that there are several steps of extinction over 50,000 years or more, and they are far too drawn out to be the result of an instant event. A catastrophist, on the other hand, would see instant extinction occurring in as short a time as 1–1000 years, and would argue that the slightly stepped pattern is the result of incomplete preservation, inadequate collecting or mixing of the sediment by burrowing organisms which churned up the mud before it hardened into rock.

What is to be done? Will detailed studies of mass extinctions always lead to this kind of stand-off, where catastrophists and gradualists see what they want to see? Many groups of organisms do indeed seem to have gone extinct geologically rapidly during mass extinction events, perhaps in spans of time less than 0.5 myr., but that is still far too long an interval for proper biological interpretation since it could encompass gradual or catastrophic kinds of causes. The dispute always comes down to problems of sampling.

Sampling the death of the dinosaurs

Fossil sampling is a key issue. Even if a palaeontologist can prove that dinosaur fossils suddenly disappear from the rock record at a particular horizon, it cannot simply be assumed that the disappearance is the result of extinction. There might have been an environmental change at that point, and the animals trotted off elsewhere, or there might have been a substantial break in deposition, or depositional processes might have changed in such a way that bones are no longer buried and preserved.

Dinosaur fossils, after all, are preserved in continental sediments (sediments deposited on land, in rivers or in lakes). Continental sediments are nowhere near as continuously deposited as are sediments in the deep ocean. Specimens are large and rare, so that detailed bed-by-bed sampling is fraught with difficulties. Is there any kind of sampling regime that might allow a palaeontologist to resolve these problems?

Two separate teams have attempted to resolve the question of the timing of dinosaur extinction once and for all. This might seem unnecessary, since most people accept that the meteorite impact at the KPg boundary was a sudden event, as discussed in the last chapter. However, it would be wrong then to assume that the dinosaurs died out overnight as a result. That is a separate claim that has to be demonstrated. So, both teams attempted large-scale controlled field sampling in Montana, and they amassed thousands of specimens from the last 5–10 myr. of the Cretaceous rocks of the Hell Creek Formation.[10] Needless to say, one team, led by David Archibald of San Diego State University, found evidence for a long-term die-off, and the other team, led by Peter Sheehan of Milwaukee Public Museum, found evidence for sudden extinction. Each sampling exercise had involved teams of dozens of people, logging in one case an estimated 15,000 person-hours of field prospecting, and in the other case a total of 150,000 identified specimens: how much more intensive does the programme have to be in order to establish what really happened?

Surely palaeontologists can at least answer that one simple question: did the dinosaurs die out instantly or over a long time? There are two main problems, however, which prevent a clearcut solution: dating and selective preservation. First is the problem of precise dating. In some places, radiometric dates at, or close to, the KPg boundary have been measured, but they have an inbuilt error of at least a few hundred thousand years. And then there is the problem of correlation, the matching of rock ages from place to place. Can we really be sure that the last rocks with dinosaurs in Montana, say, are the same age as the last rocks with dinosaurs in Wyoming? Or, more significantly, are the last dinosaurs in Montana the same age as the last dinosaurs in France, Romania or Mongolia?

Until these questions can be answered unequivocally, it will be impossible to say with certainty that the dinosaurs died out instantaneously everywhere, or over the course of half a million or a million years. Selective preservation is also a critical issue.

Taphonomy

Selective preservation of fossils is a fact of life. The preservation of fossils depends on the biology and ecology of the organisms, the vicissitudes of taphonomic processes (everything that happens between death and fossilization) and the style of collecting. These factors can be thought of as a series of filters that an organism has to pass through before it can be treated as a known part of the fossil record. Indeed, such is the severity of the filters that it is a wonder that any fossils are preserved at all!

Biology and ecology first. The building blocks of an organism determine critically whether it can ever become a fossil. If there is a skeleton of some kind – bones, a calcareous shell, woody tissue in a plant – then the organism may well be fossilizable. Soft tissues, such as guts and muscles and skin, may be preserved in unusually good chemical situations, but that is rare. So, jellyfish and worms do not have good fossil records since they lack skeletons. Ecology is important too. Organisms that live on the sea bed are more likely to be preserved than fliers or tree-dwellers as more sediment is deposited on the sea bed. Fast-breeding organisms with short life cycles are more often preserved, since there are more of them to find.

Taphonomic filters are many and varied. A skeleton is not a guarantee of preservation. Bones and shells are usually ground up by physical processes, say transport in a river or being rolled among boulders. Or scavengers and microbes may completely eat up all body parts of a carcass. Once in the sediment, bones and shells can be dissolved by weak acids in underground waters. Even if the potential fossil survives all these onslaughts, the rocks in which it becomes entombed may subsequently be crushed, heated, eroded or otherwise mauled by earth movements.

Human factors also come into play. Scientists are not automata. They do not, robot-like, record every fact that is to be recorded, and brave every difficulty to obtain those facts. If you look at palaeontologists' field maps, their finds occur near roads, near towns, near facilities. I don't wish to suggest that palaeontologists are slackers – the majority are enormous enthusiasts, who work unspeakable hours in the field. None the less, there are lots of fossils, and not so many palaeontologists, so the known fossil record represents only a part of the whole fossil record that is locked up in the rocks. Is the known fossil record seriously biased? If it were, then palaeontologists would have to tread with caution.

A biased fossil record?

Fortunately, a number of studies now suggest that published accounts of the fossil record are not seriously biased. For example, Des Maxwell, once my research student and now at the University of the Pacific, in Stockton, California, and I looked at how palaeontological knowledge had changed over the past 100 years.[11] We selected some key palaeontology textbooks published in 1890, 1933, 1945, 1966 and 1987, and drew up charts of the diversification and extinction of tetrapods. We found that, although the total number of known families had doubled since 1890, obviously as a result of the huge amount of fossil collecting during the twentieth century, the overall patterns had not really changed (Fig. 21). Had he wished to, the palaeontologist in 1890 could have identified the same mass extinctions and times of rapid diversification that we could see in 1990.

In a similar study, Jack Sepkoski looked in detail at how his interpretations of the marine fossil record had changed in the 10 years between his two 'Compendia of marine animal families', published in 1982 and 1992. The number of families had increased by perhaps 10%, but the increase was random through time. All that had changed was that the mass extinctions had become sharper. This is strong evidence that palaeontologists

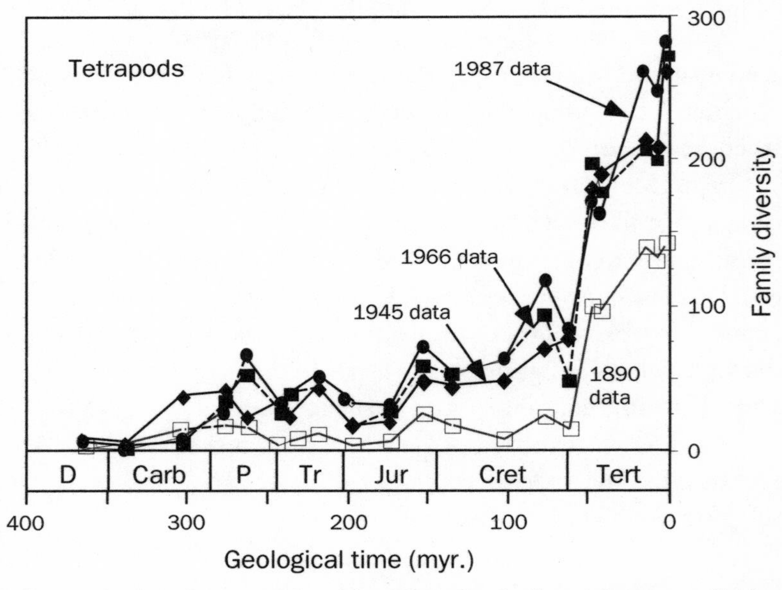

21 *The known fossil record is a good sample of the entire fossil record. Intense collecting efforts by palaeontologists over the past 100 years have doubled the numbers of families recorded, but the overall patterns of diversification and extinction have not really changed. Palaeontologists in 1890 could have identified the same events as their counterparts in 1987.*

have a good understanding of the fossil record. I would make the prediction that another 100 years of fossil collecting will not really change things very much. I'd be amazed if a fossil human is found in Mesozoic rocks, or a dinosaur in the Silurian.

These were global-level studies. What about the smaller scale, the local-level study? Peter Ward of the University of Washington, Seattle, showed how his records of ammonite extinction at the KPg boundary[12] changed with more field collecting. He chose as his field area the famous KPg boundary section in marine limestones at Zumaya in northern Spain. In two field seasons, in 1984 and 1985, Ward crawled all over the beach and collected every ammonite specimen he could find. Specimens were rare, but he identified 16 species, and drew up a chart of their occurrence. Some of the species seemed to last through only a few metres of sediment, while others spanned tens of metres of thickness, and hence, presumably, longer amounts of time. Based on his 1984–85 field collecting, Ward thought he had found a gradual pattern of ammonite extinction. All species were gone well below the KPg boundary, and they had begun disappearing many metres below.

Ward revisited the Zumaya section in 1987 and 1988, and spent longer, searching more intensively. He managed to find two new species and he extended the ranges of all the ammonites he had found by 1986. One species now reached the KPg boundary. In other words, by collecting harder, he had filled some gaps. Further collecting, over the whole region, from 1988 to 1990, filled more gaps, and now 10 out of a total of 31 ammonite species reached the KPg boundary. By intense effort, Ward had converted a gradual extinction into a catastrophic one. Was the extinction catastrophic? Probably. And here was a case where human effort was important in establishing the true picture.

The dilemma of the fossil record

Two apparently opposed viewpoints seem to be emerging. On the one hand, it is obvious that the fossil record is not complete, that many organisms that once lived are not preserved. And it must be clear that the fossil record becomes worse and worse the further back in time one goes. On the other hand, common sense says that palaeontologists know the fossil record well. The recent studies by Maxwell and by Sepkoski seem to prove that. The point is, of course, that the fossil record – all the fossils we can find in the rocks – is merely a sample of the diversity of the life of the past. So we know the fossil record well, but perhaps the fossil record does not really tell us the true history of life? That's a disturbing thought – for a palaeontologist at least. It might suggest that palaeontologists can know the fossil record, but they cannot say much about the history of life.

This thought has worried me for some time. I am a palaeontologist, and I would like to think that I can say some useful things about the history of life. In particular, I would like to think that the fossil record can tell us about major diversifications and major extinction events of the past. How can the problem be resolved? How can anyone compare the known fossil record with the true history of life, unless one has a time machine, or can read the mind of God?

The solution was found in the early 1990s. Palaeontologists are fortunate in having more than one source of information about the history of life. There are of course the fossils in the rocks, but there is also an entirely independent source of data from phylogeny. Phylogeny is the shape of the evolution of life, the evolutionary trees produced by taxonomists when they try to bring order into the diversity of organisms.

Until the 1980s, most taxonomists used some principles of stratigraphy and fossil dates in making up their phylogenies. They would arrange the fossils in order, and essentially join the dots. Older fossils were ancestors of younger fossils, and so on up to the present day. This method works well when the fossil record of a group is very good, when all the ancestors and descendants are preserved. But that is often not the case. Usually the fossil finds are patchy, and they show only the outlines of the pattern of evolution, not the detail. Alternative methods were needed.

Drawing trees

Two new methods for drawing up evolutionary trees were developed in the 1980s and 1990s: cladistics and molecular phylogeny. Both had had their origins in the 1950s and 1960s. Cladistics grew out of the need to formalize taxonomic methods in some way. Willi Hennig, a German entomologist, realized that taxonomists at the time were using all kinds of characters to make their phylogenies and classifications. They might study the shapes of skeletons and teeth, wings and spines, colours and patterns. This was fine, but there was no strict test of which characters might be helpful or not. In trying to draw up a phylogeny of humans and apes, it is no use to note that they have hair and legs. All mammals have hair and legs. It is probably more useful to look for finer features of the skeleton that are shared by two or more species of humans and apes, and then to consider whether any of these characters indicates a unique acquisition in evolution.

The methods of cladistics,[13] proposed by Hennig in the 1950s and 1960s, provided a technique for discovering evolutionarily significant characters – features that indicated a single acquisition event, a branching point. In the human-ape case, significant characters include the relative brain size, the broad chest, the lack of a tail, the shape of the molar teeth: all of these are shared by apes and humans, but not by monkeys. By trial and error, now usually carried out by computer, all possible patterns of relationship are assessed, until the one which best explains the data is arrived at. The pattern is shown as a branching diagram that demonstrates just how humans, chimps, gorillas and gibbons are related. And there has been *no* reference to fossils or stratigraphy. Fossils can be included, but their geological age does not affect the shape of the phylogeny.

The methods of molecular phylogeny reconstruction stemmed from the great discoveries of molecular biology also in the 1950s and 1960s. For the first time, biologists could determine the precise structure of biological molecules. They discovered that organisms share biological molecules, but that their structure can vary subtly from species to species. At first, studies focused on proteins. The molecular biologists discovered that, for example, the haemoglobin of chimps was identical to that of humans, and this confirmed a very close relationship (perhaps too close for some). But human haemoglobin was slightly different from that of a monkey, more different from that of a cow, and very different from that of a shark. The amount of difference in the proteins was probably proportional to the time since any pair of organisms shared a common ancestor. This is the principle of the Molecular Clock: proteins change at a fairly constant rate, and can thus give a timescale of change and a pattern of resemblances. In other words, the chemical differences could be calibrated against a timescale, and phylogenetic trees could be drawn up.

Since 1990, attention has switched from proteins to the nucleic acids, DNA and RNA, the key components of the genetic code.[14] It is possible to automate the sequencing of DNA and RNA, and long chains, often containing many genes, can be compared among species. These sequencing methods lie behind the enormous data-handling efforts of the Human Genome Project, the Tree of Life programme and many other large-scale international projects. There are now hundreds of molecular biologists around the world who specialize in turning sequence information into phylogenetic trees. These phylogenies, like the cladistic ones, are obviously entirely independent of stratigraphy. This is the key.

The test: confronting ages and clades

The first comparisons of information from the independent sources of stratigraphy ('ages') and phylogeny ('clades') were made in 1992 by Mark Norell and Mike Novacek, both from the American Museum of Natural History in New York.[15] At the outset there is no claim that either the phylogeny or the stratigraphy is correct: both are being compared to see whether they agree or not. If they do not agree, then either phylogeny or stratigraphy could be wrong (there is, after all, only one true history of life). If they agree, then they *must* be telling the true story. Couldn't they equally both be agreeing on the *wrong* story? That is not likely, since any biases that affect the quality of the fossil record (soft parts, poor preservation, loss of specimens) cannot affect the phylogenies, many of which are based entirely on living organisms in any case. What came of these studies?

The 1992 comparisons showed good agreement between phylogeny and stratigraphy in 75% of cases. Palaeontologists breathed a sigh of relief. The fossil record was telling a good part of the true history of life, and the missing bits, the soft-bodied organisms, the beasts that lived in places where fossils weren't formed, did not overwhelm the known parts of the fossil record. Various groups of palaeontologists have since pursued the idea, and the largest study, which I carried out in Bristol with Matthew Wills and Rebecca Hitchin,[16] contained comparisons of 1000 published phylogenies, cladistic and molecular. The groups ranged from basal plants to birds, trilobites to turtles, fungi to fishes. Many showed no real congruence, but the majority did agree. More importantly, and the point of the study, there was apparently no change in the quality of the fossil record through time (Fig. 22).

How can that be? Fossils from Cambrian rocks surely cannot be as well preserved as those from the Cretaceous or the Miocene. The older ones must have been crushed, mangled, heated, covered or eroded much more than more recent ones. Both points of view are correct – the damage to older forms and our results – and the resolution is in terms of scope or focus, a point made above. At the level of individual specimens or species from a locality, there will obviously be a much poorer collection of Cambrian fossils than Cretaceous fossils. But at the higher levels of genera and families, and over longer time spans, you are likely to find equivalent proportions of the true diversity from Cambrian and Cretaceous alike. True, you may have only 5 specimens representing one of the Cambrian families, and 10,000 for an equivalent Cretaceous family. But that's sufficient.

These results are highly controversial. Have we really found a way to compare the fossil record with the true history of life? And that without a

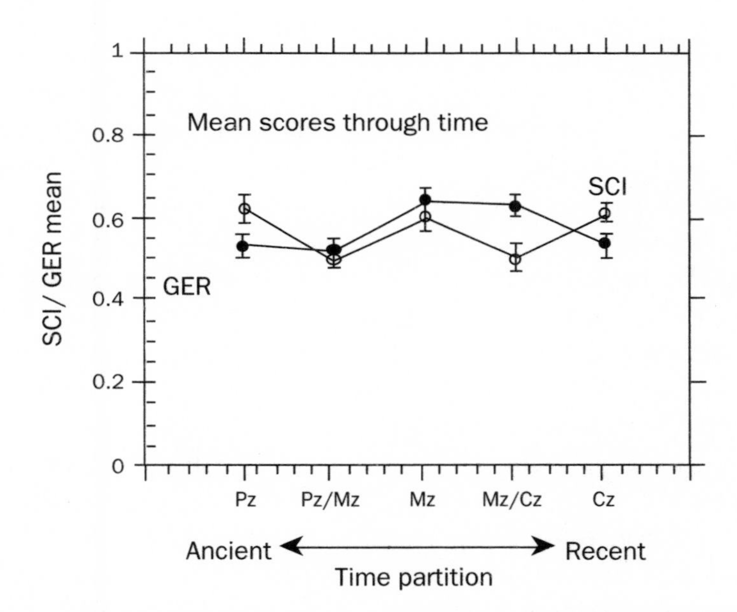

22 *Evidence that the quality of the fossil record has not changed through time. Two measures are used to compare the sequences of fossils and the shapes of phylogenies, SCI (stratigraphic consistency index) and GER (gap excess ratio). In both cases, low values indicate poor matching, and high values good matching. This study is based on a sample of 1000 molecular phylogenies and morphological cladograms, and both SCI and GER remain essentially constant through time.*

time machine? We think so. If the results are accepted, then we can conclude that the fossil record may not be good, it may be full of holes, but it is evidently adequate to tell the true story of the history of life.

Selectivity of mass extinctions

If the fossil record is adequate, then we can look again at mass extinctions for shared patterns, for evidence that might be useful in understanding how they happened, and, perhaps, to contribute to the debate about the modern extinctions. It is hard to establish the timing of mass extinctions, as we saw earlier, but it should be possible to look at some biological aspects. There are two obvious questions: are mass extinctions selective, and what happens after the extinction?

Selectivity first. The second defining character of mass extinctions was that they should be ecologically catholic, that there should be *no* selectivity. This is a somewhat counter-intuitive proposition, since most biologists would predict that some kinds of plants and animals are much more likely to become extinct than others. Surely elephants or pandas, or rare, isolated tropical trees are teetering on the brink of extinction at all times? The commonest question I am asked about dinosaurs is, 'why did they go extinct?' It is natural to try to find reasons. Surely extinction means failure, and the dinosaurs must have been lacking some key faculty?

Numerous studies by palaeontologists, however, have turned up relatively little evidence for selectivity during mass extinctions. The KPg event certainly killed the dinosaurs and some other large reptiles, but much smaller animals died too, marsupial mammals, birds, some groups of crocodiles – so large body size was not the sole weakness. Also, a large number of microscopic planktonic species became extinct. It is not clear that niche breadth (the range of habits and diets of a species) was a strong factor either, since whole clades containing generalists and specialists disappeared at the same time. There is also no apparent bias towards extinction of top carnivores.

The best evidence of selectivity during mass extinctions has been against species with limited geographic ranges. David Jablonski and David Raup of the University of Chicago[17] surveyed all the bivalve and gastropod species and genera of the latest Cretaceous and earliest Tertiary in North America and in Europe, and they found that genera which were geographically restricted were selectively killed off, when compared to those with wider species distributions. So, if you want to survive mass extinctions, make sure you are a species within a geographically widespread genus. Body size and ecological adaptations seem to be unimportant.

Precisely where you live might also be thought to be significant in selectivity during mass extinction events. For example, it has long been suspected that tropical species are more extinction-prone than are those with more polar distributions. This idea was based on the observation that some mass extinction events are associated with an episode of cooling. During such a cooling phase, temperate-belt forms could migrate towards the tropics, tracking their ideal temperature regimes, and they could perhaps survive. But the tropical species have nowhere to go, as global temperatures plummet, and they are squeezed out. Another study by David Raup and David Jablonski,[18] however, has shown no evidence for latitudinal differences in extinction intensities of bivalves during the KPg event.

Recovery

What happens after a mass extinction? Clearly the survivors are a small subset of the full pre-extinction diversity. If there is very little evidence of selectivity, then the missing species will come from all over. There will be holes in the food webs on land and in the sea. In one place a key herbivore might disappear, in another a top carnivore. The survivors will evolve eventually to plug the gaps, or at least the whole ecosystem will eventually evolve back to full complexity.

After mass extinctions, the recovery time, the time for diversity and ecosystem complexity to rebuild, is proportional to the magnitude of the event. Biotic diversity took 10 myr. to recover after major extinction events such as the Late Devonian, the Late Triassic and the KPg. Recovery time after the massive end-Permian event was much longer: it took some 100 myr. for the diversity of families of plants and animals around the world to recover to pre-extinction levels.

It is possible to examine the recovery phase after mass extinctions in more detail. One of the most-studied examples is the replacement of terrestrial tetrapod faunas after the KPg event. With the dinosaurs and other land animals gone, vertebrate communities were much impoverished. The placental mammals, which had diversified a little during Late Cretaceous times, and which ranged in size up to that of a cat, underwent a dramatic radiation. Within the 10 myr. of the Palaeocene and Early Eocene, 20 major clades evolved, and these included the ancestors of all modern orders, ranging from bats to horses, and rodents to whales. During this initial period, overall ordinal diversity was much greater than it is now: it seems that during the ecological rebound from a mass extinction, surviving clades may radiate rapidly, and many body forms and ecological types arise. Half of the dominant placental groups of the Palaeocene became extinct soon after, during a phase of filling of ecospace and competition, until a more stable community pattern became established 10 myr. after the mass extinction. We shall explore recent work on the nature of the recovery of life from the end-Permian mass extinction in Chapter 12.

Lessons from the past

Much of the material in this chapter is addresses large questions, and the solutions in many cases are still tentative. I often think that the more we study the fossil record, the less we really know. But serious study of mass extinctions is a very new science. It need not have been – after all, John Phillips had already documented some of the key aspects over 150 years ago (p. 48) – but the fear of being labelled a catastrophist ruled extinctions and mass extinctions out of bounds for all reasonable scientists until 1970 or 1980. So what do we know now?

- There have been several mass extinctions in the past, at least five.

- Mass extinctions are characterized by a loss of 20–65% of families and 50–95% of species.

- There is some evidence that mass extinctions may be set apart from normal extinctions as a separate class of phenomena, but the evidence is limited.

- There is little evidence for selectivity during mass extinctions, whether by size, diet or habits, though geographically widespread groups seem to be insulated to some extent against the effects of mass extinctions.

This may seem surprising because studies of living animals show that large body size and restricted diet are important threats to survival (think of pandas and elephants). But, during mass extinctions, the culling seems to be relatively indiscriminate.

- Durations of mass extinctions seem to range from virtually instantaneous to 10 myr. for some multiple events.

- Life has always recovered after mass extinctions, but it took some 10 myr. after most of the mass extinctions for global levels of biodiversity to reach their pre-extinction levels, and as much as 100 myr. after the huge end-Permian event.

Many of these ideas about mass extinction had been established by 1990, and yet at that time almost nothing was known about the biggest of all mass extinctions, the end-Permian event. Considerable advances have been made since then in teasing out just what happened 252 million years ago. The best fossil record exists for life in the sea, so we will first explore just what happened in the marine realm during the biggest crisis for life of all time.

7
HOMING IN ON THE EVENT

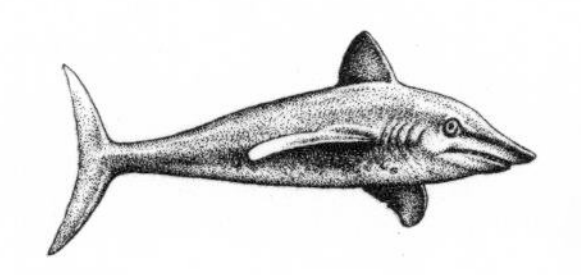

In 1990, the American geologist Curt Teichert,[1] in reviewing the end-Permian mass extinction event, wrote:

> The way in which many Palaeozoic life forms disappeared towards the end of the Permian Period brings to mind Joseph Haydn's Farewell Symphony where, during the last movement, one musician after the other takes his instrument and leaves the stage until, at the end, none is left.

This was the common view in 1990, that species had died out piecemeal through the Mid and Late Permian, a span of some 10 million years. Extinction rates were higher than normal, but perhaps an observer at the time would not have detected that they were actually in the middle of the biggest crisis life had ever faced. The sum total of the disappearance of species one-by-one is complete annihilation.

But is this the correct picture? Was the mass extinction at the end of the Permian so different from the geologically instantaneous KPg event? It might seem unusual if this were the case. After all, the Late Permian event was so much more serious in scale than the KPg. Perhaps Curt Teichert, like many palaeontologists at the time, was merely being cautious, and did not wish to point to a catastrophic model?

In fact, this was largely not the case. The problem arose through difficulties in dating the rocks immediately below and above the Permo-Triassic boundary. Indeed it was not exactly clear where the boundary should be placed. Refinements of dating can be traced through the estimates made by Doug Erwin of the Smithsonian Institution in Washington, a leading expert on the fate of marine molluscs through the crisis. In his 1993 book,[2] he estimated that the end-Permian event lasted from three to eight million years. Only five years later, he and co-workers suggested a span of less than one million years,[3] and by 2011 the timing of the extinction was less than 200,000 years, and perhaps even instantaneous.[4]

The shift of opinion from gradualism to catastrophism in nearly twenty years was not merely the replacement of old-fashioned whims and prejudices. By 1990, work on the KPg boundary had advanced enormously, and the likelihood of a major impact and a rapid extinction at that time were widely accepted. In fact, the tightening up of the end-Permian event, from 10 million years to a geological instant was mainly the result of ever-more refined study of some of the geological sections that span the Permo-Triassic boundary.

But surely, by 1990, the Permo-Triassic boundary was well understood? After all, Roderick Murchison had named the Permian in 1841, and geologists had had 160 years in which to pin down exactly where the boundary with the Triassic lay. Surprisingly, they had failed to do so, and wrangling continued until 2000, when the 'golden spike' was finally driven in, as we shall see, at a section at Meishan in south China.

Wrangling for a hundred years

The story of the argument over defining the Permian is pretty arcane, and may seem pointless, but it is worth persevering in order to appreciate that there is a method at work, that international agreement is essential and that in this case the turbulent political history of Russia hindered this agreement. Clearly, it is essential to make sure everyone is talking about the same thing before one can begin to focus on a particular event, such as a mass extinction, that might have lasted for just a geological instant.

Murchison had initially included only the middle and upper parts of what we now call the Permian in his definition in 1841 – he in fact left out all the lower part of the Permian, calling it generally Permo-Carboniferous, since he was uncertain where to draw the line. So, Murchison's[5] Permian comprised what is now assigned to the Kungurian, Ufimian, Kazanian and Tatarian stages in the Russian system, just over half of what is now included (Fig. 23).

Both the top and bottom boundaries of the Permian had continued to be disputed long after Murchison's time. In 1874, the Russian geologist A. Karpinsky included the underlying Artinskian and Sakmarian units in the

<table>
<tr><td rowspan="9">PERMIAN</td><td rowspan="2">Lopingian (Upper)</td><td>Changhsingian (= Dorashamian)</td></tr>
<tr><td>Wuchiapingian (= Dzhulfian)</td></tr>
<tr><td rowspan="3">Guadalupian (Middle)</td><td>Capitanian (= Maokouan)</td></tr>
<tr><td>Wordian (= Chihsian)</td></tr>
<tr><td>Roadian</td></tr>
<tr><td rowspan="4">Cisuralian (Lower)</td><td>Kungurian</td></tr>
<tr><td>Artinskian</td></tr>
<tr><td>Sakmarian</td></tr>
<tr><td>Asselian</td></tr>
</table>

23 *Timescale of the Permian, showing the international divisions. The Lower Permian is based on the Russian succession, the Middle Permian on the North American (two Chinese equivalents are noted), and the Upper Permian on the Chinese (two Russian equivalents are noted).*

Permian.[6] He showed that Murchison had identified these units wrongly when he trekked down the west side of the Urals. Murchison thought the marine sandstones and limestones around the town of Artinsk were similar to the English Millstone Grit, a familiar rock unit he had seen in northern England, and definitively part of the Upper Carboniferous. Karpinsky's studies of the fossils, however, proved that the Artinskian rocks had more in common with the overlying Kungurian, and were clearly younger than Late Carboniferous. So they were assigned to the Permian.

The true base of the Permian became clearer only after more detailed studies by Russian geologists and palaeontologists around one hundred years after Murchison's work in Russia. V. E. Ruzhentsev restudied Karpinsky's Sakmarian rocks and he realized that they fell naturally into two units, each distinguished by its own suites of fossils. So he divided the Sakmarian into two, naming a lower Asselian stage in 1950,[7] and retaining the name Sakmarian only for the upper portion.

During the nineteenth century, North American geologists evolved their own system for dividing their Permian rocks, and last century, the Chinese did just the same. So now there are three independent schemes available (Fig. 23), and evidently this does not help in identifying particular time lines on a worldwide scale. How is one to choose among these schemes?

The 1937 meeting in Leningrad

Why should Russia, North America and China have their own systems of naming rock units? Obviously, nationalism is at play here. Geologists desperately want to have their national names used worldwide – just as Murchison did in the 1840s, when he had argued that it was clearly right and proper that British systems should be extended worldwide for the universal benefit and improvement of all foreigners.

There is also a geological reason why it is difficult to come to international agreement on matching rock units. Many different kinds of rocks are being deposited at any instant in time. For example, right now desert sands are being deposited in the Sahara, coral-rich limestones in the Caribbean, river sands and muds in the Mississippi, soils and lake sediments in northern India, and glacial sands in northern Canada. How will the future geologist be able to tell that all these sedimentary rocks are of the same age?

The problem is the same for Permian stratigraphers. The Lower Permian rocks of the Ural Mountains in Russia were laid down mainly in the sea, and they contain typical marine fossils; at the same time, continental red beds were accumulating in Texas and New Mexico, complete with their faunas of amphibians and reptiles. Inevitably, the first geologists who arrive in a new district have to choose the marker beds and fossils that are available. It is only now, some 160 years after Murchison's first work in Russia, that the correlations between Russia and China and North America can be achieved with accuracy. In this case, politics played a part in delaying the necessary comparative work.

The key task of the stratigrapher is to define the *base* of his or her rock units. (As mentioned above, unit tops are ignored, since they are automatically defined by the base of the next overlying unit – there might be even more disputes if tops were also defined, and did not somehow correspond with the superincumbent unit base.) In naming the Permian System in 1841, Murchison should have defined the base of the new rock succession. This he believed he had done in including the red beds and salt deposits between Oka and Kazan, and around Perm, on the western flanks of the Urals. We now know, however, that the huge thicknesses of fossiliferous marine limestones and sandstones lying below these units, and above definitive Carboniferous rocks, are also Permian; he had omitted over 3 kilometres of rock accumulation, corresponding to perhaps 25 million years!

This omission was not merely a national affair, of interest solely to Russian geologists. It affected opinion around the world. American stratigraphers were also puzzled, since they had to compare the fossils of their extensive Permo-Carboniferous rock successions with what they could find out about the Russian rocks. Until the Russians decided where to fix the boundary, the Americans could not move.

It was only in 1937 that scholars from around the world were able to examine the Russian Permian seriously. This was at the height of Stalin's rule in Russia, but the Seventeenth International Geological Congress met in Leningrad (St Petersburg) that year, and a long field trip was organized so that overseas geologists could see the whole of the classic Russian Permian.[8] This was a unique experience for the American and European visitors, familiar with their own Permian successions, but desperate to examine the rocks that had given Murchison the evidence to name the system. Equally, for the Russian geologists, unable to leave their country, this was a chance at last to discuss their work with scholars from other nations.

The trip clearly made a huge impression on Carl O. Dunbar, Professor of Paleontology and Stratigraphy at Yale University, and Chairman of the Permian Subcommittee of the National Research Council's committee on stratigraphy. He wrote[9] in 1940 of 'the never-to-be-forgotten days of the Permian excursion' in offering his detailed overview of what he had seen, really the first detailed account in English since Murchison's work one hundred years before. But he noted many areas of disagreement. He reported how the Russian geologists still disputed which rocks were to be correlated with the European and North American Carboniferous, and which should be called Permian. What we now call Asselian he, and they, still assigned to the Carboniferous.

The golden spike

Dunbar's subcommittee was one of many that had been set up to try to pin down international standards of definition of geological units. These committees passed to an international level when they became part of the International Union of Geological Sciences in 1961, and of the International Geological Correlation Programme in 1972, under the umbrella of UNESCO. The role of these committees, with input from interested geologists of all nations, is to hammer in 'golden spikes' at the agreed boundaries of geological units. 'Golden spike' is a figurative term; the boundary is fixed precisely, once and for all, in a particular location that can be visited and studied – a spike made from gold is not actually hammered into the rock.

In Murchison's day, it was enough to refer in general terms to the rocks one wished to include in a newly named unit. By 1940, geologists were seeking something more precise, and they realized that the task would be difficult, since they had to survey the whole world and choose the best available rock section. To do this, they had to master the intricacies of many sequences of rocks, and the local nomenclatures from different lands. They had to attempt to *correlate*, or match, exact rock horizons from place to place, from Russia to Argentina to China. They then had to choose the geologically most appropriate section to be designated as the global *stratotype*, that is, the single reference point that all geologists would use for ever more. Well, that's the theory of how it should work.

By agreement, golden spikes are driven into boundaries in marine successions, since fossils are more abundant and sedimentation is usually more regular in shallow seas than on land. The stratotype section must be thick and continuous, and without any obvious major gaps. Fossils should be abundant and, ideally, of different kinds, such as ammonoids (coiled mollusc shells) and conodonts (microscopic tooth-like fossils that are used extensively for dating marine rocks at that time). Abundant fossils allow the palaeontologists to fix the golden spike immediately at the first appearance of a distinctive fossil. Then, palaeontologists around the world can seek the first appearance of this fossil species in their local successions, and thereby make the correlation with some confidence. If ammonoids are absent, then perhaps they will be able to find the equivalent conodont that marks the same age of rock.

With the work by the Russian stratigraphers, and by western geologists like Dunbar, it might seem that the Carboniferous-Permian boundary could have been settled. But, after the Second World War, contacts between Russian and western geologists became strained, and very few serious exchange visits were possible. Many westerners tried to arrange further fieldwork in the 1950s and 1960s, but they rarely got further than Moscow, where they would be politely taken on tours of the Kremlin and Red Square, but desperate requests to visit the countryside and see some rocks were blandly refused, or paperwork was misplaced.

The situation improved after 1990 and, following many further meetings and debates, the golden spike for the base of the Permian was finally agreed in 2000, and it was placed in a stream section on Aidaralash Creek, in the Aktöbe (formerly Aktyubinsk) region of northern Kazakhstan, lying some way south of the Ural Mountains. It is marked by the first occurrence of the conodont *Streptognathodus isolatus*. So, the base of the Permian was finally agreed. But what of the top, of much greater interest for understanding the end-Permian mass extinction event?

Seeking the base of the Triassic

It became clear early on that the Russian rock successions would not be appropriate for fixing the base of the Triassic (and hence the top of the Permian). The uppermost units of the Tatarian stage in the Urals, and between Moscow and Kazan, were largely red beds, sandstones and mudstones laid down in ancient rivers and as soils between the river channels. Fossils were present, but these were mainly the bones of ancient amphibians and reptiles, with associated plant fossils and small pond-living creatures. Murchison may have been impressed by his finds of reptile bones (see Chapter 1), but modern stratigraphers were not. The search was on for a marine succession that spanned the Permo-Triassic boundary.

Much of western Europe was ruled out. The Permo-Triassic rocks of Britain, France and Germany, for example, were well enough known, but there had been a major switch in deposition from the marine Zechstein to continental red beds in most areas. Indeed, there was a strong suspicion that quite a span of time might also be missing just at the boundary. In southern Europe, on the other hand, there were some successions that spanned the Permo-Triassic boundary entirely in marine rocks. Could these provide the key?

In northern Italy, for example, in the Dolomites, part of the southern Alps, the rock successions show that sea waters were generally deepening.[10] The latest Permian *Bellerophon* Formation consists of dolomites – altered limestones – that were deposited in lagoonal, subtidal conditions. Above these are oolitic limestones of the Tesero Oolite Horizon. Oolites (the word literally means 'egg stones') are small limestone balls, made up from thin shells of limestone that precipitated around a sand grain or shell fragment, and they are formed generally in warm-water lagoons. The overlying Mazzin Member consists of thinly bedded dark grey limestones containing abundant pyrite, evidence of further deepening of the water. The Permo-Triassic boundary lies within the Tesero Oolite Horizon. Below it are abundant latest Permian fossils; above it, these fossils have apparently all disappeared.

Not only did the water become deeper, but oxygen levels also fell. Iron pyrites, commonly called 'fool's gold', a mineral composed of iron and sulphur which forms only in the absence of oxygen, is seen abundantly in rocks of the Mazzin Member. Lack of oxygen, 'anoxia', means that nothing can live in such waters (even the meanest of sea creatures requires oxygen to survive).

Similar successions are found in the Sosio Valley in Sicily. However, neither of these European sections seemed appropriate as an international marker for the base of the Triassic since some critical fossil groups were missing, and it was not possible to tell how complete the sections were. There appeared to be similar gaps, perhaps even greater, in much of North America, where the Late Permian appeared to be largely absent. During the 1960s, stratigraphers turned their sights on Asia.

Geotourism in Asia

In 1961, two American geologists, Curt Teichert and Bernhard Kummel, decided to grasp the problem of the Permo-Triassic boundary seriously. They read all the papers that had been published and decided they should visit every available section. They identified key sites in southern China, Kashmir, northern Pakistan, the northern border between Iran, Armenia and Azerbaijan, and northeast Greenland, and they visited all, or nearly all, of them, in an impressive example of geotourism with a purpose.[11] Sadly, however, they could not visit southern China, largely for political reasons.

In the Iranian-Armenian-Azerbaijani rock successions, Teichert and Kummel encountered many problems. Not least of these was that geologists had been arguing for years over the exact location of the boundary bed. They found the same situation, if not even worse, in the Guryul Ravine in Kashmir, lying between India and Pakistan. Since 1909, palaeontologists had confidently placed the Permo-Triassic boundary at no fewer than *seven* different levels, ranging through a rock thickness of 400 metres in all. With such confusion and dispute, Kummel and Teichert had to sift through masses of contradictory evidence. They opted for the lowest position of all, below the first appearance of the bivalve *Claraia.* However, gaps in the succession of rocks suggested to them that the Guryul Ravine section would not make an ideal global stratotype.

On to Pakistan. Kummel and Teichert visited the classic sections in the Salt Range mountains of northern Pakistan. These had been identified as a richly fossiliferous Permo-Triassic succession in the nineteenth century by the German ammonite expert W. Waagen and others. In the 1950s, Otto Schindewolf and his students returned to the Salt Range, in their efforts to establish what had happened during the end-Permian mass extinction, at a time when his 'neocatastrophist' views were distinctly unfashionable (see Chapter 4).

The Salt Range sections were revisited in 1991 by two British sedimentologists, Paul Wignall from Leeds University and Tony Hallam from the University of Birmingham. They initially believed that there was a major gap in sedimentation just below the Permo-Triassic boundary in the Salt Range, but they reversed their view after further study of the conodonts.[12] The latest Permian Chhidru Formation consists of sandy limestones containing abundant fossils – bellerophontid gastropods, bryozoans, foraminiferans, crinoids, echinoids, brachiopods and algae – evidence for a rich and diverse shallow marine fauna.

The overlying Kathwai Member, which spans the Permo-Triassic boundary, begins with sandy limestones that contain Permian brachiopods and bellerophontids, while higher up there is a switch to thinly bedded sandy limestones, probably indicating deeper water. The main extinction level coincides with this switch in sedimentation style. The overlying Mittiwali Member is also generally thinly bedded, and black shales make their appearance in places. Wignall and Hallam interpret this as evidence for deepening of the water and a reduction in oxygen levels.

So, Wignall and Hallam had confirmed a switch from thick limestone units in the latest Permian to more thinly bedded limestones and black mudstones in the earliest Triassic. But, faced with problems of identifying exactly where the Permo-Triassic boundary lies in the Salt Range, it has been harder to pinpoint the precise pattern of the extinction. If their interpretation of the conodont evidence is correct, it would mean that the end-Permian mass extinction actually happened in the earliest Triassic in Pakistan.

The golden spike is driven

The best evidence for events at the Permo-Triassic boundary seems to come from the north shore of the great Palaeotethys Ocean (Fig. 24). The sections in northern Italy, in Iran-Armenia-Azerbaijan, in Pakistan, in Kashmir and in south China all lie in this vast tract of land. And the boundary sections, from Italy to Kashmir, all appear to show a switch from normal limestone beds in the latest Permian to thin-bedded pyritic limestones at the boundary.

The studies by Kummel and Teichert, and by other geologists, in these Asiatic sections had been devoted mainly to sorting out the stratigraphy, and not much can yet be said about the exact pattern of what happened. Surely geologists must be queuing up to do such important studies? Sadly not at the moment. As Doug Erwin pointed out in his 1993 book,[13] most of these areas coincidentally lie in the midst of territorial disputes and civil wars. The difficulty of access of some of the classic central Asian sections means that we have to wait for essential studies – detailed bed-by-bed sampling to record how the different fossil groups died out, allied with fine-scale dating and geochemical studies.

Kummel and Teichert had been unable to visit China during their global tour of the Permo-Triassic boundary in 1961. Here again politics held up progress, since fieldwork was impossible for anyone in China

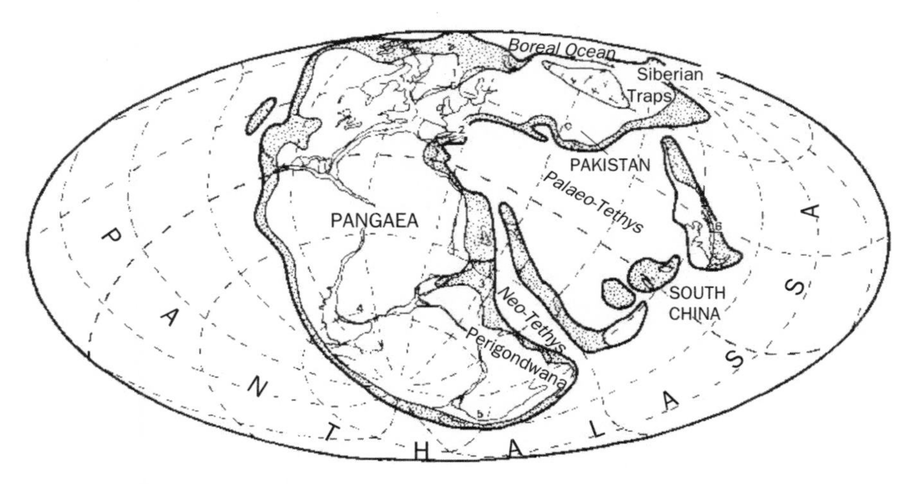

24 *The Late Permian world, showing the distribution of the continents and sites of important rock successions that span the Permo-Triassic boundary.*

during the Cultural Revolution (1966–76) and it took some time for Chinese geologists to recover. Liberalization of the political regime in China since 1980 allowed proper scientific exchanges, and intensive field-work has been completed on many Chinese rock successions that span the Permo-Triassic boundary.

Finally, after much lobbying and campaigning, the section at Meishan, in Zhejiang Province, south China, was selected as the global stratotype for the base of the Triassic in 2000, with the golden spike driven in at the first appearance of the conodont *Hindeodus parvus*.[14] Not only is the top of the Permian defined in China, but so too are the other subdivisions of the Late Permian (Fig. 23), because of the good quality of the marine successions there. So, 159 years after Murchison named the Permian, the limits of the system had finally been pinned down.

The Meishan section

The Meishan section is exposed in five quarries, all close together. Combining information from the different quarries gives a section some 50 metres thick, with the critical boundary bed occurring 40 metres above the base (Fig. 25). The lower part of the succession, termed the Baoqing Member, consists of thick units of limestone, with thinly bedded limestones in between. These limestones were laid down on a marine slope, as shown by evidence of movement under gravity and abundant fossils such as foraminiferans, brachiopods and conodonts. Rarer fossils include cephalopods (coiled molluscs), echinoderms (sea urchins and starfish) and ostracods (small crustaceans), again all typical of shallow seas.

The sediments tell the story of the environments in which they were deposited, according to Paul Wignall and Tony Hallam.[15] The thick and thin limestones attest to warm-water shallow seas, with some water currents that washed the shells and other animal remains around on the sea floor before they were finally incorporated into the rock. The thin beds include some black muddy limestones, the black colour coming from organic matter that remains in the sediment. This indicates anoxic ('no-oxygen') conditions, since organic matter is usually completely decomposed by scavenging organisms which can only survive if oxygen is present. The thin-bedded limestones lying between the black mudstones do not contain any burrows, further evidence of an absence of life on the

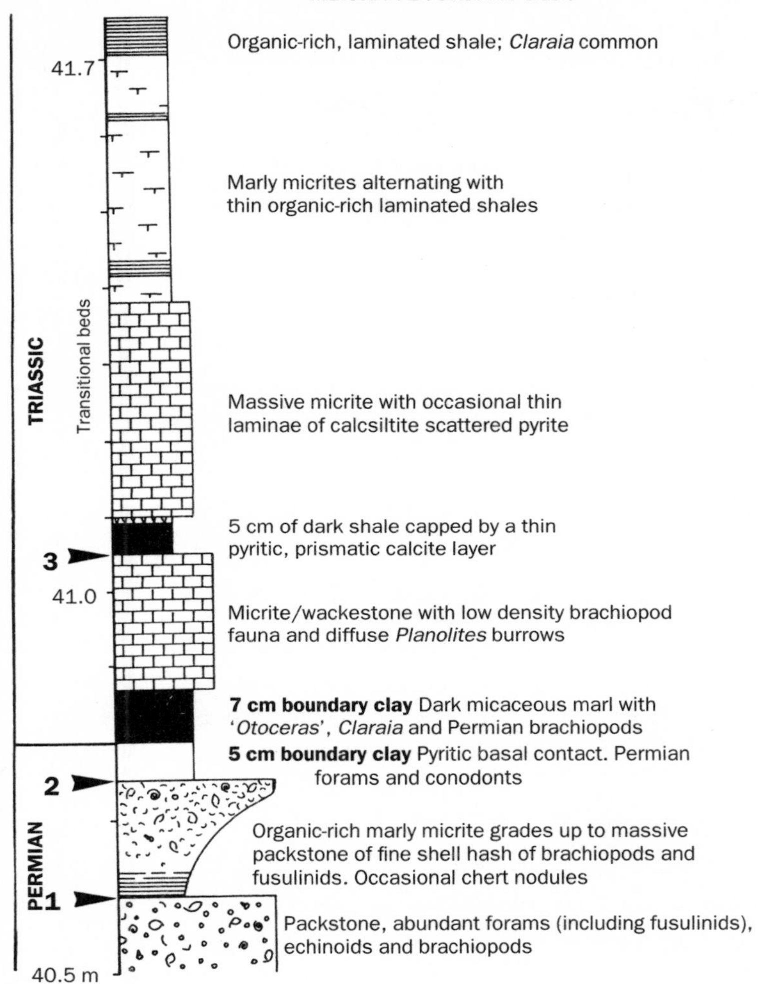

25 *The Meishan section, including the critical Permo-Triassic boundary horizon, as defined by international agreement in 2000. Extinction levels are marked with arrows (1, 2, 3).*

sea floor, since there are always worms and shellfish that creep about on the sand and burrow into it in search of food or safety.

On top of the sediments of the Baoqing Member are those of the Meishan Member, the last to be deposited in the Permian. The rocks are limestones again, and depositional conditions were apparently similar to those of the Baoqing. Near the top, there is extensive burrowing in the limestones, indicating a return to conditions of full oxygenation. Suddenly, everything changes. The thick, burrowed limestones have gone, and so too have the abundant fossils.

The highest limestone of the Meishan (bed 24) is followed by 29 centimetres of clays and more limestone. First in the clays comes a pale-coloured ash and clay bed, then a dark organic-rich mudstone, followed by a muddy limestone. In the Chinese system these are numbered as beds 25, 26 and 27 respectively. Above bed 27 comes a long succession of thin limestones and black shales, termed collectively the Chinglung Formation, containing only rare, small burrows. Here is a major succession of low-oxygen beds spanning one or two million years. Layers of pyrite crystals scattered abundantly throughout the Chinglung sequence provide further confirmation of the low-oxygen conditions. Beds 25, 26 and 27 are where it was all happening, so let's look at them in more detail.

The boundary beds

In the Meishan quarries, the rocks that mark the biggest mass extinction of all time do not look particularly unusual, just a succession of grey and black limestones and mudstones (Fig. 26). But a millimetre-by-millimetre dissection tells a strange story.

The last typical unit of the Permian, the limestone bed number 24 of the Meishan Member, contains Permian fossils, highly worn specimens of brachiopods and foraminiferans. At the top of bed 24 is a mineral-rich layer of pyrite and gypsum, the pyrite certainly suggesting anoxic

26 Photograph of the Permo-Triassic boundary in the Meishan section, northeastern China. Beds 24 to 28 are numbered. The end-Permian mass extinction happened in three closely spaced phases here, at the base and the top of bed 24, and at the top of bed 27. The global stratotype for the base of the Triassic is defined in the middle of bed 27, with the first appearance of the conodont Hindeodus parvus.

conditions. The meaning of the gypsum has been debated. Gypsum is normally found as a salt that has been produced by evaporation of sea water, but Wignall and Hallam have suggested that here it may simply indicate an interaction of the pyrite and limestone during modern-day weathering.

The so-called lower boundary clay, bed 25, is a thin band of pale-coloured clay, only 5 centimetres thick, which contains scarce Permian foraminiferans and conodonts. Under the microscope, this clay contains small iron-rich pellets and decayed pieces of quartz that indicate that it was modified from an acidic tuff, an amalgam of volcanic fragments and ash from an explosive volcanic eruption.

The upper boundary layer, bed 26, consists of 7 centimetres of dark, organic-rich limey mudstone which contains a mixture of Permian and Triassic fossils – Permian brachiopods and goniatites and Triassic bivalves (clams) and ammonoids. Goniatites and ammonoids are cephalopod molluscs, distant relatives of the modern squid and octopus. Conditions during deposition of bed 26 were low in oxygen, but not anoxic, based on the relatively diverse fossils, and on geochemical evidence.

These two mudstone beds, 25 and 26, form a distinctive black-on-white marker band, called an ash band. It has been detected in many localities throughout south China and is useful for geologists wishing to make correlations from location to location. The double deposition event that produced it must have occurred over as much as 1 million square kilometres of China. What kind of process could have produced such a thin ash band carpeting such a huge area?

Bed 27, a 17-centimetre thick limestone, contains pyrite crystals here and there, but it is also full of burrows, showing that bottom conditions were not particularly low in oxygen. The unit contains rare Permian brachiopod fossils near the base, in the so-called units 27a and 27b. Then these disappear, and the conodont *Hindeodus parvus* is found for the first time at the base of subdivision 27c. Here, 5 centimetres above the base of bed 27, Chinese geologists have now convinced the world that the golden spike that marks the base of the Triassic should be placed.

Dating the end of the Permian

Older geological time charts give ages for the Permo-Triassic boundary anywhere between 225 and 250 million years ago, while publications since 1980 have homed in on dates of 245, 248 or 250 million years ago. These dates were not very precise, however, being merely *interpolations*. Radiometric dates were available for the middle of the Triassic (238 myr.) and for the base of the Artinskian in the Lower Permian (268 myr.), and something from 245 to 250 myr. sounded about right for the Permo-Triassic boundary. This was a broad guess, and the span of 30 million years between the two radiometrically dated fixed points was far too long.

The Chinese sections offered fantastic new opportunities for dating – not only could the boundaries be determined by means of the fossils, but there were also clay/ash bands in bed 25, close to the boundary, and further clay bands both below and above the Permo-Triassic boundary and these could be dated scientifically. Geologists descended on the section, and bags of clay were whisked off to laboratories in China and elsewhere in the world.

The first dates, published in 1991 and 1992,[16] were assessed using the uranium-lead method, and these gave measures of 250 ± 6 myr. and 251.1 ± 3.4 myr. The uranium-lead method is based on the change of the uranium-238 isotope to lead-206, a transition with a half-life of 4500 million years. The plus-or-minus (±) figure is the assessment of experimental error – the range of ages that were found by repeated analyses in the laboratory. It is not a measure of global error, in the sense that no one knows what the maximum range of possible age estimates might be. Repeated age determinations by different laboratories, and using different isotope series and different equipment, are the best test of the accuracy of any published radiometric date.

Such a test came in 1995, when an American group[17] used the new argon-39 to argon-40 technique to achieve a date of 249.9 ± 1.5 myr., well within the range indicated by the uranium-lead dates. In a further, even more detailed study, an American-Chinese team led by Sam Bowring of MIT[18] reverted to the uranium/lead method, but they used two isotopic series, the decay of uranium-238 to lead-206, and the decay of uranium-235 to lead-207, which has a shorter half-life of 700 million years. By using the two isotope series, this team was effectively cross-checking every measurement they made. The ash bands in bed 25 were dated at 251.4 ± 0.3 myr. and bed 28, immediately above the Permo-Triassic boundary, yielded a date of 250.7 ± 0.3 myr. So, beds 26 and 27, the latter of which contains the official base of the Triassic, fall between those two dates, representing up to 700,000 years (0.7 myr.). Splitting the difference gave a date for the Permo-Triassic boundary of 251.0 myr.

Despite the thoroughness of their study, it has been claimed that Bowring and colleagues had not allowed for errors coming from unaccounted loss and inheritance of lead, and from mixing of grains of slightly different ages within their samples. These dates were revised upwards to about 253 myr. by Roland Mundil in 2001. Finally, consensual work by Shu-Zhong Shen and colleagues, using new uranium-lead methods, dates the ash-bearing beds 25 and 28 as 180,000 years apart, and the Permo-Triassic boundary as 252 ± 0.06 myr. We use the 252 myr. date here. The three apparent pulses of extinction, at the bottom and top of bed 24, and at the top of bed 27, may, in total, span about 1 million years. But what was going on in this relatively short interval?

Carbon isotope shifts

Isotope geochemistry can provide evidence about ancient environments as well as rock dates. In the case of environments, geochemists focus on isotopes of carbon and oxygen. Carbon isotopes show a sharp negative excursion (see Fig. 27), dropping from a value of +2 or +4 parts per thousand to -2 parts per thousand in the pale-coloured mudstone, bed 25, which corresponds to the main extinction level. What does this mean?

The impressive-looking term $\delta^{13}C$ is essentially the ratio of the two stable isotopes of carbon, ^{13}C and ^{12}C (the 12 and 13 are the atomic weights) measured against a standard, and calibrated as parts per thousand. When plants photosynthesize, they take up the isotope ^{12}C from the soil (land plants) or from seawater (the floating phytoplankton) by preference, and this has the effect of increasing the proportion of ^{13}C left behind. A geologist in the future who measures the $\delta^{13}C$ ratio from the soils or sea-bed sediments being deposited today will note relatively high values, and will interpret these as an indication of high levels of biological activity, sometimes termed 'productivity'.

A negative shift of 4 to 6 parts per thousand in the $\delta^{13}C$ ratio might seem fairly minor, maybe just a local effect. This negative excursion, however, seems to be a global phenomenon, having been found in Permo-Triassic sections worldwide. It is also actually large. At the KPg boundary, the negative excursion is smaller, a mere 2 or 3 parts per thousand. The

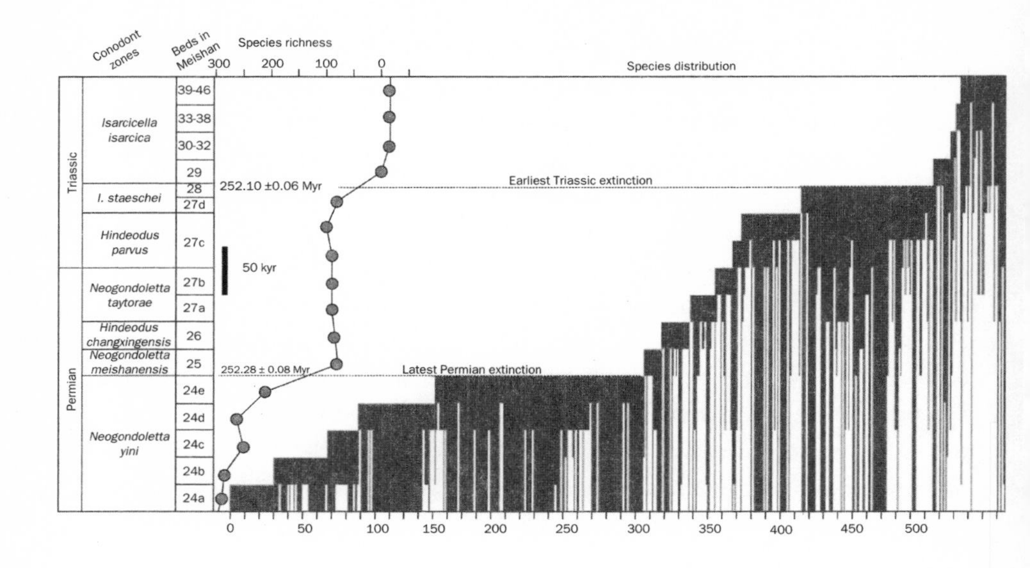

27 *The extinction of life at the end of the Permian in China, showing radiometric ages, the carbon isotope curve, and the ranges of 560 species of fossils identified from 90 metres of rock in the Meishan quarries. The three apparent pulses of extinction are indicated.*

drop in the $\delta^{13}C$ ratio at the Permo-Triassic boundary could be interpreted as evidence for a major reduction in biological productivity, just as would be expected at a time of mass extinction. But, as we will see later, even the most astonishing level of extinction could not produce such a drop.

Oxygen isotopes in limestones show a similar negative shift, from a high value of -3 to -1 parts per thousand in the Late Permian, to -7 just at the boundary. The oxygen isotope ratio, $\delta^{18}O$, is often used as a palaeo-thermometer, and a drop of 4 parts per thousand could indicate a global increase in temperature of around 16°C.

So, in summary, the carbon isotopes could suggest a major drop in productivity (but there must be much more to it than simply that), while the oxygen isotopes suggest that temperatures rose dramatically at the same time. These are the simplest interpretations of the geochemical data, however, and there are many complicating factors that we must explore later in trying to work out exactly what happened to the world 252 million years ago (see Chapter 11). We have seen evidence of some startling changes in marine environments in the Meishan section. But what of life in them?

The biggest extinction anatomized

Chinese and western palaeontologists have collected fossils intensively through the Meishan section in an attempt to discover exactly what happened. Early reports suggested that there had been three levels of extinction, perhaps spanning half a million years altogether. This clearly had to be assessed in the greatest detail possible, so a Chinese-American team set about the task. Y. G. Jin and his colleagues from the Nanjing Institute of Geology and Palaeontology, and Doug Erwin from the National Museum of Natural History in Washington, undertook a huge sampling programme.[19] This work was revised and extended by Haijun Song and colleagues in 2013.

These teams collected fossils from 64 levels through the 90 metres of rocks that span the Permo-Triassic boundary in the Meishan quarries, from bed 1 to bed 46 of the standard numbered succession. In all, 537 species were identified, belonging to 15 marine fossil groups – microscopic foraminifera, fusulinids and radiolaria (all microscopic organisms, some of them living on the sea floor and others floating at the sea surface), rugose corals and bryozoans (both of which form colonies made from numerous individual animals living in a shared structure), as well as brachiopods, bivalves, cephalopods and gastropods (shellfish that lived on, or near, the sea floor), ostracods (swimming, shelled shrimp-like creatures), trilobites, conodonts, fishes and algae (seaweeds) – and they plotted their exact ranges against the stratigraphic section (Fig. 27).

The plot shows a number of minor extinctions low in the section. Some 150 species died out at different points through bed 24, but these were 'normal' extinctions, part of the regular turnover of species. There seems to have been a major step at the top of bed 24e, and then a series of small stepwise extinctions up to the top of 28, where a further major loss of species seems to have occurred. What does all this mean: was there a single mass extinction event, or two, or are these steps of extinction just reflections of incomplete recording of the data?

The Signor-Lipps effect and forward smearing

It is important to determine whether the end-Permian mass extinction happened at one level – in a geological instant – or whether it occurred in several pulses. Does the Meishan section show a single extinction event, the main step in the fossil ranges at 252 myr. (Fig. 27: B), or two levels of extinction? Even though the extinction step at the top of bed 28 is comparatively small, it is still there.

But palaeontologists are always cautious about reading the rocks literally, as they may mislead the unwary investigator. Imagine that extinction rates are constant, with a species disappearing, on average, every 100,000 years. If sedimentation is not relatively constant, the regular background extinction pattern may be disrupted. A gap in sedimentation, representing half-a-million years would automatically create an apparent sudden extinction of five species. But that's not all.

Palaeontologists may be sure of one thing: they will never find the last surviving member of a particular species. This observation is based on the low probability that any individual organism will actually become a fossil, and the likelihood that the last stragglers of a species may be holed up in some obscure location that never fossilizes, or is never found. Therefore, when a mass extinction occurs, and 100 species die out in an instant, the best a palaeontologist can hope for, even after months of assiduous rock smashing, is a pattern where only 20 or so actually disappear in the extinction level, and the remaining 80 seem to drop out in the levels below. So, palaeontologists expect to find backwards smearing of a mass extinction event, a superficially gradual dropping out of species one-by-one. This is the Signor-Lipps effect, named after the American palaeontologists, Philip Signor and Jere Lipps, who first presented the idea in statistical terms.[20]

Smearing can occur forwards as well. Fossils may be found by chance *above* their extinction level. This happens commonly when sediments are burrowed by worms and shrimps, which can churn the sands and muds to a depth of a metre. In a section like Meishan, a metre of rock in the Permo-Triassic boundary beds can represent several hundred thousand years. Fossils in the burrowed rock may be moved up and down by the burrowers. Careful study of the rocks should show whether fossils are in place, or whether they have been moved by burrowers, but upwards smearing of a mass dying level has to be considered.

In their 2000 paper, Jin and colleagues assessed the reliability of what they called extinction levels A and C, at the tops of beds 24d, 24e and 28 (Fig. 27). Could they simply be backwards and forwards smearing of a single mass extinction, level B, at 251.4 million years ago? Steps A and C are certainly much smaller than B, which immediately suggests the possibility that they do not mark real events. Following their careful calculations, the investigators found that the lower level A, at the top of bed 24d, simply disappeared: some of the species that apparently died out at this level are almost certainly false last records, evidence of the Signor-Lipps effect.

As for the top of bed 28, there is evidence for burrowing in bed 27, for example, but the fossils have not been shifted up through 17 centimetres of limestone here. About 110 species apparently disappear between the top of bed 24e and the top of bed 17c, and further species drop out in levels above. The sedimentological observations, and the statistical tests, do not suggest that these have all been smeared forwards from the true mass extinction level. The picture is of two distinct mass extinction phases, one at the top of bed 24e, and the second, perhaps 180,000 years later, at the top of bed 28, these accounting for the seemingly instantaneous loss of 56% and 71% of species. These pulses of extinction lie on either side of the Permo-Triassic boundary, between beds 27b and 27c.

Is this, however, merely a local pattern, found at Meishan, but not elsewhere? If it is, then it cannot tell us so much about what happened 252 million years ago. But, if the same pattern of geochemical changes, and dying, can be found all over the world, then geologists can begin to look for the global-scale processes behind the end-Permian mass extinction.

Events in China and Pakistan

In 1991, after visiting Meishan, Paul Wignall and Tony Hallam went on to examine other sites in south China.[21] The Hushan section, 200 kilometres to the north of Meishan in Anhui Province, 'records a virtually identical series of earliest Triassic events to those seen at Meishan'. There are the white and black clay bands marking the boundary, and containing a mixture of Permian and Triassic fossils. The same chemical changes are documented, with a major drop in oceanic productivity, a rise in temperature and a progressive shift to anoxic conditions in the earliest Triassic. Numerous other Chinese sections show the same pattern, and the extinction event seems to happen at the same time in each section.

In Pakistan, however, as we have seen, the end-Permian mass extinction may have happened rather later, in the earliest Triassic. This possible difference in timing of the mass extinction in China and Pakistan is far from certain. Admittedly, south China and northern Pakistan are some distance apart today – about 4000 kilometres – and they were probably as far apart in the Late Permian. At that time (see Fig. 24), both regions were under the sea, as the sediments show, Pakistan on the north shore of the great Tethys Ocean, and China part of a microcontinent to the east.

The sedimentary successions in Pakistan and China accumulated on the shores of separate continents, but a difference in timing of the mass extinction would lead to a very different model for what happened 252 million years ago from one in which events were exactly coincident. What of the rest of the world?

The Panthalassa Ocean

Not much is known about what happened in the vast Panthalassa Ocean. This ocean (see Fig. 24), surrounding the continents, corresponds to the modern Pacific, but it was much wider, since the Atlantic did not exist. In the Mino-Tanba Belt in southwest Japan, on the west shore of the Panthalassa Ocean, the boundary rocks consist of thicknesses of red cherts: hard silica-rich, almost glassy rocks, made up from the skeletons of radiolaria, minute planktonic animals that fell to the sea floor and accumulated slowly for millions of years. The monotony of these cherts is broken by the Toishi Shale that spans the Permo-Triassic boundary. Exactly at the boundary is a black shale. Why the change in sedimentation?

A red colour in sediments indicates the presence of abundant oxygen, just as a black colour usually indicates anoxia, the absence of oxygen. Grey is something in between. Colours are a useful diagnostic tool for geologists, and they are usually reliable since the colours are produced by the same properties of the rocks in most cases. Red colour in a sandstone, a mudstone or a chert, often indicates the presence of iron oxide in the form of haematite. If haematite is deposited in a sedimentary rock in low-oxygen conditions, much of the oxygen is lost, and the haematite transforms into other iron minerals that are grey or green in colour. A black colour usually means that there is a great deal of organic carbon – the remains of dead plants and animals which have not been scavenged by detritus-eaters or combined with oxygen to form carbon dioxide.

Simply interpreted, then, the Japanese section shows well-oxygenated marine conditions in the Late Permian (red cherts), after which oxygen levels fell (grey cherts), falling further to a very low level in the lower Toishi Shale (grey shales), to almost none at all at the boundary (black shales). Then, in a precise reversal of the sequence, oxygen levels built up in the Early and Middle Triassic, a span of 15 million years, from black shales, to grey shales, to grey cherts and finally to red cherts again.

The Japanese section is not yet well enough dated for palaeontologists to be sure exactly how the end-Permian extinctions happened. Late Permian radiolarians died out as oxygen levels fell, but that is all we know. Whether the Japanese extinctions happened at the same time as the main mass extinction horizon in China, or before or after it, cannot yet be ascertained. Events in northern continents may have been rather more drawn out, however.

Greenland

Geologists, from Kummel and Teichert onwards, have also turned their eyes northwards, to Greenland. In the Late Permian, a narrow seaway extended down from the great northern Boreal Ocean into Laurasia (see Fig. 24), bringing shallow marine sedimentation and faunas into the north-eastern coast of that continent.

The Greenland succession consists of a series of rocks called the Schuchert Dal Formation, essentially latest Permian in age, overlain by the Wordie Creek Formation, dated as Triassic. The Schuchert Dal Formation consists of a mixture of marine limestones, shales and gypsum containing many fossils and it is heavily burrowed. The Wordie Creek Formation is composed of dark grey sandstones and siltstones, some of them containing pyrite. Fossils are rare, especially in the lower portions, and there is little sign of burrowing. So here we see the usual switch from limestones and mixed sediments with rich fossil faunas to deeper-water anoxic mudstones, with little sign of life, after the end-Permian event. However, the fossils in the Greenland successions contain a conundrum.

Examples of Permian fossils – brachiopods, corals, foraminifera, crinoids, echinoids and bryozoans – are found as much as 20 metres into the 'Triassic' Wordie Creek Formation. One explanation could be that the final extinction of Permian organisms happened slightly later in the Boreal Ocean than in equatorial regions. Curt Teichert and Bernhard Kummel could not accept that. Surely the Permian fossils should have disappeared at the end of the Permian, in other words, at the top of the Schuchert Dal Formation? But how then could one account for the presence of abundant, and undamaged, Permian fossils so high in the Wordie Creek Formation?

Armoured mudballs.[22] Teichert and Kummel argued that the delicate Permian fossils in the Wordie Creek Formation had been reworked in the Early Triassic. The beasts had indeed died at the end of the Permian, leaving their skeletons in the Schuchert Dal Formation. Erosion by seabed currents in Wordie Creek times supposedly cut into the Schuchert Dal beds, releasing the shells and other remains. But how could these remains have been transported without being damaged? Teichert and Kummel suggested that great balls of mud had passed over the older rocks, picking up delicate shells and fragments on their surfaces, and that these had then come to rest, dissolved somehow, and had left the older fossils in younger sediments.

Hallam and Wignall gave their succinct view of '[t]his frankly ridiculous scenario'. It was clearly nonsense, they suggested, since no examples of giant dissolving mudballs have ever been reported, and the Permian-style fossils in the first 20 metres of the Wordie Creek Formation are preserved more or less in life position. In other words, the shells are sitting on their backs as they would have when alive; the corals and bryozoans stand upright like little trees. No rolling mudball could do that.

So were extinctions in Greenland later than in China? Was there a latitudinal, perhaps climatic, effect? Probably not.

Timing of the end-Permian event

The clue to timing is in the Meishan section. There the end-Permian mass extinction happened essentially in a geological instant, but the initial event was followed by a period of thousands of years when species continued to disappear (see Fig. 27). Perhaps the seemingly later extinctions in Pakistan and Greenland correspond to part of this later decline. It is striking that the highly condensed sections in China (some 50 metres document 6 myr.) show the same patterns as the extended Greenland sections, where the equivalent time span is represented by over 500 metres of sediments.

Work published in 2001 gave some insights into the timing of events. Richard Twitchett, then based at the University of Southern California, and Cindy Looy from the University of Utrecht, and their colleagues,[23] had been studying the Greenland sections in intense detail. They found that in fact the marine record there documents a collapse of ecosystems within 50 centimetres of the section. At the top of the Schuchert Dal Formation, life was normal: a typical latest Permian assemblage of brachiopods, corals, ammonoids and foraminifera is preserved. The sediments are intensely burrowed, indicating a well-oxygenated environment.

Then, through the next 50 centimetres of green and grey muddy siltstones, life is devastated. The burrows disappear and there are almost no fossils. All the seabed life that existed just a few centimetres below has gone, and only a few species reappear above the boundary beds (the late survivors at Meishan and the 'armoured mudball' fauna in Greenland), but these Permian survivors rapidly disappear. At the base of the Wordie Creek Formation are black anoxic mudstones containing virtually no fossils of any kind. Based on their calculations of rates of deposition, Twitchett and Looy estimated that the 50-centimetre death bed represents something between 10,000 and 60,000 years.

It is too soon, however, to be sure whether this timescale holds true worldwide. Only the Meishan and the Greenland sections have received the sort of detailed stratigraphic study that is necessary to plot the precise course of events. Until geologists and palaeontologists return to the sections outside China and Greenland to collect the fossils centimetre-by-centimetre, and obtain some good radiometric dates using modern analytical machines, we shall be uncertain.

And until such detailed results are available, it will be most sensible to read the patterns from China and Greenland as indications of what happened. We have seen that diverse groups of brachiopods, corals, bryozoans, foraminiferans and others died out. It is important now to look at those groups of marine animals to try to understand the seriousness of the event. Ecosystems were ripped to shreds, but what were those ecosystems?

8
LIFE'S BIGGEST CHALLENGE

Doug Erwin of the Smithsonian Institution in Washington called the end-Permian event 'the mother of all mass extinctions', a reference to the remark made by Saddam Hussein, President of Iraq, in 1990 that the Gulf War would be 'the mother of all battles'. Erwin characterized the biotic effects of the extinction:[1]

> Killing over 90% of the species in the oceans and about 70% of vertebrate families on land is remarkably difficult. The end-Permian mass extinction was the closest metazoans [animals] have come to being exterminated during the past 600 million years. The effects of the extinction are with us still, for it changed the structure and composition of marine communities far more than any event since the Cambrian radiation.

This is dramatic stuff, but no exaggeration. The quotation highlights three topics that we will consider in this chapter.

First, it is necessary to review the different groups of animals in the sea, how they lived and their role in ecosystems in the Late Permian, just before the event. In each case, it is important then to determine what happened to each group. Did they all die out at once, or were some more resistant to extinction than others?

Second, it will be critical to estimate just what was the magnitude of the event. Palaeontologists bandy all kinds of figures around for the percentage of species that were lost: 75%, 80%, 90%, or 96%. In a sense, it doesn't really matter, since such losses mean that only 4–25% of species survived, a tiny fraction of the total. Equally, however, it is important to get the figures right, for comparison with the KPg event and others of that magnitude – and indeed with present-day rates of species loss.

Third, it is necessary to try to visualize just which plants and animals managed to creep through the crisis. Which groups survived, and which were entirely wiped out? The broader-scale recovery after the end-Permian event will be considered later (Chapter 11), but death and survival will be considered here. In addition to determining the victims and the survivors, it is important then to enquire whether there is any evidence for selectivity. Did the survivors share any features that set them apart from the victims?

Before and after

Comparison of the scene before and after the end-Permian event illustrates the severity of the mass extinction (Figs 28, 29). The 'before' scenes, from the latest Permian, show reef communities with rich faunas of corals, bryozoans and crinoids, with ammonoids and fishes swimming above. Brachiopods, sea urchins, snails and foraminifera rest, or move slowly, on the bottom. These are rich and complex ecosystems, whether on the shores of the Tethys Ocean, the Chinese example, or on the edge of the Boreal Ocean, the Greenland example.

In the earliest Triassic the reefs have gone, and all the organisms that were specialized to form reef structures have died out. The rich accumulation of shelly debris is missing, since everything has died, and each scene is dominated by a single kind of bivalve, *Claraia*. The earliest Triassic Greenland fauna seems to be richer than that of China, with a small shark and other small fishes, a couple of ammonoids and a conodont animal swimming over a sparse fauna of *Claraia* and some small snails.

These impressions of life before and after the great extinction show just how severe it was. A global survey of the major groups of organisms in the sea confirms the enormity of the changes.

Microscopic floaters

Some of the key organisms in the sea are invisible to the naked eye. These are the plankton, the microscopic plant-like and animal-like organisms that float near the surface. The plant-like plankton capture energy from the sun by photosynthesis, just as green plants do on land, and they are fed on by the animal-like plankton. Planktonic organisms keep afloat by an

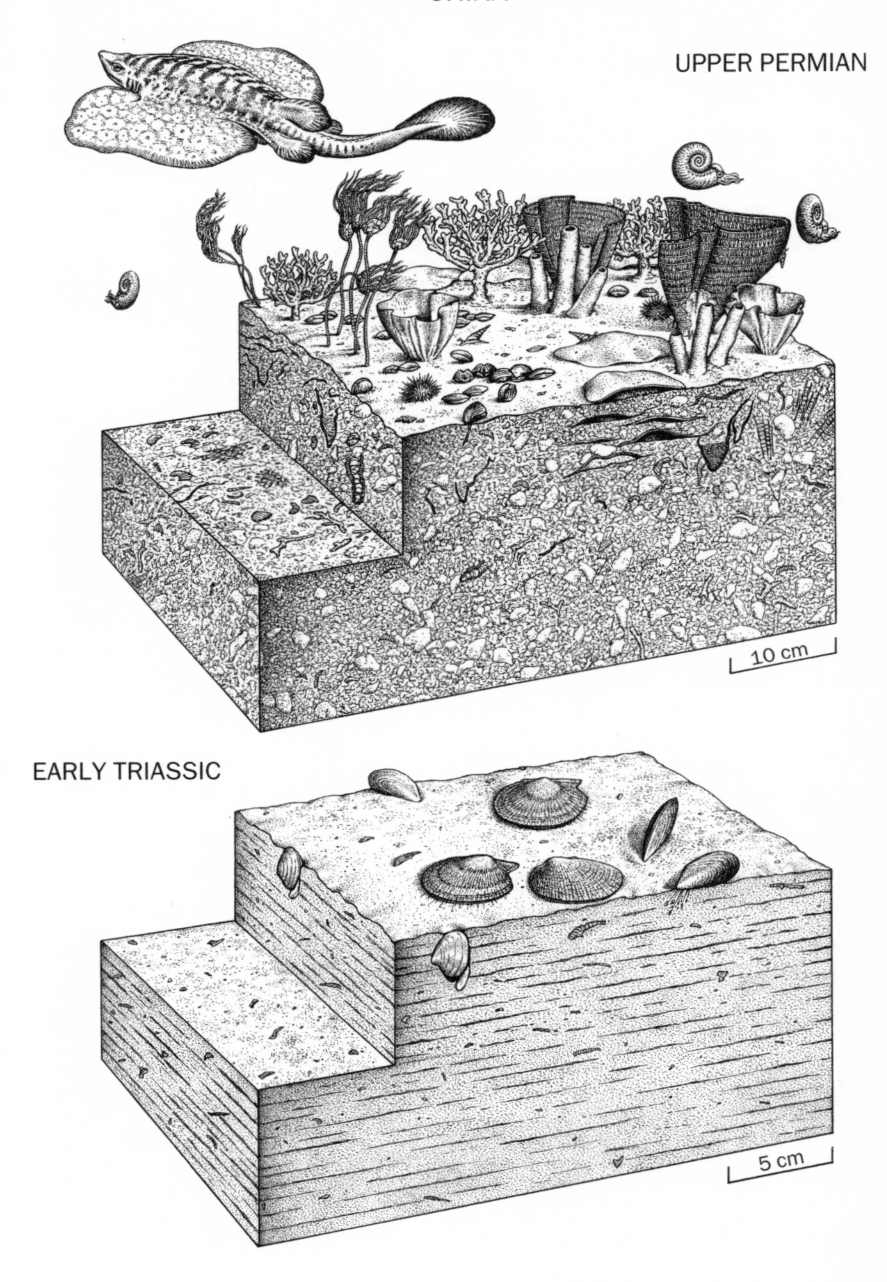

28 *Before and after: the tropical ocean. A reconstruction of the latest Permian sea-bed (above) and the earliest Triassic sea-bed (below), immediately after the catastrophe; based on information from the Meishan section, China.*

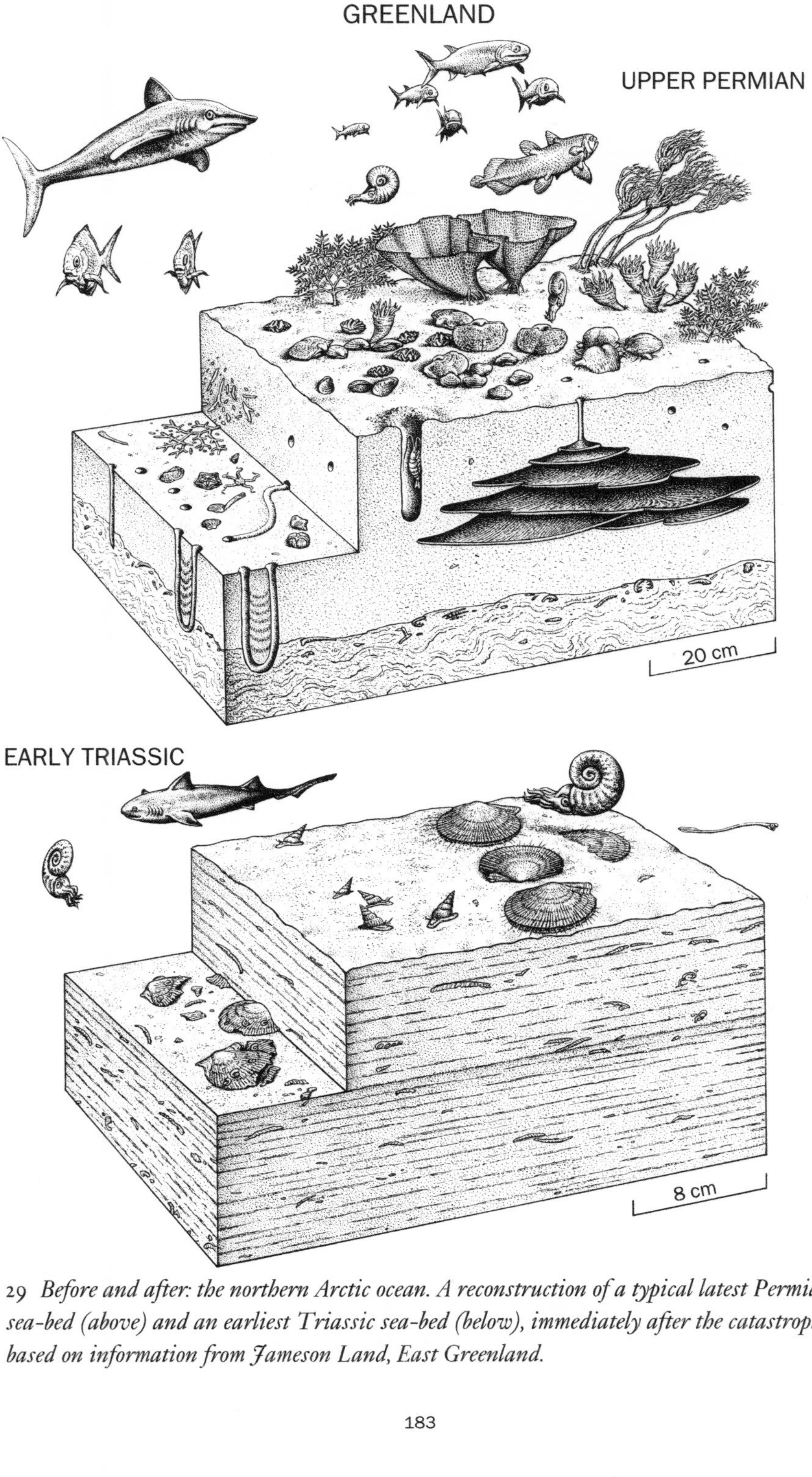

29 *Before and after: the northern Arctic ocean. A reconstruction of a typical latest Permian sea-bed (above) and an earliest Triassic sea-bed (below), immediately after the catastrophe; based on information from Jameson Land, East Greenland.*

amazing array of special devices – broad flanges, spikes, gas-filled balloons – that stop them from sinking; some have spiral body shapes so that they spin slowly. They include many unique groups that spend their entire lives in that form, while others are the larvae of typical marine animals, such as crabs, sea urchins or corals, that will eventually metamorphose into their adult forms. The plankton form the base of all food chains in the sea. They are eaten by shrimps, fishes and other larger animals, and these in turn form the diet of larger fishes, sharks, seals and whales. Kill the plankton, and you kill all life in the sea.

The radiolarians, delicate net-like little organisms with a light skeleton generally made of silica (silicon dioxide, the main component of sand and of flint), today feed on bacteria and plant-like plankton. Their skeleton is made up of tiny spicules, or needles, of silica, forming perforated spheres, some with spikes, just like miniature Christmas decorations. Others are like tiny string shopping bags, suspended from a single corner. All are perforated with numerous regular holes. When they die, the flinty little skeletons of radiolarians rain down on to the deep ocean floor where they accumulate slowly, at only 4 or 5 millimetres per 1000 years. Nevertheless, over millions of years, radiolarian oozes have solidified into cherts, glassy pure-silica deposits, that currently make up about 3% of the modern ocean floor.

In Late Permian deep marine rock successions, radiolarian cherts are found quite commonly in China, Japan and Canada, but then disappear completely at the end of the Permian, only to reappear in the Mid Triassic, some 8 million years later. This 'chert gap' is matched by the virtual annihilation of all species of radiolarians at the end of the Permian – an event that is particularly striking since otherwise the radiolarians had had a singularly uneventful history for the previous 450 million years. It takes some enormous environmental shock to kill off such minute planktonic organisms that must have been present as millions of millions of individuals all over the world. Far easier to kill off large and less numerous animals such as dinosaurs.

Another element of the plankton today are the foraminifera. Foraminifera look like tiny spiral or coiled snails, with a shell made from calcium carbonate (calcite) or glued sand grains, enclosing their single-celled soft parts. The shell may form a tall spiral, a flat coil or disc, or a mass of small globules, in each case being divided into a number of internal chambers. In the Permian all foraminifera were sea-bed dwellers, feeding themselves on the rain of organic matter and plant-like plankton that sank from the surface. The dominant foraminiferan group in the Permian were the fusulines. Their shells were made from many tiny crystals of calcite. Some of them reached as much as 10 centimetres in length – most unusual for such a single-celled animal. The fusulines flourished in the Permian, evolving fast and giving rise to over 5000 species. Indeed, they evolved so fast, and achieved such diversity, that they are used as important guide fossils for dating Permian marine rocks. Then they all disappeared.

Until recently, the die-off of the fusulines was thought to have lasted for most of the second half of the Permian, some 20 million years or more. It was said to have been a gradual die-off, perhaps linked to long-term climatic deterioration. But new work has shown that the early studies had been too limited, and had relied on a literal reading of the record. The problem was backward smearing, the Signor-Lipps effect, as noted above (Chapter 7).

A study of the Permo-Triassic boundary in Austria[2] by Michael Rampino and Andre Adler, both of New York University, has shown on the basis of very detailed collecting that most fusuline species did indeed go extinct right at the Permo-Triassic boundary. Other fusuline species disappeared from the rock record as much as 16 metres below the boundary, which could be taken to imply a long-term die-off. However, these species were rare forms, known only from small numbers of specimens. Rampino and Adler argued that these were misleading datum points: the disappearances do not indicate extinctions. If these rare forms had been commoner, they would probably have been sampled up to the Permo-Triassic boundary as well. As mentioned before, with a patchy fossil record, it is unlikely that palaeontologists will ever find the very last member of a particular species to have lived on Earth. If a hundred species died out in an instant, the fossil record might still paint a picture of long-term gradual disappearances. Rare forms are especially liable to this phenomenon of false early disappearance, the Signor-Lipps effect.

Not all foraminifera died out during the end-Permian crisis. The generally large fusulines had completely crashed out of sight. But survivors included mainly smaller forms and some flattened species that burrowed in the sea-floor sediments feeding on detritus. A characteristic feature of these, and other survivors, is that they may have been adapted to living in conditions of low oxygen. Perhaps this is a clue to the nature of the environmental stresses at the end of the Permian.

Reefs

Reefs were common in Permian shallow tropical waters. For example, huge reefs developed over much of west Texas and New Mexico. During the Mid Permian, as the tropical seas became deeper, reefs built up around the edge of the ancient Delaware Basin, reaching a height of some 600 metres. Just like modern coral reefs, the living corals and other animals, were in the upper parts of the reef, keeping in close touch with the sea surface which allowed the associated plant-like organisms to photosynthesize. Deeper parts of the reef were formed from overgrown, dead coral skeletons, as well as shells and other reef rubble.

Reef life then was hugely diverse, with hundreds of species living in close proximity. The framework of the reef was built from sponges, corals and bryozoans, animals that secrete a stony skeleton in which they live. Living on the dead corals were various clinging molluscs and worms. Creeping among the coral fronds were snail-like molluscs, sea urchins, starfish and shrimps. And swimming above were jet-propelled nautiloids and ammonoids (relatives of squid and octopus), swimming arthropods and fishes of various kinds. Just as today, Late Permian reefs were diversity hotspots – locations of unusual species richness.

The seas have retreated now of course, but the west Texas landscape has not changed much in the past 250 million years. The visitor today can essentially stand on the Late Permian sea bed and look around at the towering Guadalupe Mountains, made up of reef limestones, termed the Capitan Limestone Formation. The vast size of the reef, hundreds of metres thick and 400 kilometres long, is immediately clear, and the richness of Mid Permian tropical reef life is evident. Such large and richly diverse reefs are also known from China right to the end of the Permian.

Reefs were entirely wiped out by the end-Permian event. Like the 'chert gap', there is a 'reef gap' lasting for some 7 or 8 million years in the Early Triassic. Many sponge groups survived apparently unaffected through the end-Permian crisis, but others, especially those associated with the tropical-belt reefs, were decimated.

The corals were even harder hit. Through the preceding 200 million years, limestone deposits of tropical zones teem with the skeletons of rugose and tabulate corals. Any novice fossil collector will have accumulated dozens of specimens of these corals – the tiny ice cream cones of small, solitary rugose corals and the fist-sized, rounded tabulate coral colonies composed of dozens of regular, honeycomb-like or star-shaped chambers. Some are even shaped like a bursting sun – hence the name of the coral, *Heliolites*, the 'sun rock'. Solitary corals built themselves tubular houses from calcite which they laid down round and round their soft bodies as protection from predation. Their tubular houses range in size from a few millimetres to the vast metre-long cones of *Caninia* in the Carboniferous. Mostly they were fixed upright on to rocks or other hard materials on the sea bed.

Colonies of rugose or tabulate corals are among the most beautiful of fossils. Colonies arose when one progenitor coral animal cemented its skeleton down to the seafloor, and then set about building its stony dwelling chamber. By endlessly splitting, the original coral animal formed numerous identical clones of itself, each of which constructed a little chamber. There is economy in such an arrangement, since each subsequent coral animal has to build only a few side walls and can use the pre-existing parts of the colony. In the end, most colonies formed bulbous, cabbage-shaped structures or broad plates on gradually expanding trumpet-like stalks. Each coral animal kept itself free of the others, and fed by capturing food particles on sticky tentacles. If danger threatened, the coral animals withdrew deep into their stony houses.

The rugose and tabulate corals, which had been the mainstay of reef formation worldwide for 200 million years of the Palaeozoic, all died out at the end of the Permian. It seems that they *had* undergone a long-term decline before the very end. First to go were the massive colonial forms, and at the end it was the turn of the colonies made from less intimately intergrown tubes and the solitary forms. The early losses of some coral groups seem to relate to changing habitats. For example, the warm tropical seas that had covered Texas and New Mexico had withdrawn in the Late Permian. This was not part of the crisis, merely a change in sea levels and continental positions. Where coral reefs are found in the latest Permian, however, such as in South China, the corals survived right to the end.

Other reef-builders are less well known. The bryozoans, or 'moss animals', form regular colonies of tiny habitations – less than a millimetre across – each housing an individual animal. Some of these colonies grow as vertical sheets from the seafloor, while others encrust corals or shells as a fine meshwork. Of the four major Permian bryozoan groups, one became extinct during the crisis and the other three suffered heavy losses of species.

After the 'reef gap', which marked a time when there were only a few simple, low patch reefs formed from sponges and serpulid, the first reefs to form in the Mid Triassic were small patches on the sea floor, known from shallow-water sediments of the Dolomites in northern Italy. But it took many millions of years more before large-scale framework reefs like the great Capitan Reef of the Guadalupe Mountains in Texas reappeared. The first Triassic reefs were formed from some surviving species of bryozoans, stony algae (distant relatives of seaweeds), sponges and the first members of a new group, the scleractinian corals. Scleractinian corals dominate modern reefs, with their multicoloured fleshy bodies and tentacles enclosed within stony chambers, and they owe their origin to the after-effects of the end-Permian mass extinction.

Sea lily forests and bottlenecks

The echinoderms ('spiny skins') include sea urchins, starfish, sea cucumbers and sea lilies. They are all characterized both by a skeleton made from calcite plates and their five-rayed symmetry, whether the five legs of a starfish (ten in some) or the ten-panelled structure of a sea urchin. Echinoderms were hit particularly hard by the end-Permian crisis – indeed, most had disappeared several million years before.

Most of the fixed echinoderms disappeared. Sea lilies (called more properly crinoids) were hugely successful in the Palaeozoic, forming parts of the great reefs, either growing on and around the corals and sponges or forming huge crinoid forests on their own. Typical crinoids look like plants: a long flexible stalk fixed down by a root-like structure, with a blob at the top of the stalk and tentacle-like arms shimmering in the currents above. Crinoids feed on small organic particles in the water, which they capture on their sticky arms and then waft down a central groove on the upper surface of the arm into the mouth, located in the centre of the body. The body is the blob on top of the stalk. Two subclasses of crinoids died out at the end of the Permian, and the recovery of crinoids in the Triassic was probably founded on a single genus that survived through the crisis. There are still some stalked crinoids today, but most are free-swimmers, rolling and crawling along the sea floor, propelled by their slender, multi-coloured, billowing arms.

Other stalked echinoderms that had once dominated the Palaeozoic sea floors were also wiped out. Prominent among these were the blastoids, which looked like stone tulips or gooseberries on sticks. Lower than most crinoids, they formed large spiky forests around reefs in the Carboniferous and Permian. They disappeared without trace at the end of the Permian.

Both these losses effectively destroyed the very strange echinoderm forests of the Palaeozoic sea floor. Catastrophic for the crinoids and blastoids of course, but also for all the little creatures that lived in, on and below the forests. In some ways, cutting down the echinoderm forests on the sea floors was like cutting down a true forest on land. Dozens of other organisms depend on the forest for their livelihood, and the same was true for the crinoid-blastoid forests of the Carboniferous and Permian. Their loss marked the end of a unique habitat that has never been reconstructed on the ocean floor.

Starfish and sea urchins were also nearly wiped out. Indeed, the end-Permian crisis marked something of a bottleneck for both groups. A bottleneck in evolutionary terms is a time of extreme reduction in diversity of a group. The group flourished before and after the bottleneck, but what comes after clearly can represent only a small sample of the original population. So, post-Palaeozoic sea urchins and starfish were quite different from their Palaeozoic forebears. Among starfish, their most delightful specialization, the ability to turn their stomachs inside-out, engulf their prey and begin digesting it before even eating it, arose only in the Triassic and Jurassic. Permian starfish had to content themselves with a more discrete and concealed stomach.

Only one genus of sea urchin, *Miocidaris*, is known to have survived the end-Permian crisis. This was an extreme bottleneck, cutting global diversity from 20 or 30 species down to one or two. From this one genus, seemingly, arose all the diversity of later sea urchins. In the cases of starfish and sea urchins, the bottleneck actually had a positive effect. Both groups successfully multiplied in numbers through the Triassic and rose to become important components of the seabed fauna, as they are today.

Of course, had this not happened – if, for example, the tiny numbers of species that squeezed through the bottleneck from the Permian to the Triassic had actually all died out – the effects would have been negative. But then it would have been a total wipeout, pure and simple.

Shellfish

First-year geology students always complain about having to learn the groups of fossils. One of the key facts they have to grasp is the difference between brachiopods and bivalves. These are two distinct groups of shelled animals which have different ancestors, but which look superficially similar. A brachiopod consists of two shell halves, more properly called valves, that enclose and protect the animal inside. The shell is fixed to the seabed by a tough thread that emerges from the tip of one of the valves. The two valves are joined along the hinge line, and they may be opened by muscular activity to allow food particles to be sucked in and waste material expelled.

It is easy to tell a brachiopod from a bivalve: brachiopod valves are different in dimensions, while those of bivalves are identical mirror-images. One valve of the brachiopod is larger than the other, and it is often shaped like a Roman oil lamp – teardrop-shaped, with the extension at the hinge-end often perforated by a large circular hole for passage of the attachment thread (just like the hole for the wick in the Roman oil lamp). The other valve is circular and smaller. Bivalve valves, on the other hand, generally fit exactly over each other, being identical in size.

The brachiopods and the molluscs of the Permian were hit hard by the mass extinction. Molluscs, such as clams, oysters, mussels, whelks, octopus and squid, dominate the seafloor today, while brachiopods are relatively rare, being found only in rather deep waters and confined to certain parts of the world. However, the situation was the reverse in the Palaeozoic, and the end-Permian crisis perhaps has a large part to play in engineering the switchover.

To human eyes the brachiopods may seem pretty limited in their potential – all they really do is sit on the seabed opening and closing their valves. They feed by sucking water, plus food particles, into their shells, passing it over a looped filtering organ, the lophophore, and blowing water out the other side. However, the Permian was a time of astonishing innovation in the group. The cone-shaped richthofenids copied the corals, cementing themselves to a rock or another shell with the tip of the cone and standing upright in tight clusters to form mini-reefs. The smaller valve had become simply a small lid, like that of a pedal bin, which could be opened to allow feeding. The fat, and often large, brachiopods also flourished in the Permian, when remarkable new spiny forms appeared. The spines were delicate tubular structures sprouting wildly all over the base valve, and these extraordinary brachiopods must have used them to anchor themselves in soft, muddy seafloors. So these two successful groups had conquered new habitats and modes of life, and who knows where the brachiopods might have gone but for the mass extinction.

The brachiopods were devastated by the end-Permian event. At the level of superfamilies, 16 out of 26 disappeared, which is bad, but not awful (it equates to a loss of 62%). However, at the level of families, 40 out of 55 died out (73% loss). It has been estimated that some 95% of genera of brachiopods were hit by the extinction, and that equates to about 99% of species. So all but a tiny handful of this hugely diverse and abundant group bit the slime.

In contrast to this collapse of the brachiopods, some of the molluscs weathered the end-Permian crisis much better. There are three main groups of molluscs: the bivalves ('two valves'), gastropods ('stomach foot') and cephalopods ('head foot'). The origins of the last two of these names are rather startling. Gastropods do indeed have a stomach in their foot, but their foot is actually almost their whole body. Technically, the soft slimy portion of a snail or whelk that creeps over the ground is the foot, but obviously the whole body of the animal, from eye stalks at the front to anus at the back, is enclosed in the foot. Cephalopods include the octopus and the squid, as well as the fossil ammonoids with their coiled shells. The 'head' is the front portion with its huge eyes and ring of massive tentacles that haul food towards the mouth. The 'foot' consists of the tentacles and a siphon that can squirt water or black ink for rapid jet propulsion and confusion of an enemy. So, technically, the head and foot are closely associated, and the 'body' of the ammonoid, or of the octopus, is a bag-like structure containing the stomach and guts.

Most families of bivalves passed through the event relatively unscathed, with only three out of 40 disappearing. Bivalves in the Late Permian were rarer elements of the seabed faunas than were the brachiopods, but they occupied a range of habitats, including living on the surface of sandy and muddy seabeds, and burrowing in the sediment. The extinctions did not eliminate any of these modes of life, but species diversity was, as always, hit hard.

Gastropods include some rather fiendish predators that prey on other shelled seabed animals such as bivalves and arthropods. They attack them by an array of mechanical and chemical means, breaking and piercing through the shelly armour of their prey and sucking out the soft flesh. Others creep about on rocks, scraping off the green algae with a tough rasping radula, something like a woodworker's mechanical sanding belt. Yet others hoover up organic detritus from the sea floor. Gastropods were diverse (i.e. lots of species) but rare (i.e. not many individuals in any location) in the Late Permian, and they were apparently hit harder than the bivalves, but they recovered well after the mass extinction. It has been estimated that perhaps 90% of gastropod genera died out, and this scales up to the loss of three out of 16 families. Most of the losses were species that had limited geographic ranges or specialized diets: the more cosmopolitan forms survived, as did the generalist detritus-feeders.

The third major mollusc group, the cephalopods, were severely affected by the mass extinction. The ammonoids, the free-swimming forms with coiled shells, were nearly wiped out. Ammonoids have always shown boom-and-bust patterns of evolution. In good times they evolved rapidly and diversified into hundreds of species worldwide. Then, when an environmental crisis came along, they always died out. In the latest Permian Chinese sections, 20 out of 21 genera, and 102 out of 103 species, disappeared at the very end of the Permian. The few species that crept through into the Triassic then radiated rapidly, but the ammonoids virtually disappeared during the next mass extinction at the end of the Triassic. After another huge and successful recovery from the brink, in the Jurassic, the ammonoids were finally and definitively killed off during the KPg event. Other cephalopods were much less affected by the end-Permian crisis.

The post-extinction world retained only sorry remnants of the brachiopods and bivalves, and they were fairly similar worldwide. An assemblage consisting of the brachiopods *Lingula* and *Crurithyris*, both leftovers from the Permian, is found everywhere in the Early Triassic. *Lingula* in particular is a well-known disaster species, able to survive in all kinds of conditions, including low oxygen and low salinity. Indeed, *Lingula* is one of the most astonishing 'living fossils', known virtually unchanged through most of the past 500 million years. It might be expected that such longevity would be associated with some remarkable features, but it is not. *Lingula* is a brachiopod shaped roughly like the nail on your little finger that lives buried in the mud, held down by a long fleshy foot and quietly pumping seawater through its body cavity so that it can extract food particles. The secret of its success is presumably a wide environmental tolerance, and the ability to survive on very little food.

The post-extinction bivalves were a similarly restricted group. In the Early Triassic, four or five genera occur everywhere. Although these supposedly contain about 100 species, most belong to *Claraia* and *Eumorphotis*, both thin-shelled, scallop-like paper pectens (they are called paper pectens because they are both pectens, like modern scallops – the symbol of the Shell oil company – and they are also paper-thin) that were attached by fine threads to irregularities they found in the black anoxic Early Triassic muds. Cephalopods were rare in the Early Triassic, but microgastropods are suddenly, and bizarrely, hugely abundant worldwide. Then everything changed.

In the Mid Triassic, the uniform and cosmopolitan disaster species of brachiopods, bivalves and microgastropods were supplanted by bursts of evolution in all groups, but the bivalves and other molluscs radiated explosively. New species and life modes appeared, many of them with distributions in individual ocean basins. By the end of the Triassic, the molluscs had recovered to Permian diversity levels and had even begun to surpass them. The brachiopods could not respond fast enough it seems, and they never recovered their Permian diversity. This episode has formed the basis of a long-standing debate about the relative success of the brachiopods versus the bivalves, which exemplifies a wider clash in our views of evolutionary history.

The search for pattern

Does history follow rules or not? In matters of human affairs, some historians see a horrible inevitability in the rise and fall of nations. Countries and peoples can be said to go through cycles of rising fortunes, successful conquest, then decadence, followed by collapse. Marxist historians argue that there is a pre-ordained plan, driven by socio-economic forces and the tension between the ruling classes and the ruled. Others would use militaristic or martial metaphors: nations rise and fall as a result of their fighting power. When confidence is high and the steel is sharp, that nation prevails.

In 1923, the Russian astronomer and archaeologist A. L. Tchijevski published his 'Index of Mass Human Excitability'. He argued that humans behave excitedly and violently at regular intervals, nine times each century. Each cycle of excitability lasts for 11.1 years, and the maximum level of violence is associated with sunspot activity. In 1943, the American psychologist Raymond P. Wheeler of the University of Kansas argued that civil wars follow a 170-year cycle. Every third wave is supposedly more dramatic than the others, giving phases of extreme violence every 510 years. The cause in this model is supposedly periods of drought, which apparently have happened every 170 years. Other historians view such notions as utter nonsense, entirely without substance.

The search for pattern and meaning in history is an incredibly attractive pursuit. A quick search of the internet reveals hundreds of websites devoted to showing how history follows patterns. The theories range from the crackpot to the seemingly sane. Why should this search for pattern be so popular? If you look hard enough you can find cycles, patterns, regularities in any series of historical events. Such claims can only be justified if they can be turned into predictions. So, the millennialist has to show by calculating the birth dates of Genghis Khan, Napoleon and Hitler, or whatever, that the rise of the next nationalist conqueror can be predicted.

Pattern in history could all be in the mind. Humans are tidy creatures who like to file information away in meaningful boxes. It is comforting to be able to catalogue all the kings and queens of England, and classify them as either good or bad, not something in between. Far easier to see patterns in history than to have to accept the alternative – a raw and uncontrolled series of changes of immense complexity, and devoid of meaning. The same is true of evolution.

Evolutionary history was a comfortable phenomenon for the pre-evolutionists, and indeed for modern creationists. God planned everything, and He imposed a clear pattern. Every fossil had its place in the scheme of things, and the apparent coming and going of different groups of plants and animals was all part of a parade towards perfection. Whether these evolutionary changes were seen as unidirectional (time's arrow) or cyclical (Charles Lyell's view), there was a goal and a measure of predictability.

Early theories of evolution, such as that of Jean Baptiste Lamarck, were no less goal-directed – what the philosophers call 'teleological'. As simple organisms evolved into more complex organisms, the next stage below moved up a step in the great chain of being. Darwin forced people to reject such comforting views. If evolution happened by natural selection, it could not be predictable. But even Darwin could not entirely reject some of the comforting pattern of Nature. He retained the pre-evolutionary concept of plenitude, literally 'fulness'. Plenitude encompassed the idea that everything that could be done in Nature was being done: the Earth was full of species of plants and animals, and that number was pretty well fixed. In a famous analogy, Darwin wrote:

> Nature may be compared to a surface covered with ten-thousand sharp wedges... representing different species, all packed closely together and driven in by incessant blows,... sometimes a wedge of one form and sometimes another being struck; the one driven deeply in forcing out others; with the jar and shock often transmitted very far to other wedges in many lines of direction.[3]

In another analogy, Darwin compared life at any time to a flotilla of apples floating on the surface of a huge barrel full of water. If a new apple is to insert itself in among the floating mass, it has to push a pre-existing apple either up or down.

So, whether wedges or apples, in Darwin's view species could come and go through the course of evolution, but a new species could not arise and find its place without first displacing a pre-existing species. This seems to make sense – or does it?

The race is not (always) to the swift

In fact there is no evidence for plenitude. The history of life through long spans of time shows that species diversity has risen time after time. New spheres for life have opened up the opportunity for bursts of diversification: for example, when life first moved on to land, when the first forests appeared, when the first insects took to the air, when the first coral reefs arose in the sea, when vertebrates evolved warm-bloodedness, and so on. A study of plants and animals that have become extinct in the past million years shows plenty of empty niches. What creature today is doing what the mastodons, mammoths and woolly rhinos of North America and Europe did until 10,000 years ago? What about the giant ground sloth of North and South America, which fed on caper and mustard plants, and which disappeared 11,000 years ago? It would be presumptuous in the extreme to assume that all potential for the expansion of life into new habitats and niches has come to an end just at the moment when we happen to be here and worrying about our place in Nature. Darwin's barrel of apples has more to do with the limits of our imagination than with reality.

So, what about the end-Permian crisis and the brachiopods and bivalves? This was for a long time a key example of plenitude in action. The brachiopods had to go in the end because the superior bivalves ousted them step-by-step through the Palaeozoic. Comparisons of modern brachiopods and bivalves seemed to support this view: brachiopods have a limited array of modes of life, while bivalves live in a wide range of habitats, swimming, creeping and burrowing on sea beds, on shores, and in rivers and lakes around the world. Brachiopods eat very little and don't do very much, while bivalves eat much more and (some at least) are very active. Here was a classic example of competition on the large scale. Not competition between two individuals for a mate, or between two species for food or space, but between brachiopods as a whole and bivalves as a whole. There was some sort of justice about the rise of the mighty clam, and the demise of the miserable brachiopod.

In a classic study in 1980, Stephen Jay Gould and Jack Calloway challenged this comfortable vision.[4] First, they asked, how could the arms race between brachiopods and bivalves have sustained itself in the Palaeozoic seafloor ooze for 300 million years? When ecologists study competition between modern organisms, they expect to measure selective advantages on time scales of years or tens of years at most. Over hundreds of millions of years, the selective advantages would be so minute as to disappear in the statistical variation of normal populations. Gould and Calloway then made counts of the numbers of genera of brachiopods and bivalves through the Palaeozoic and post-Palaeozoic. It turned out that the expected steady rise of bivalves and steady decline of brachiopods had not happened (Fig. 30). Both groups remained at pretty well constant diversity through the

Palaeozoic, the brachiopods with 150–200 genera, and the bivalves with 50–100. Then came the end-Permian crisis. Both groups were hit hard, as we have seen, and only about 50 genera of each survived through into the Triassic. What was critical was the post-extinction recovery.

In the Triassic, the bivalves made it and the brachiopods did not. Bivalve diversity climbed steadily from 50 genera to over 100 by the end of the Triassic, and continued rising to more than 400, while brachiopod diversity remained firmly at 50 or so. So, concluded Gould and Calloway, brachiopods and bivalves were not involved in continuous hinge-and-lophophore struggle throughout the Palaeozoic. In fact, their modes of life were often quite different, and competition between an average brachiopod and an average bivalve would be as ludicrous as suggesting that rabbits and snails compete, just because we find them together and they both eat plants. In Gould and Calloway's words, brachiopods and bivalves were 'ships that passed in the night', groups that lived side-by-side but did not seriously interact. The competitionists were incensed by Gould and

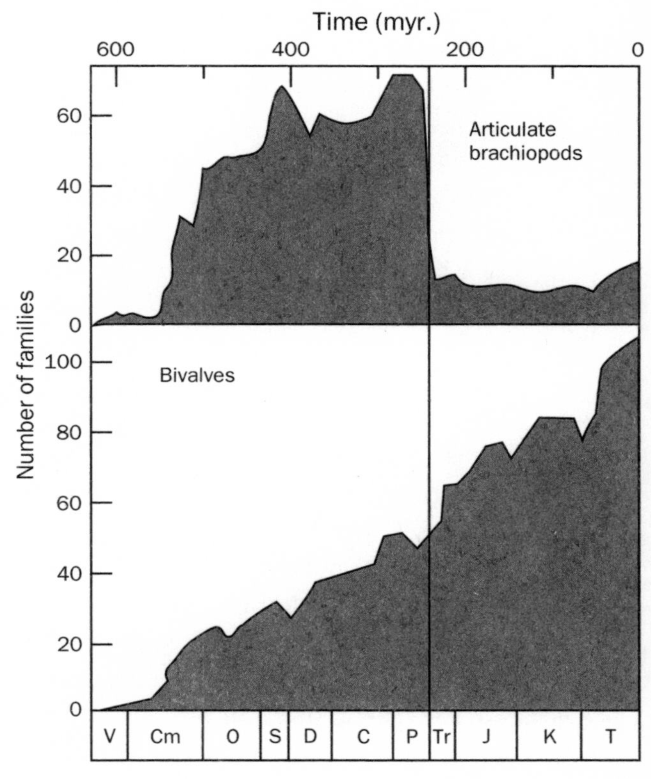

30 *A classic example of supposed competitive replacement. The steady rise of the bivalves through the past 500 million years was once thought to have caused the demise of most brachiopod groups. However, it seems that the brachiopods and bivalves never interacted in a major way. It was the end-Permian crisis that hit both groups hard, and only the bivalves were able to recover.*

Calloway's analysis, but the facts have to lead to the rejection of the comfortable competition theory.

'Comfortable competition' may seem an odd juxtaposition of words. But, in evolutionary biology there has been a pervasive view that competition (that is, the interaction between any pair of organisms in which one gains a benefit at the expense of the other) guides the large patterns of evolution. Such a viewpoint leads to explanations of larger-scale patterns, such as the rise and fall of dynasties of plants and animals, migrations, the sequence of groups of organisms through time, and extinctions, by competition. The mathematics of competition are relatively tractable. So competition *is* comfortable for many biologists and theorists.

Competition can also be seen as part of the desire for predictability and pattern. Just like Tchijevski and his cycles of human excitability, we find it very hard to let go of the reassurance of numbers and meaning, whether in the history of human societies or in the history of life. Far bleaker to have to accept that nation states come and go, animal and plant groups appear and die out, and there is no pattern, no innate system guiding their relative fates.

Perhaps the author of Ecclesiastes (9:11) got it right long ago:

> I see this too under the sun: the race does not go to the swift, nor the battle to the strong; there is no bread for the wise, wealth for the intelligent, nor favour for the learned; all are subject to time and mischance. Man does not know his hour; like fish caught in the treacherous net, like birds taken in the snare, so is man overtaken by misfortune suddenly falling on him.

Trilobites: did they or didn't they?

All classic accounts of marine life of the Palaeozoic say that the trilobites were one of the most striking groups to have died out at the end of the Permian. Other, more detailed, overviews say the exact opposite: the trilobites were not involved. Both views are right. But, how can such contradictory statements be made?

Trilobites[5] had indeed dominated the sea-floor since Cambrian times. With their three-lobed bodies, consisting of a midline body portion and two lateral areas that bear the eyes at the front and cover the legs and gills further back, trilobites were the most active, and seemingly most complex, animals of the Palaeozoic. They belong to the Phylum Arthropoda, a huge group of animals that includes insects and spiders, as well as marine forms such as crabs and lobsters. Arthropod means 'jointed limb', an apt name for animals which have a tough outer skeleton like a suit of armour with flexible joints all along the legs. Trilobites ranged in size from the microscopic to over 30 centimetres long. Many of them had complex eyes and, presumably, excellent vision, as well as electrical sense organs. They lived largely on the sea bed, ploughing through the surface mud feeding on detritus. Many of them could swim, and some may have hunted actively.

Trilobites were common and diverse especially in the Lower Palaeozoic, and they survived through to the Permian, but had disappeared before the Triassic. It is true, then, to say that they became extinct in the Permian, but the class was already much reduced before the beginning of the Permian. Only three small families of trilobites are known from the Permian. One or two species are known from the latest Permian sections in China, and these indeed disappeared at the Permo-Triassic boundary. The final disappearance of the trilobites was a fairly feeble little blip in their history, and it followed a long-term decline. So, although the rule of the trilobites finished at the end of the Permian, closer analysis shows that the group was already very much on the way out.

While other arthropod groups in the sea, including ancestors of modern crabs and lobsters, were apparently little affected, an unusual arthropod group, the ostracodes, were hit relatively hard. Ostracodes are generally tiny, a millimetre or less across, and they look like tiny animated beans. Their bodies are enclosed in two tightly fitting valves, and they swim and feed by opening the valves and extending feathery little legs and gills out into the water. The only fact about ostracodes that my students ever remember is that the males have a penis that is often longer than their bodies. Shallow-water ostracodes suffered some major extinctions, while the deeper-living, more cosmopolitan groups were apparently little affected by the end-Permian crisis.

Fishes: extinction or Lazarus taxa?

As with the arthropods, the fossil record of fishes has been interpreted in two ways. Some commentators say that the fishes simply swam unconcernedly through whatever was happening at the end of the Permian, just as they were seemingly unaffected by most other mass extinctions. Others claim that they suffered extinctions, like other groups. Who is right?

Certainly the supporters of fish survival can make a case. They note that all the major fish groups from the Permian are found in the Triassic. But their opponents argue that the sharks show major changes in size. The diverse array of medium- and large-sized sharks of the Late Permian was reduced to a fauna of only small sharks in the Early Triassic (see Figs 28 and 29). Whether it was only small forms that managed to survive, or whether the survivors, variable in size, then became dwarfed because of some evolutionary pressure is uncertain. Also, among the bony fishes – the group that today includes everything from haddock to goldfish, and tuna to seahorse – two out of eight families disappeared. But did they really disappear?

The extinction apologists point to the possibility of Lazarus taxa. Lazarus taxa are species or genera or families that seem to disappear, and then miraculously reappear. Well, not 'miraculously'. Their temporary absence may just be a failure of preservation. The term comes from the famous biblical story of Lazarus (St John, 11: 17–44):

> On arriving, Jesus found that Lazarus had been in the tomb for four days already. . . . The dead man came out, his feet and hands bound with bands of stuff and a cloth round his face. Jesus said to them, 'Unbind him, let him go free'.

In a palaeontological example, it is much more likely that a group that apparently disappears and then reappears later has actually been there all the time, and the apparent absence, or death, merely represents a gap in the fossil record.

So, in the case of the end-Permian fishes, there was a substantial diversification of bony fishes in the Early Triassic, when many new families and a major new group, the Neopterygii, which includes all modern bony fishes, radiated in steps. New work by Carlo Romano and colleagues[6] shows the patterns in some detail. Fish diversity was simply lower in the Middle and Late Permian, and several lines went extinct. Some groups recovered a little in the Early Triassic, but the explosion of neopterygians in the Middle Triassic was real and dramatic.

Survivors and victims

So, in the end, despite much debate about individual groups, current studies suggest that there were major extinctions at the end of the Permian among virtually all marine creatures. Groups, like the fishes, thought to have been little affected, turn out, on closer study, to have suffered just as much as the others. Older statements that some groups survived unscathed often turn out to have been based on incomplete and perhaps rather general examinations of the data. The more detailed, site-by-site, bed-by-bed, studies that are now being done show us more exactly what was going on.

A comparison of the survivors and the victims can provide clues about the crisis. One might hope to find evidence that, say, carnivorous animals were harder hit, or perhaps tropical dwellers, or large animals, or specialists, or whatever. Actually, it is very hard to find evidence for selectivity of this sort. In general, of course, geographically widespread species survive better than those restricted to small areas. In addition, those that have wide tolerance – animals that can survive in a wide range of temperatures or on a variety of different food types – might also be among the survivors. Any plant or animal that is adapted to a very narrow habitat and that feeds on a small range of foods is likely to be vulnerable during any crisis.

In a detailed survey of the victims and survivors of the end-Permian crisis in the sea, Tony Hallam and Paul Wignall[7] argue that the survivors do share one key feature – an ability to live in low-oxygen conditions. This is true of the surviving foraminifera, brachiopods (at least *Lingula* and *Crurithyris*), bivalves (the paper pectens *Claraia* and *Eumorphotis*) and ostracods.

Dwarfing was another feature of the survivors, most notably the gastropods and fishes. In the Late Permian, the gastropods and the fishes display a broad range of sizes. The same is true for the Mid Triassic. But in the Early Triassic, a new rock type is found worldwide: microgastropod grainstone. This is a limestone made up solidly of billions of millimetre-sized shells of gastropods. Perhaps less dramatically, sharks and bony fishes of the Early Triassic were all a size or two smaller than either before or after that period.

Survival in low-oxygen conditions and dwarfing are both perhaps responses to low productivity seas. The crisis wiped out so many organisms – 90% or more of all species had disappeared – that established food webs and food chains broke down. Normal processes of transmitting energy and carbon from the plankton up through ever-larger predators were destroyed. Perhaps those small and undemanding organisms that did not need much food, or much oxygen, were the only ones that could eke out a living in the post-disaster conditions.

Rarefaction – rarefiction?

The approximate figures for the scale of extinctions during the end-Permian crisis were given at the beginning of the chapter. Some 50% of marine families died out, and this scales to 90–96% of species. The higher estimate of 96% was made by the University of Chicago palaeobiologist David Raup[8] in 1979. He used a mathematical method called rarefaction, which is a way of estimating how low-level effects, for example the loss of individual species, feed up to higher-level phenomena, such as the extinction of families. We saw this scaling effect in considering how the extinction of a relatively small proportion of brachiopod superfamilies implied the extinction of a very high proportion of genera and species.

Critics suggested that Raup had perhaps been a little over-enthusiastic. Toni Hoffman, a Polish palaeobiologist, suggested a more conservative 75% species loss, arguing that the rarefaction approach could not be regarded as reliable. Mike McKinney, another critic, accepted the validity of the rarefaction approach, but he came up with a revised figure of 90% species loss. This revised figure followed from a detailed reconsideration of the rarefaction methods. He found that Raup had overestimated the species extinction figure since he had assumed a random distribution of species extinctions across the evolutionary tree. But this, argued McKinney, is not correct.

McKinney noted from his studies of sea urchin evolution that the patterns of extinction of species within genera were not uniform. In other words, certain genera, each containing several species, were more likely to go extinct than others. If species loss is not randomly distributed, then some genera will have a higher chance of going extinct than others. Equally, others will survive better. Overall, this non-randomness means that a fixed level of family or generic extinction can be founded on a lower species extinction rate than an entirely random distribution.

So the best estimate for species loss at the end of the Permian is 90%. This is less dramatic perhaps than a loss of 96% of species, but it is still by far the largest mass extinction of all time, and certainly the closest that life has ever come to complete annihilation.

Palaeontologists have always accepted that the end of the Permian marked a major crisis for life in the sea. But their views about the extinctions on land have been much more mixed. Indeed, some claimed that nothing much happened at all. New research has overthrown that idea, as we shall see in the next chapter.

9
A TALE OF TWO CONTINENTS

There seems to be little doubt about the scale of the end-Permian crisis in the sea. Ever since the work by John Phillips in the 1840s, palaeontologists have accepted that this was a time of major turnover in the history of marine life. But what of the land? The story has been very different. As we saw earlier (Chapter 4), many, perhaps most, palaeontologists once happily denied that there had been any more than a hiccup in the history of the reptiles and amphibians.

The same was also true for experts on fossil plants and fossil insects; they saw their favourite fossil groups in Permian rocks and in Triassic rocks, and did not recognize any major differences. Closer study now shows that the crisis on land was just as severe as in the sea. Why this apparent failure to recognize the truth?

Part of the problem has always been the notoriously patchy terrestrial fossil record. Marine palaeontologists rejoice in great thicknesses of rock, sometimes hundreds of metres, even kilometres, of limestones and mudstones laid down year-by-year on ancient continental shelves, and stuffed with fossils. The same is rarely true for the continental fossil record. The term 'continental' is used as a proper equivalent to marine, meaning the combination of terrestrial (ancient soils, screes, landslides), lacustrine (lake) and fluvial (river) deposits. Even if the plants and animals of interest lived completely on land, their fossils are commonly found in sediments deposited by ancient rivers or lakes, since most of the land is not associated with deposition. In fact quite the opposite. Mountains and hills are sites of erosion, so rocks and fossils are rarely found there. Lowland soils are rare also, since most of them have a transient existence, deposited, rained on, washed away, redeposited and finally carried into the sea. Rivers pick up sediment and inexorably wash it downhill until it finally enters the sea at an estuary, when it may be transported further across the continental shelf by slumps and turbidity flows until it may finally enter the deep oceanic abyss.

There might seem, on reflection, to be little hope of finding any continental fossils. Not so, fortunately. There are some settings on land where sediments can accumulate in considerable thicknesses. Most notable are ancient lake systems in subsiding basins. For example, rift valleys are sites where a continent unzips. A good modern example is the Great Rift Valley running halfway down the eastern half of Africa, from Ethiopia to Mozambique. The continental plate carrying Africa is in the process of pulling apart, and a great ocean may some day flood into the split. Over the past 20 million years, however, the rift has been filling with sediment, to a depth of kilometres in places. Palaeontologists can follow the evolution of their terrestrial fossil groups of interest, whether they be snails, grasses or humans, through short-term, even annual, layers.

The great Karoo Basin

In the Late Permian, southern Africa lay near the south pole. The base rock of that part of the continent had been formed from molten rocks thousands and hundreds of millions of years earlier. During the Permian, Africa was part of the supercontinent Gondwana. This was a single, fused landmass with no divisions, and to draw the modern continents over it is somewhat misleading. None the less, it is impossible to visualize the nature of Gondwana without making a mental jigsaw reconstruction using the familiar shapes of the modern continents. South America snuggled up close against the western margin of Africa, with the eastern bulge of Brazil fitting neatly into the Congo Basin. Land extended from the eastern margins of Africa through Madagascar, now an island, and across to India. The triangular shape of India today relates to its ancient Gondwanan position, when it fitted, like a wedge, into the northern margin, squeezed up against the eastern coast of Africa and Madagascar. To the south, Antarctica wrapped itself around the southern tips of South America, Africa and India. The eastern margin of Gondwana was made from Australia, rotated so that its south coast faced west and fitting neatly against the eastern side of India and Antarctica.

A great basin, the Karoo, forms the heart of South Africa today. The scrub-covered land of this great desert-like bowl offers poor returns to the farmers, but it is rich territory for geologists and palaeontologists. In the Permian it was ringed by mountains to the south, lying on what is now Antarctica. The basin is 1500 kilometres across, and up to 10 kilometres of sediment accumulated from the Early Permian to the Early Jurassic. During this time, the whole basin was subsiding, partly as a result of ancient fault systems that remained active and allowed the underlying basement rocks to sink ever deeper, and partly from the accumulating weight of sediment, which added to the downward pressure. Sediment thicknesses are greatest in the south of the Karoo Basin, good evidence that that was the main direction of supply. In geology, the proximality principle applies: the closer you are to the source of some rocks, the thicker they will be, whether they are volcanic lavas or river-deposited sediments. So, even though the mountains that supplied rocks, sand and mud to the Permo-Triassic rivers of South Africa have long gone, there is no doubt that they existed, and that they lay to the south.

The sediments of the Karoo Basin span the Permo-Triassic boundary, and they are full of plants, insects and reptiles. Ideal territory to study the nature of the end-Permian event on land.

Professor Owen and the Scottish engineer

The Karoo reptiles first came to public notice in 1845. In 1844, Henry De la Beche, 'foreign secretary' of the Geological Society of London (and creator of Professor Ichthyosaurus; see Fig. 8) had received a long letter from Grahamstown in South Africa from Andrew Geddes Bain (1797–1864), a Scottish engineer who was employed in military road building in what was then known as Cape Colony.[1]

Bain had been an enthusiastic geologist ever since he had read Charles Lyell's *Principles of Geology* in the 1830s, and he never ventured out of doors without his hammer and collecting bag. In the course of his work in the southern parts of the Karoo Basin, Bain had stumbled across dozens of reptile specimens, the first in 1838. He had extracted them from the rocks with care and had made a collection of many different species, some as small as a mouse and others as large as a rhinoceros. In his letter to De la Beche, Bain described the reptile finds, and also reported on the geology of the rocks in which they had been found. The fossils arrived at the same time as the letter, shipped in several packing cases from Grahamstown.

The committee of the Geological Society immediately sent Bain the Wollaston Fund, a sum of 20 guineas (£21), as a reward for his work, and they encouraged him to continue to collect. Bain's fossils were handed to Richard Owen for description. Shortly after, a paper about the geology of the Karoo Basin, extracted from Bain's letter, was read to the Geological Society, followed by Owen's consideration of the reptile specimens.[2]

Owen recognized, as Bain had, that most of the new South African specimens belonged to a single group which Owen termed the dicynodonts. These creatures had a short, high skull, and apparently a horn-covered turtle-like jaw (Fig. 31). They typically had only two teeth (Bain had called these beasts 'bidentals'), and Owen's term *Dicynodon* meant 'two-dog-tooth', in reference to the fact that the pair of teeth were the canines, the sharp teeth that are particularly long in flesh-eating mammals today and in humans are more pointed than the neighbouring incisors in front and cheek teeth behind. The dicynodonts had rather massive bodies, with stumpy legs, and a short, one might say disappointing, tail.

Owen's considerations of the Karoo reptile specimens came before his description of the isolated reptile bones from Russia (Chapter 1), in March 1845. He looked at Bain's bones in late 1844, and read his first account to the Geological Society of London in January 1845. Little did he know it, but Owen had in his hands, from Russia and from South Africa, at the same time, evidence for a major new group of reptiles – evidence that could have revolutionized palaeontological understanding. But the Russian fossils were too incomplete for him to spot the similarities, and the link was made only a few decades later.

Bain sent back more collections of fossil reptiles to London, and Owen was able to publish further descriptions. Indeed, Owen made sure that Andrew Bain, and later his son Thomas, were well remunerated by British government sources. For example, in 1877, Thomas Bain was

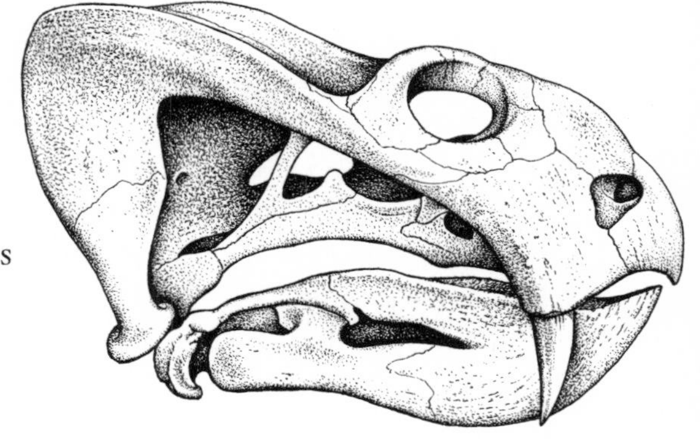

31 *The skull of the plant-eating synapsid* Dicynodon.

sent £200 to carry out a large-scale fossil-collecting expedition in the Karoo Basin by ox cart. The trip coincided with a severe drought, and Bain was hard pressed to find enough water to keep his men alive. Nevertheless, he reported to Owen that he had come back with 'almost 280 heads' of Permo-Triassic reptiles, which were all sent to London for Owen to describe and name.

Later in the nineteenth century, the younger British palaeontologists Thomas Henry Huxley and Harry Govier Seeley (1839–1909) also began to describe and name new South African specimens. Seeley went to South Africa in 1889 to look at specimens in the museums there for himself, but he found that the best ones had already been sent to the British Museum in London. Nevertheless, Seeley did collect some further specimens, and brought them back to London.

Wing collars and addiction to fossils

Everything changed in the twentieth century. It became less common for colonial fossils to be sent back to Europe in general, whether Britain, France or Germany, and South Africa gained independence from Britain in 1902, following the Boer Wars. In addition, the arrival there of an energetic palaeontologist by the name of Robert Broom (1866–1951), a Scottish-born physician, meant that there was no longer any need to seek advice from palaeontologists based in Europe. After training as a doctor in Glasgow, Broom[3] went first to Australia to build his career as a country doctor. In his spare time, he studied fossil and modern marsupials in his adopted country. Then, on a visit to London in 1896, he saw the South African fossil reptiles in the British Museum, and he was hooked. He decided not to return to Australia, emigrating instead to South Africa where he became a country doctor again, but turned his attention to the burgeoning collections of Permo-Triassic reptiles. He was always properly dressed, and conducted his field work in a formal coat and hat, and a stiff wing collar, as was the Edwardian fashion, even in the blistering African sun. However, it is said that sometimes it became too hot even for him, and he would strip naked to cool down.

By 1900, collectors in South Africa were so overwhelmed with fossil specimens that they concentrated simply on picking up skulls; the skeletons were left to erode away. Broom was soon sending descriptions and reviews of the South African Permo-Triassic reptiles to scientific journals in London, and increasingly to those of the scientific societies in his adopted country. In the early 1900s, he managed to write a mere four or five scientific papers each year, but this soon rose to a steady twenty or so. By the time of his death in 1951 he had written a total of over 400 scientific papers, as well as several books.

Clearly, Broom worked fast and he had a prodigious memory – he is said to have been able to remember everything he had ever written and seen, and indeed everything he had ever read. Also, he approached his God-given duty to find, name and describe new fossils with zeal and immense speed. Typically, he would write a description and make freehand drawings of his new finds in a matter of days, and send off the manuscript to the press within a week. Despite this haste, not uncommon at the time but quite a contrast with the slow and painstaking work that is more normal today, Broom's observations were generally accurate. His one failing, and again this can only really be said in retrospect since he was following the late Victorian norms he had learnt from his predecessors, was that he named far too many new species and genera. He saw a new form in every half-decent fossil, and named over 200 new species of Permo-Triassic reptiles.

Precursors of the mammals

The majority of reptiles named by Owen, and later by Broom, from the Karoo Basin were mammal-like reptiles. At first, however, Owen had been unclear just what they were. He toyed with the idea that *Dicynodon* was a relative of turtles, based on its generally toothless jaws, pointed beak and solid skull bones. Then he thought it might be somehow related to modern lizards, although his logic was somewhat arcane in that case. But in the end, he realized that these reptiles were related in some way to mammals. In his words, these Karoo predators 'made a most singular and suggestive approach to the mammalian class.'

The most obvious clue is in the teeth. Instead of the typical reptilian arrangement of essentially identical teeth from fore to aft, the jaws of the synapsids, as the group is termed (see Chapter 1), show a clear flesh-eating, mammal-like complement of teeth: small spiky incisor teeth at the front, then a long curved canine tooth and a number of cheek teeth behind

(Fig. 32). Other mammal-like features include a dog-like face, a single lower opening in the skull behind the eye socket, various features of the skeleton, and even hair. Synapsid hair is not preserved in any Late Permian fossils, but some early cynodonts have masses of small nerve openings in the snout region, which imply that they had whiskers, presumably for sensory purposes – and if they had whiskers, they also had body hair.

The early synapsids are often called pelycosaurs, and their most famous representative is the iconic *Dimetrodon. Dimetrodon* has a smiling face, sharp flesh-eating teeth and a vast sail on its back. Commonly misclassified as a dinosaur, *Dimetrodon* graces every child's book of prehistoric beasts or collection of plastic models. The sail along the ridge of its back is made from skin stretched over the enormously extended spines attached to its backbone. Whether the sail was used for thermoregulation or not (and that is what every book says), most pelycosaurs lacked a sail. By Mid Permian times, perhaps 260 million years ago, the pelycosaurs had given way to the new breed of synapsids, termed the therapsids ('beast arches'). The ancient Greek *ther*, *therion* means 'beast', and is interpreted as equivalent to mammal in more correct modern terminology.

The Karoo therapsids include a basal, and varied, group of meat-eaters and plant-eaters, the dinocephalians, some of them quite large. The plant-eating *Moschops* in particular was an extraordinary animal, equipped with massive shoulders and neck and tiny limbs and head. *Moschops* also had a massively thickened skull roof, and it has been suggested that it engaged in stentorian head-butting contests in seeking mates. The meat-eating dinocephalians were more lithe and superficially dog-like.

The commonest herbivores were the dicynodonts, with either no teeth at all or merely a pair of tusks (see Fig. 31). They sliced vegetation with their sharp, horn-lined beaks, and ground it with a circular jaw motion. The dicynodonts and large plant-eating dinocephalians were preyed upon by the first sabre-tooths, the gorgonopsians. Gorgonopsians were powerful sleek animals

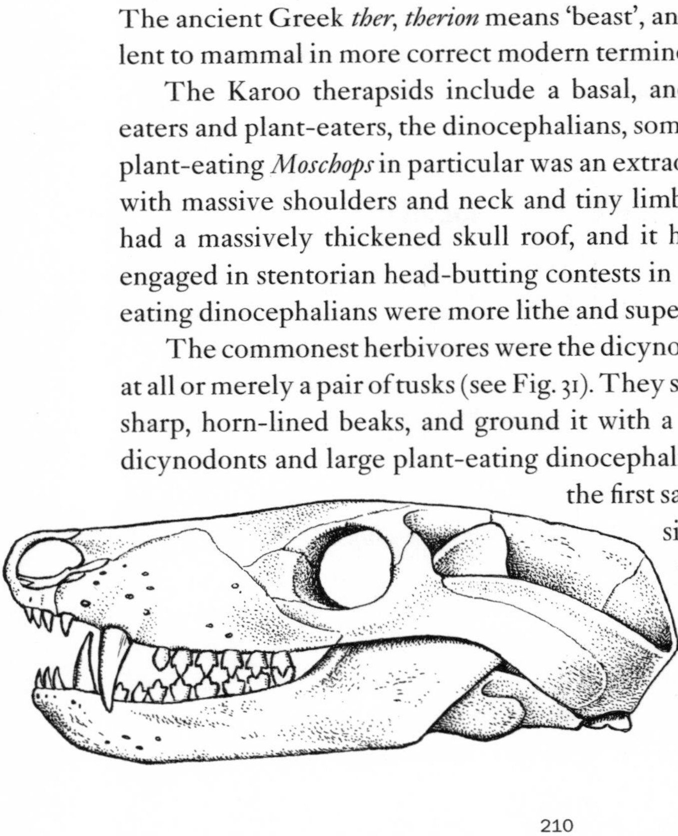

32 *The skull of the flesh-eating synapsid* Thrinaxodon *from the Early Triassic of South Africa.*

with massive tusks. They could drop the lower jaw clear of the fangs, and then leap at a herbivore with the teeth exposed and pierce their thick skins.

The light dawns . . .

Unknown to Owen, synapsids had already been independently described from the Late Permian Copper Sandstone beds of the Ural Mountains in Russia by a number of palaeontologists there – F. Wangenheim von Qualen, S. S. Kutorga, J. G. F. Fischer von Waldheim and E. I. von Eichwald, all of whom were introduced in Chapter 1. Owen is not really to blame, however, for not making the link at that time. These impressively Germanic gentlemen published rather limited accounts, some of them in somewhat obscure journals in St Petersburg and Moscow. More importantly, in the 1840s, neither they, nor Owen, had any concept of synapsids.

Enlightenment was rather slow to follow. Owen saw all the new South African materials first-hand, but he was never able to visit Russia to confirm that the Permian reptiles there were essentially the same as those from the Karoo. The first light was shed in 1866 when Hermann von Meyer (1801–69), the senior German palaeontologist of the day, published a large monograph on the amphibians and reptiles from the Urals, based on a collection of von Qualen's fossils that had been sent to Germany.[4] The connection was completed ten years later when the British Museum acquired its own collection of Copper Sandstone reptiles and Owen was able to compare them with the more complete Karoo specimens. He then concluded that there was a firm connection between the reptiles of the Russian Copper Sandstone and those from South Africa.

Another English researcher, William Harper Twelvetrees, published several articles in 1880 to 1882 on vertebrate remains and other fossils from the Kargala mines in the Urals, where he had worked as a mining engineer for a number of years. Twelvetrees emigrated to Australia in 1891 and was employed as director of the new Tasmanian Geological Survey which was launched in 1899. But it was Harry Seeley who was probably the first person to visit both South Africa and Russia, and to see the Permo-Triassic rocks, and their enclosed reptiles, himself. Having visited the Karoo Basin in 1889, he published[5] an account of the collections of Russian Permian reptiles in the St Petersburg Mining Institute and those of Kazan University.

Copper mining in the Urals had become uneconomic by the mid-nineteenth century, largely because of competition from new and more efficient enterprises in Africa and Australia. The last copper mine in the South Urals closed by 1900. New discoveries of vertebrates from the Copper Sandstones almost completely stopped, except for occasional finds – which can still occasionally be made – of scraps of bones in the spoil heaps of the old mines.

A timescale of the crisis

With all these reptile specimens from both South Africa and Russia, it should perhaps have been straightforward to follow through time exactly what had happened at the end of the Permian period. Surely the palaeontologists could readily track, bed-by-bed, the collapse of the Late Permian faunas, and the rise of new forms after the cataclysm?

In reality, of course, things are not at all so simple. A great deal of groundwork had to be done during the twentieth century before geologists could begin to disentangle the exact nature of the huge catastrophe that had taken place on land. The puzzle was, had all these creatures been living at the same time, or did they constitute a sequence of faunas that in fact spanned millions of years? Bain, Broom and others had begun to divide up the sequence of rocks in the Karoo into 'zones', each representing a unit of time, associated with and named after particular reptiles. But there were problems.

The synapsid skulls and skeletons had been extracted from sites all over the huge area of the Karoo Basin. Precise levels in the rock sequence could not be determined clearly, mainly because the Permo-Triassic succession looked very similar throughout – red sandstones and mudstones. There was no single, simple vertical reference section in which the horizons could be identified. So two skulls found 1000 kilometres apart could well be from the same species, and from rocks of the same age. Equally, two finds from the same farm could turn out to be utterly different, and separated in time by 20 million years.

A second major problem was that no one in 1900 had a clear view of the timescale involved in the Permo-Triassic, nor of how to match the continental sequences from South Africa, Russia, or indeed from England or Germany, with marine successions known from the Alps.

The South African Karoo sequence was known to consist of a long run of continental sediments, ranging in age from the Carboniferous to the Jurassic. At the base of the sequence is the Dwyka Tillite, a deposit made from the ground-up rocks deposited beneath glaciers. Here was evidence for the huge south-polar glaciation of the Late Carboniferous and Early Permian. The Dwyka Group also includes, above the tillite, Early Permian beds with fossil plants. Above the Dwyka Group (700 metres thick) comes the Ecca Group (2000 metres thick), largely Mid Permian in age, and this is followed by the Beaufort Group (up to 4500 metres thick), ranging in age from Late Permian to Mid Triassic. The famous Karoo fossil reptiles virtually all come from the Beaufort Group. Above this lies the Stormberg Group (500 metres thick), a Late Triassic to Early Jurassic succession of continental red-bed sediments. In some horizons, the remains of early dinosaurs have been found, in the form of both skeletons and footprints. The sequence is capped by the Drakensberg volcanics, basalt lavas that poured out in the Early Jurassic as Gondwana began to rift into its constituent modern continents. In 1900, geological understanding of the Karoo Permo-Triassic rock succession was well ahead of that in Russia. But by 1950, the situation in Russia had greatly advanced, largely due to the work of one man.

The father of the Russian Permo-Triassic

Ivan Antonovich Efremov (1907–72) was the right man at the right time (Fig. 33).[6] With a brilliant synthetic mind, he was able to draw together the seemingly unconnected records of Permo-Triassic vertebrates from the vast Russian territories, and cajole them into a meaningful stratigraphic scheme. Not only this, but Efremov is credited with the invention of a new branch of palaeontology, which in 1940 he called taphonomy, the study of the preservation of fossils. He was also a writer of science fiction novels, which were highly popular in Russia, both for their imaginative story lines and for the mildly pornographic images of women which he used as illustrations.

Efremov's scientific activity began in 1925, when he was 18 years old. He was employed as a fossil preparator in the Mining Museum in Leningrad and he began straight away to lead fossil-hunting expeditions south to the Caspian Sea, east to the classic Urals Permo-Triassic and northeast to the

33 *Portrait of Ivan Antonovich Efremov, who worked out the stratigraphic scheme for the Russian Permo-Triassic vertebrates.*

Dvina River. The North Dvina River is a huge waterway that begins high in the Ural Mountains and runs for 1000 kilometres, heading first southwards, then east and northeast to the icy port of Arkhangel'sk (often called Archangel in English), where it enters the Arctic Ocean. Efremov took over the major expeditions that had been begun on the North Dvina by V. P. Amalitskii (1860–1917), professor of geology at Warsaw in Poland. Working against considerable obstacles, and with very little finance at first, Amalitskii had proved the richness of the Late Permian beds along the North Dvina. Over years of return visits, he and his wife extracted hundreds of skeletons of a whole new array of amphibians and reptiles, some of them unique to Russia, such as the amphibians *Dvinosaurus* and *Kotlassia*, and the sabre-toothed gorgonopsian *Inostrancevia* (see Fig. 35). Efremov collected more amphibians and reptiles along the Dvina River, and made sure that Amalitskii's unfinished work was completed.

Nearer to home, Efremov excavated sites north of Moscow, where he found some of the first Early Triassic vertebrates in Russia, including skulls of the fish-eating amphibian *Benthosuchus*. He also reinvestigated the copper mines in the South Urals, crawling about in some of the old shafts in the hope of making new finds. While he was disappointed in this aim, he was able to provide an overview of what had been found, and the conditions of their preservation. More promising were the Permo-Triassic red beds that lay around the South Urals. Murchison had visited these in 1841 and had collected from them the reptile bones he passed to Owen. Very little had been done since. But Efremov and his colleagues began to identify numerous localities on both sides of the Ural and Sakmara rivers. They noticed how certain species tended to occur together regularly – that there appeared in fact to be a succession of quite distinctive faunas.

Efremov recognized seven zones:

Zones I, II: Dinocephalian (Late Permian)
Zone III, IV: Small reptile and pareiasaurian (Late Permian)
Zone V: 'Neorachitome' (Early Triassic)
Zone VI: *Capitosaurus* (Early Triassic)
Zone VII: *Mastodonsaurus* (Mid to Late Triassic)

The division between Zones IV and V marks the line of the end-Permian crisis. Each zone is defined by a characteristic reptile or amphibian, part of a typical assemblage. The Late Permian Zones I–IV are named for particular reptiles, while the Triassic Zones V–VII are founded on aquatic amphibian species. These zones seemed to work throughout Russia, from Moscow to the Urals, and even for more remote localities in Siberia.

Efremov rose through the ranks in the Soviet Palaeontological Institute. In 1937 he superintended its removal to Moscow when Stalin shifted his seat of power from the old imperial capital of St Petersburg, where Murchison had been so lavishly entertained by Tsar Nicholas nearly one hundred years before. As well as his Permo-Triassic researches, Efremov used his position to initiate the first of a long series of Soviet and Russian expeditions to explore the dinosaur-rich lands of Mongolia.

When Efremov died in 1972, the Palaeontological Institute was still located in the Neskuchny Palace, an architectural monument of the eighteenth century in the centre of Moscow. Efremov had planned for a larger and more modern museum, and in 1987, 15 years after his death, the palaeontologists took over a new building on the outskirts of Moscow. The new museum boasted extensive public displays and a staff of over 100 – an impressive legacy of Efremov's influence and authority. I was able to visit it in the 1990s, at the beginning of my Russian adventures (Chapter 10).

The first complex ecosystems

At the same time as Efremov and his students were sorting out the stratigraphy of the Russian Permo-Triassic, South African geologists were working on that of the Karoo, and by 1995 a detailed scheme had been established[7] that allows us to dissect the end-Permian crisis on land in some detail. We will not explore all that is known about the life of each of the zones of the Beaufort Group in the Karoo Basin, but let us look at the *Dicynodon* Zone, just before the end of the Permian. In retrospect, of course, we know that this scene of burgeoning life was not to last, but there can have been no presentiment of the impending cataclysm.

If we could visit the *Dicynodon* Zone we would experience a time of tropical climates and monsoonal rain. Great rivers cut across the flat plains, sweeping masses of sandstone down from the Antarctic mountains to the south. Occasionally the rivers spill over their banks, forming floods of finer sediment. Here and there, lakes provide a home for fishes. Trees and bushes around the margins of the lakes drop their leaves into the waters, where they sink and are entombed in the dark muds at the bottom. Typical plants around the margins of the lakes and rivers are mosses, horsetails and ferns. Here and there bush- and tree-like conifers provide some higher foliage. A particularly common plant is *Glossopteris*, a 'seed fern'. Usually a tree with a woody trunk and standing about 4 metres tall, some species are lower and more bushy. *Glossopteris* leaves are tongue-shaped smooth structures with a central stem, arranged in star-like bunches of 20 or so at the ends of branches.

Scurrying in the undergrowth are centipedes, millipedes, spiders, silverfish, cockroaches and beetles. Bright mayflies and dragonflies skip and glint in the dappled sun over the lakes, while bugs and flies feed on the leaves and on animal dung. There are no termites, ants, bees, wasps or butterflies – all of these come much later. Snails and worms crawl in the mud at the edge of the lakes, and leave their trails pressed in the wet sediment.

During the dry season, the lakes dry up and water-living animals and plants die off. The bugs and beetles can move elsewhere. The dragonflies and mayflies die, but their line is maintained by a few eggs and larvae in the much-reduced ponds. Fishes die too, except for those that happen to be in rivers whose water supply comes from the high, cool mountains to the south. Other rivers dwindle to a trickle and disappear, leaving behind an arid, cavernous wadi.

Some animals had strategies to survive the hardships of the dry season, burrowing deep in the cool mud and sealing themselves in. It is likely that lungfishes of *Dicynodon* Zone times were able to do this. They certainly do it today, and fossils have been found of lungfishes apparently sealed into burrows. Sensing that its pool or pond is drying fast, the lungfish fattens itself with all the food it can and then constructs a chamber as much as a metre beneath the muddy bed. It then makes a watertight cocoon from mucus and turns its body systems down to stand-by. This process of aestivation ('over-summering') is analogous to hibernation ('over-wintering'). Some of the Karoo reptiles seem to have done the same thing: burrows have been found on river banks containing clutches of skeletons curled up and evidently sheltering from the heat of the sun. Such fossilized examples show us the victims: they had starved to death, or perhaps their burrow was flooded by monsoonal rain before they could escape.

It is the vertebrates for which the *Dicynodon* Zone is famous, however. At present, 74 species are recognized, consisting of two species of fish-eating amphibians in the rivers and 60 species of reptiles. At the base of the plant-eating food chain (Fig. 34) are two species of procolophonids. These reptiles have triangular-shaped skulls and rows of blunt peg-like teeth that were used to crush and tear their plant food. The procolophonids are probably distant cousins of modern turtles. At the base of the flesh-eating food chain are the millerettids, of which two species are known from the *Dicynodon* Zone. The millerettids look superficially like little lizards, but they are probably relatives of the procolophonids. Their tiny pointed teeth suggest a diet of insects. Other insect-eaters include a species each of

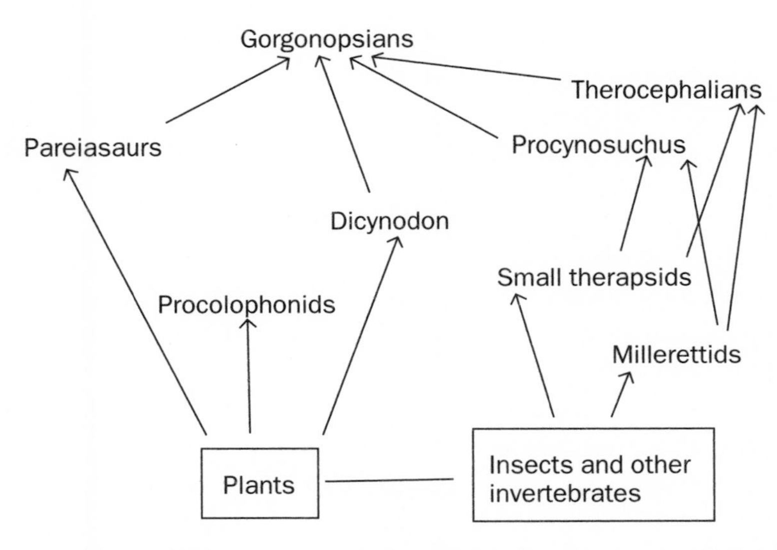

34 *A simplified food web of the main reptiles of* Dicynodon *Zone times, one of the first complex ecosystems on land and entirely destroyed by the end-Permian mass extinction event.*

Youngina and *Saurosternon*, both of which also looked like lizards but were actually much more primitive. There are also three species of small, possibly insect-eating therapsids.

Next up in size are the therocephalians, of which 15 species have been reported so far from the *Dicynodon* Zone. The therocephalians are cat-sized and have longish skulls with sharp, piercing teeth concentrated near the front. In the same dietary category are the early cynodonts *Procynosuchus* and *Cynosaurus*, which are the size of terriers and have a full set of teeth along the length of their jaws. These were all probably second-level predators that hunted the smaller plant-eating procolophonids as well as the insect-eating anapsids, diapsids and synapsids.

The *Dicynodon* Zone is named, of course, for *Dicynodon*, of which nine species are recognized; a further nine species of other dicynodonts are also known. These toothless herbivores, of the type first described by Owen from South Africa in 1845, had clearly undergone a great diversification. The different species varied somewhat in size, as Andrew Geddes Bain had noticed when he made his first collections. Doubtless each species fed on a slightly different plant diet, allowing such a diverse array of 18 closely related species to co-exist, just as numerous species of apparently similar antelope do on the grassy savannahs of Africa today.

The largest herbivores include two species of pareiasaurs. Hefty animals, each about 2 metres long and with massive, barrel-shaped bodies carried on rather feeble-looking sprawling limbs, these reptiles are close relatives of the much smaller procolophonids. Pareiasaurs are renowned for their uncomely looks, having broad mouths, a roughly triangular skull and numerous prominences and excrescences all over the head. Their unfortunate plainness of visage was matched by low intelligence, as indicated by the relatively small size of the head in comparison to the body, and the even smaller size of the braincase within the skull. They doubtless munched placidly on *Glossopteris* leaves, and resisted attack with their formidable size and knobbly head.

The dicynodonts and pareiasaurs are preyed on by a fiendish array of gorgonopsians, some 25 species of them. These predators (Fig. 35), sometimes smaller than their prey, being generally about 1 metre in length, were by no means ineffective. They all have excessively elongated canine teeth, as well as the jaw mechanism and muscles to operate as sabre tooths, a mode of attack that no longer exists. Until the end of the last ice age, 10,000

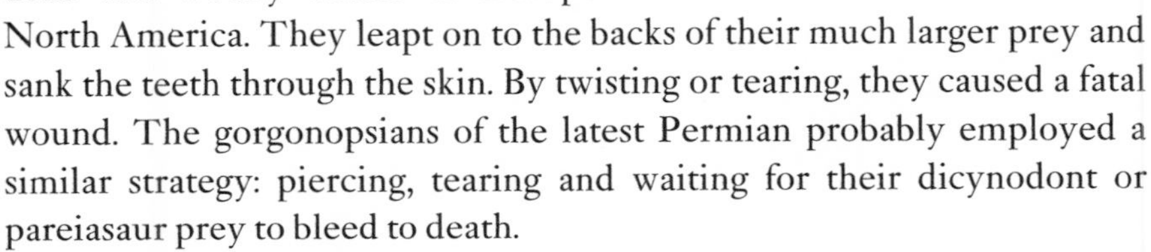
35 *The gorgonopsian* Inostrancevia *from the latest Permian of Russia.*

years ago, various sabre-toothed cats, with curved knife-like canine teeth up to 15 centimetres long, hunted the large thick-skinned mammoths, mastodons and woolly rhinos of Europe and North America. They leapt on to the backs of their much larger prey and sank the teeth through the skin. By twisting or tearing, they caused a fatal wound. The gorgonopsians of the latest Permian probably employed a similar strategy: piercing, tearing and waiting for their dicynodont or pareiasaur prey to bleed to death.

The key fact about the *Dicynodon* Zone is the ecological maturity of the vertebrate fauna. Food webs (see Fig. 34) were complex, with herbivores of various sizes feeding on a broad range of plants, and with three or more levels of predators feeding respectively on the small, medium and large herbivores. This kind of complexity and diversity of land vertebrate life is common enough today, but was first achieved only during the Late Permian. Before that time, herbivorous reptiles were rare or non-existent, and flesh-eaters existed only in a couple of size categories. The tertiary-level sabre-toothed gorgonopsians were something new altogether.

Collapse of stout pareiasaur

All this complexity was wiped out by the end-Permian crisis. At the moment, we can only see the picture partially, as if through several layers of frosted glass. But the astonishing finding from all the current work is that perhaps the end-Permian crisis was actually more severe for life on land than for life in the sea. This is a striking turn-around from the view of only 15 years ago, that essentially nothing out of the ordinary had happened on land at all.

It is hard to work out the precise patterns of collapse of the plants and insects because of their poor fossil records in the Karoo. However, worldwide, 11 families of true land plants fell to three across the Permo-Triassic boundary. Species numbers fell from about 140 to 50 or fewer: the quality of the evidence is just not good enough to be sure. The story appears to be clearer for the vertebrates.

The 62 species of amphibians and reptiles from the *Dicynodon* Zone in the Karoo contrast with only 29 in the subsequent *Lystrosaurus* Zone, the earliest phase of the Triassic. But only two species of reptiles have been discovered that survived through the Permo-Triassic boundary, the dicynodont *Lystrosaurus* (species undetermined) and the therocephalian *Moschorhinus kitchingi*. The other 27 species of the *Lystrosaurus* Zone had all arisen during the intervening period. So, at species level, the end-Permian crisis apparently accounted for 60 of the 62 vertebrates, a loss of 97%. But we know this is an overestimate.

A number of reptile groups are known to have survived into the Triassic, such as the procolophonids, the basal diapsids, the dicynodonts and the cynodonts, so at least a further three species must have passed through the Permo-Triassic boundary. Perhaps their fossils have yet to be found, or they may indeed have died out in the Karoo Basin but survived elsewhere, and then recolonized South Africa in the earliest Triassic.

Worldwide, the picture seems clearer.[8] Of the 48 families of amphibians and reptiles that were present in the last 5 million years of the Permian, 36 died out, representing a loss of 75%. This is higher than the generally accepted figure of a 50% loss of families in the sea. So, for the vertebrates at least, the end-Permian crisis was apparently more severe than for marine animals. If a 50% family loss scales to 90% species loss in the sea, then the loss of 75% of vertebrate families must scale to something like 95% species extinction on land. Whether in South Africa, or worldwide, the pareiasaurs, millerettids and gorgonopsids, as well as numerous other groups, had gone for good.

New work in the Karoo Basin has helped to identify just what the environmental changes were on land. Roger Smith, a sedimentologist from the South African Museum in Cape Town, and Peter Ward, a palaeontologist from the University of Washington, Seattle, pinpointed the exact position of the Permo-Triassic boundary in the Beaufort Group sequence.[9] The boundary between the sediments of the *Dicynodon* and *Lystrosaurus* zones was marked by a 20-metre thick succession of channel sandstones and finely bedded thin sandstones and shales. The end of the *Dicynodon* Zone is marked by the last appearance of fossils of *Dicynodon* and of *Glossopteris*, and this also marks the beginning of the *Lystrosaurus* Zone. There is not a single level at which all these things happen. In fact, the last *Dicynodon* occurs at the top of the section, while the first *Lystrosaurus* occurs at the base, so there is an overlap. However, the two dicynodonts are generally found in different beds.

The last specimens of *Dicynodon* are found preferentially in green- and olive-coloured mudstones and fine sandstones, while *Lystrosaurus* is found in red-coloured sediments. In the larger picture, the green-coloured sediments of parts of the *Dicynodon* Zone are replaced by red beds. So, Smith and Ward interpreted this to mean that there was a step-by-step change in the conditions of deposition of the sediments, which is seen in detail in the 15-metre transitional zone. Just what was going on?

Plant die-back and catastrophic erosion

The red and green sediments of *Dicynodon* Zone times were deposited by broad meandering rivers, carrying sediment at relatively steady rates from the southern mountain ranges. Sediments in the lower parts of the *Lystrosaurus* Zone are quite different, consisting of more sandstones and less mudstones. The sandstones are not seen in discrete channel fills but form instead thicker, more complex packets of sediment, evidently deposited by braided rivers. Braided rivers, where the individual channels twist and turn dramatically, are typical of alluvial fans, formed when fast-flowing, high-energy upland parts of river systems come hurtling down the mountains and meet a lower-angle slope.

Smith and Ward argued that a major change in sedimentation patterns was happening in the Karoo Basin just at the Permo-Triassic boundary. They believed that the pace of sedimentation increased dramatically. In the 15-metre transitional section, the *Dicynodon*-bearing green beds of the lowlands are progressively overstepped by higher-energy red sandstones with associated *Lystrosaurus*, until the latter association of sediments and fossils prevails.

What caused the increase in the rate of sedimentation? One obvious cause might be that the mountains that supplied the sediments had been uplifted by some major tectonic activity. However, there is no evidence for anything of the kind. So, Smith and Ward proposed that the increase was caused by a major die-back of plants. Plants prevent erosion on land by binding rocks with their roots, and building up layers of soil. If all plants are removed from any typical continental area, rates of erosion multiply by 10 or 20 times – as seen in the catastrophic effects of deforestation on the foothills of the Himalaya in recent years. After the trees had been removed for building huts and for firewood, rainfall washed huge amounts of soil and rock from the slopes, causing catastrophic floods in Bangladesh every year or so.

Smith and Ward only had good evidence for this major change in sedimentation in the Karoo Basin, but they believed that other continental Permo-Triassic boundaries show a similar phenomenon. So perhaps this was a worldwide aspect of the end-Permian crisis. Their hypothesis depends on the dramatic loss of plant life over wide areas. What evidence is there?

Detailed studies of ancient soils, and of plants, across the Permo-Triassic boundary seem to support the Smith-Ward hypothesis. Greg Retallack of the University of Oregon studied numerous Permo-Triassic soil successions in Gondwana (Antarctica, Australia, New Zealand), and he found a shift in soil and plant types.[10] Latest Permian soils include coals and rooted sands, while earliest Triassic soils are mainly root-filled mudstones. This indicates a shift from cold-temperate broad-leaved swampy floras, dominated by *Glossopteris*, ferns, horsetails, mosses and the like, to cool-temperate conifer forests. Temperatures and levels of rainfall rose slightly across the Permo-Triassic boundary, and this was associated with increased erosion and runoff.

The disappearance of swamp vegetation for a time at the beginning of the Triassic is associated with the famous 'coal gap'. Nowhere in the world did coals form during the 20 million years of the the Early and Mid Triassic, whereas coal is usually forming somewhere, wherever climates are warm and humid. Early Triassic floras were dominated by stress-tolerant, opportunistic kinds of plants – what are commonly called 'weeds' – which are adapted to spread quickly and to live in poor-quality soils lacking in nutrients. These are the plants that will grow up first on a piece of disturbed ground, such as the rose-bay willowherb and other weeds that colonize abandoned building sites. In time the weeds give way to different plants, representing the more mature successional flora.

Some additional evidence tends to support this idea of a catastrophic loss of plant cover in the earliest Triassic. When he was examining samples of sediment under the microscope, Yoram Eshet, a palaeontologist from the Geological Survey of Israel, noticed a tremendous shift in pollen and spore types across the Permo-Triassic boundary.[11] In the latest Permian, in samples taken from all over Israel, he found a typical selection of pollen and spores from a wide range of terrestrial plants. The same was true for samples taken from higher levels in the Early Triassic. But for a few metres at the very beginning of the Triassic, almost all typical plants disappeared, and all he could see were cells of fungi. In other words, the end-Permian crisis was marked by a fungal spike, a dramatic spread of mushrooms all over Israel.

Not only Israel, but the world. Eshet found that other authors had spotted the fungal spike in Europe, North America, Asia, Africa and Australia. It has not been observed in Greenland, probably because the sections there are so thick that an increase in proportions of fungi is spread through several metres. So, something had temporarily displaced the normal floras and encouraged fungi to blossom, or at least to extend their hyphae and spores widely. Eshet suggests that the sudden proliferation of fungi indicates a high-stress event that killed most other plant life. Fungi flourish in the presence of decaying organic matter, and the widespread killing of more complex plants provided a perfect environment for them. But the plants had not been entirely wiped out. Some clearly survived in isolated patches. When the stress was removed, the normal diversity of plants returned, and the fungi retreated to their usual position in the system.

Plant ecosystem collapse

In a further study, Cindy Looy of the University of Utrecht and Richard Twitchett, then of the University of Southern California,[12] were able to tease apart in some detail the step-by-step process of the plant ecosystem collapse. They showed quite convincingly that plants did not disappear overnight, and that normal ecological processes continued for some time, even during the phase of severe environmental stress. They based their work on the remarkable Permo-Triassic succession in Greenland, which we encountered in Chapter 7 – the transition from the latest Permian Schuchert Dal Formation to the earliest Triassic Wordie Creek Formation. These sediments were deposited in shallow seas and are well known for their marine fossils. But they were also close enough to shore to contain abundant pollen and spores that had been blown offshore from land plants. So there is a unique opportunity here to see what was happening in the sea and on land, and to match the timing, centimetre-by-centimetre, through the sediments.

This is the story of the plant extinctions. The first phase saw the loss of closed woodland to open herbaceous vegetation – more bushes than trees. Fungal activity increased at the same time, possibly related to the decay of the trees that were dying off. The place of the trees is taken by opportunistic, weedy species – smaller and faster-growing – which take advantage of the gaps in the once uninterrupted forests. These species of weedy club-mosses were previously kept in check by the dominance of woody trees, the seed ferns and conifers, which blocked sunlight from the forest floor. With the trees gone, the small green plants that were formerly suppressed now expanded and spread.

In a second phase of the collapse, the open space was further occupied by ferns; these had been present before, but in low abundance. In addition, some new forms arrived from the south. Cycads and other plants that had been known before only in North America, for example, suddenly appear in the crisis floras of Greenland. So, in the face of death and destruction, it seems opportunistic species colonized and took advantage of the crisis. But not for long.

Then, in the final killing phase, most of the invaders disappear. The last of the woody trees and bushes go too. All that is left is a much-reduced flora of lycopsids, club-mosses – low plants that presumably survived in damp pockets here and there. During the three-phase killing cycle, the overall volume of pollen and spores diminishes dramatically, indicating rarity of plant life at the final killing level, and for many metres of sediment thereafter in the earliest Triassic.

How long did all this take? Looy and Twitchett estimated that the three-phase decimation of floras lasted for some 500,000 to 600,000 years, spanning the Permo-Triassic boundary, corresponding to the duration of marine decline in the Meishan section in China.

Lystrosaurus the survivor

What of the reptilian survivors of the crisis on land? The most prominent was *Lystrosaurus*, the modest-sized dicynodont which we met in

Chapter 1 and which, as we have seen, overlapped for a short time with its typical latest Permian relative *Dicynodon*. The extraordinary thing about *Lystrosaurus* is not that it survived, but that it dominated the whole world for a short time. Species of *Lystrosaurus* have been described not only from South Africa, but also from South America, Antarctica, India, China, Russia and possibly Australia. The *Lystrosaurus* Zone of South Africa is matched by contemporary rock units in all those continents. The apparent absence of *Lystrosaurus* in other areas is probably due to the fact that the right rocks have yet to be found. It was truly cosmopolitan – a time indeed when 'pigs ruled the earth'.

Lystrosaurus was not only widespread, but also hugely abundant. In South Africa, palaeontologists cry with frustration when they find another *Lystrosaurus* skull. James Kitching, one of the greatest collectors in South Africa, estimates that over 2000 skulls of five species of *Lystrosaurus* have been collected in the Karoo Basin, and many more have been smashed with hammers. Collectors are desperate to find anything else. But the reptiles and amphibians associated with *Lystrosaurus* are unbelievably rare. Only a few specimens of its smaller relative, the dicynodont *Myosaurus*, have been found. The same is true for the other *Lystrosaurus* Zone animals: isolated specimens of a few species of amphibians, a couple of procolophonids and basal diapsids, and 10 species of small flesh-eating therocephalians and cynodonts. So, in all, *Lystrosaurus* makes up at least 95% of all animals found in the *Lystrosaurus* Zone.

Such dominance by a single animal is far from normal. If the 20 or so other species comprise less than 5% of the fauna, we are looking at something that is so unnatural as to resemble a product of human intervention. For example, a field of cows is a typical agricultural monoculture. True, there may be half-a-dozen rabbits in one corner of the field, and a few field mice, voles and weasels lurking in the hedgerows, but the scene is dominated by cows. Such monocultures cannot survive without human assistance. In Nature, when food is abundant, the one sure thing about life is its diversity. No one species will dominate, and there may be as many as five or ten that are considered to be abundant.

The unnatural abundance of *Lystrosaurus* confirms that it is a disaster species, something like a weedy plant. However, unlike the fungal spike and the weedy plant species in the earliest Triassic, 'normal' faunas could not re-establish themselves. Among the plants, sufficient diversity had evidently survived here and there to allow normal floras to re-establish themselves relatively rapidly in the Early Triassic. But *Lystrosaurus* was on its own. At about 1 metre in length, it was the largest animal on Earth. All the others were much smaller. This was a huge change from the latest Permian situation, when many species of dicynodonts, pareiasaurs and gorgonopsians were larger. But they had all died out.

What did *Lystrosaurus* do? Like all dicynodonts, it had two upper canines, with otherwise toothless jaws no doubt lined with a horny beak. It used the standard dicynodont feeding mechanism of grasping a mouthful of leaves at the front of the beak, perhaps using the canines as rakes to gather the plant material into a bundle, then closing its lower jaws and pulling them back, slicing the plants into bits. The roughly cut stems and leaves were then swallowed more or less whole, since *Lystrosaurus* could not chew by side-to-side movements. Doubtless much of the digestion of the tough plant material took place inside the huge gut. Like most herbivores, *Lystrosaurus* had a vast rib cage and enormous guts (plant food takes a lot more digesting than meat). The gurgles and eructations would have been an Early Triassic phenomenon.

When *Lystrosaurus* was named in 1903 by Robert Broom, he classed it as an aquatic animal, an active swimmer, perhaps something like a pygmy hippopotamus.[13] This view held sway until the 1990s, when Gillian King and Michael Cluver of the South African Museum in Cape Town pointed out that *Lystrosaurus* was actually rather like other dicynodonts, and was almost certainly a land-dweller. In opposition to this, Russian palaeontologist Mikhail Surkov of Saratov University argued that the Russian species, *Lystrosaurus georgi*, at least, and possibly other species of *Lystrosaurus*, did have adaptations for swimming. The forelimbs are relatively larger than in other dicynodonts, and the distribution of the muscles indicates that the animal used these to paddle vigorously, while the hindlimbs trailed behind. Of course, it was still as adept on land as any other dicynodont, which isn't saying much.

A superior pig-like reptile?

So why did *Lystrosaurus* do so well? It is easy to ask such a question, since there is no doubt that *Lystrosaurus* was in its way one of the most successful reptiles of all time, known from a dozen or more species from all continents, and surviving for a few million years at the beginning of the Triassic. But perhaps it is wrong to seek particular reasons why *Lystrosaurus* was so successful. It was a success *faute de mieux*. It may have been no more than pure chance that this genus, among all the others of the latest Permian, did not die out. Then, as the sole modest-sized survivor, it had an empty world all to itself. Perhaps *Lystrosaurus* was actually not particularly skilled at surviving: it just had to be good enough to scratch a living. There were no other significant plant-eaters with which to compete, and there were no large predators to pursue it.

This sort of question is very popular, as we saw in Chapter 1. But surviving an extinction is not like winning a sprinting race. In the case of the race, the athletes know in advance exactly what is expected. They train and try to hone their performance to the very best they can. And, on the day, the best athlete wins. An extinction event is like a race with unpredictable obstacles and of unknown length. The athletes have no advance warning of the obstacles, any of which could be fatal. So in this race, the athletes all have pretty much the same chance of reaching the other end of the track first – if at all. This is also evidently true of the mass extinctions in the past.

Back in the USSR

The Karoo Basin of South Africa has yielded up some of its secrets. Thanks to the work of Broom, Kitching, Smith, Ward and many others, we now know something about the nature of the end-Permian crisis on land. The rate of extinction among reptiles was as high as any extinction in the sea, and older ideas that not very much happened at the Permo-Triassic boundary among continental organisms have been thoroughly rejected.

The Russian Permo-Triassic sequences are just as good, and just as rich in some places. In the early 1990s, before Roger Smith and Peter Ward had published their work on the Karoo successions, there was an opportunity for me to go to Russia. The communist regime had collapsed, and Boris Yeltsin had come to power. Russian and western scientists were looking for contacts, and I met Akademik Leonid P. Tatarinov, a noted expert on the Russian mammal-like reptiles, but now retired, at a conference in Montbéliard in France in 1992. Tatarinov suggested that the University of Bristol should collaborate with the Palaeontological Institute in Moscow, and we signed an agreement.

This was a very exciting opportunity. We knew the older published work by Efremov and others. Surely Russia might provide a good testing ground for hypotheses based on the rock sequences in the Karoo Basin? Surely also it would be useful to look at a part of the world that in the Permian was a long way from South Africa? Are the extinction patterns comparable, or are there some regional differences? Many of the Russian reptiles are very similar to those from South Africa, but there are some unique Russian groups, not known from elsewhere. So could some of the extinction patterns be different? What of the apparent loss of plant cover and heightened erosion that seemed to occur in the Karoo: could this be confirmed in Russia as well?

It was with these questions in mind that we went to Russia in 1993.

10
ON THE RIVER SAKMARA

The dragonflies flew above our heads at dusk, swooping and weaving in pursuit of mosquitoes. They did not appear during the day, but emerged at nightfall, when their prey came out. They were shadowy forms, with their great paired wings, more like birds than insects.

'Ubivayoot kameri. Harashoa!' murmured one of our Russian colleagues, 'They kill the mosquitoes. Good!'.

The bonfire crackled in front of us, and we crowded round in an attempt to escape the biting insects. This was the routine every night. We ate during daylight, and spent the early evening trying to avoid being eaten by the voracious blood-sucking mosquitoes. They were larger than anything we were used to at our respective homes in Britain and North America, and all our lotions proved totally ineffective against them. The leader of the trip, Valentin Tverdokhlebov of the Geological Institute in Saratov, offered us some Russian anti-mosquito mixture. This was a pinkish lotion, rather like baby cream, with an unusual smell. We never found out what was in it, but it certainly worked.

We were here, by the Sakmara River, at the south of the Ural Mountains in Russia, in July 1995. The Sakmara flows southwestwards from the foothills of the Urals to join the Ural River itself, which then passes southwards to the Caspian Sea. Our camp site was not far from some of the old Copper Sandstone mines that had produced fossils in the eighteenth and nineteenth centuries. These had long been out of action, and we hoped to find specimens in the red sandstones and mudstones of the surrounding foothills and river banks.

Our aim was to explore the Russian Permo-Triassic succession in detail to see whether it might shed light on what had happened 252 million years ago.

In the galleries

Two years earlier, during our first visit to Russia, Dr Glenn Storrs and I spent a week at the Palaeontological Research Institute (known as PIN, Paleontologicheskii Instituta Nauk) in the outskirts of Moscow, on the avenue Profsoyuznaya. This was the institute Ivan Efremov had dreamed of, but never saw in his lifetime. Glenn was working then at the University of Bristol, and he is now Curator of Fossil Vertebrates at the Cincinnati Museum of Natural History.

We stayed at a hotel belonging to the Russian Academy of Sciences in central Moscow. When we arrived there we were given 10,000 roubles each, seemingly a huge sum, but worth only about £70 then. Andrey Sennikov, one of our Russian colleagues, explained the purpose of the money: 'This is for scientific purposes. You must not spend it on bad girls.' We heeded Andrey's advice, and spent the week studying the remarkable collections of fossil reptiles at PIN.

PIN is both research institute and museum. In its heyday, in the last years of the Soviet Union, and for a few years after *perestroika*, and Gorbachev and Yeltsin, PIN employed around 100 palaeontologists, both in an older building in central Moscow and in the new museum that had opened in 1987 on Profsoyuznaya. Numbers have dwindled since then, as salaries have gone unpaid and people have been forced to leave for economic and other reasons. The new museum is a striking brick edifice, with an extensive basement in which vast numbers of bones are stored, and two floors of public galleries. The public galleries are a wonder. The architect has used the structure and ornament of the building to illustrate its subject matter. There are great terracotta sculptures of mammal-like reptiles, dinosaurs and mammoths on the walls, ceramic images of pterosaurs, coelacanths and ancient humans, and a *trompe-l'œil* stratigraphic column, using mirrors, which gives a striking impression of the seeming endlessness of geological time as you look down. The galleries at PIN are full of wonderful displays of the antediluvian treasures of the Soviet Union, including some of the earliest forms of life from the Precambrian of Siberia, as well as dinosaurs and fossil mammals from Mongolia. Glenn and I were there to look at the Permo-Triassic reptiles and amphibians.

In the very first Permo-Triassic gallery we were astounded. There, as a centrepiece, was an assemblage of 20 pareiasaur skulls and skeletons. Pareiasaurs (Fig. 36) had been described first from the Karoo Basin, as was noted in the last chapter, and they were later found in abundance in Russia. From quarries near Kotlas, endless numbers of specimens of the pareiasaur *Scutosaurus* have been excavated and brought back to Moscow. Russian scholars had divided their pareiasaurs into some 15 species, but later work has shown that this was more to do with nationalistic pride than reality. There were probably in reality only two species.

Ranged around the walls of the Permo-Triassic gallery were skeletons of many other kinds of amphibians and reptiles, all from the southern Ural Mountains, from northern parts of the Urals, and from Arctic Russia, north of Moscow, especially along the North Dvina River – the fruits of Amalitskii's excavations of the early twentieth century. Most striking were the synapsids, which are very similar to those from the Karoo. These range from small insect-eating cynodonts and therocephalians, to large pig- and hippo-sized dicynodonts. The most astonishing of the Late Permian

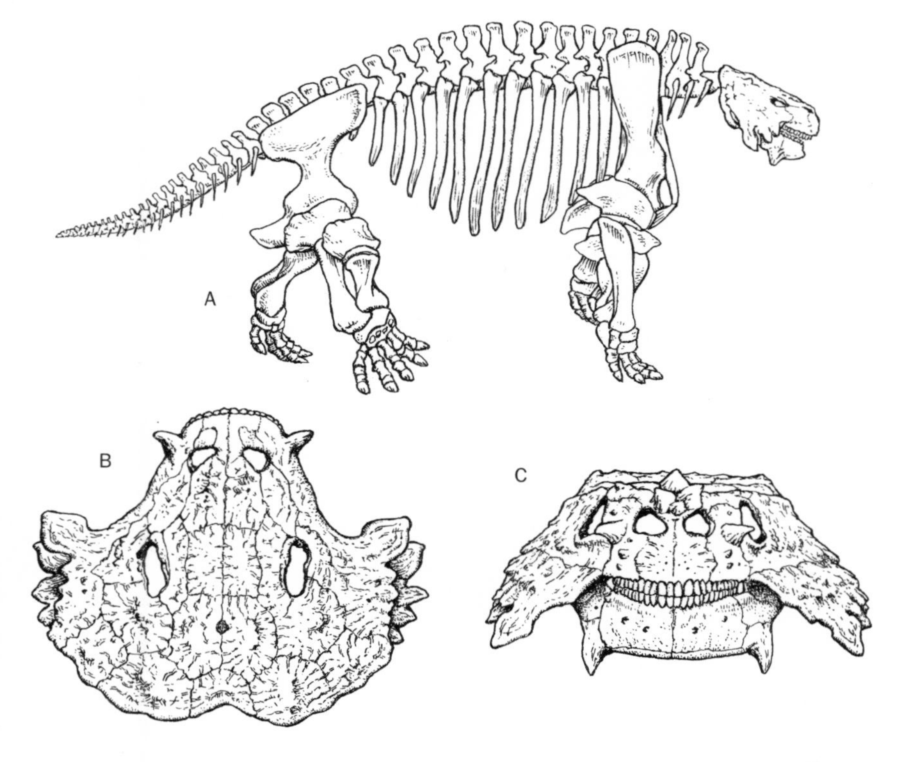

36 *The Russian pareiasaur* Scutosaurus *from the latest Permian: A, skeleton; B, C, skull viewed from above (B) and from the front (C).*

synapsids was the gorgonopsian *Inostrancevia.* Its large wolf- or tiger-like skeleton indicates a powerful runner; its long dog-like skull suggests intelligence. The canine teeth are arresting – fully 50 millimetres long – and the lower jaw could drop open to an angle greater than 90°. This gorgonopsian was similar to those known from the Karoo.

Here, in the Late Permian of Russia, is an assemblage very like that of the *Dicynodon* Zone of South Africa, which we examined in the last chapter. The latest Permian fauna of Russia, the Vyatskian assemblage (Fig. 37), known from the North Dvina river and from the South Urals, was rich and diverse: the pareiasaur *Scutosaurus*, the dicynodont *Dicynodon*, four species of gorgonopsians, including *Inostrancevia*, and two smaller carnivores, a therocephalian and a cynodont. In other localities, latest Permian reptiles include *Archosaurus*, a 1-metre long slender fish-eating reptile, oldest member of the Archosauria, or 'ruling reptiles', the group that includes crocodiles and dinosaurs, and procolophonids, small triangular-skulled reptiles, related to pareiasaurs, but superficially looking somewhat like fat lizards. At the water's edge were three or four species of amphibians.

This was a rich and complex ecosystem, with as many animals as in any modern terrestrial community. There were herbivores specializing in plants of different kinds, fish-eating amphibians, insect-eating synapsids, carnivores feeding on small prey, and the gorgonopsians, so-called top predators, feeding on the largest of the herbivores. All these animals were wiped out by the end-Permian crisis. But what happened? We had to learn more about the rocks from which the bones came.

On the banks of the Volga

During our first visit to Moscow, we were able to visit the famous Tikhvinsk site, near the town of Rybinsk. We reached Rybinsk by train. Rail travel is the main means of travelling long distances in Russia, and

37 *The latest Permian Vyatskian fauna from Russia. At the back, the gorgonopsian* Inostrancevia *looks speculatively at the plant-eating pareiasaur* Scutosaurus. *A dicynodont stands at the water's edge, while the flesh-eating synapsid* Annatherapsidus *sits on a log, with* Dvinia *below. The temnospondyl amphibian* Chroniosuchus *sits on a sand bank, with* Kotlassia *in the water. In the foreground, the little procolophonid* Microphon *is to the left, the temnospondyl* Raphanodon *to the right.*

the trains are clean and efficient, if not particularly fast. We packed into a sleeping compartment for four: Glenn Storrs, myself, Andrey Sennikov and Igor Novikov, who was Deputy Director of PIN. The trip was organized as part of a joint research scheme, funded by the Royal Society (London) and the Russian Academy of Sciences (Moscow).

The rail tickets were cheap, and we had invested in a second-class berth. Third-class passengers slept in open carriages divided by bunks into sections, sleeping people on three levels. Our compartment was very like a British or French sleeper, with a two-level bunk. We settled on the train in Moscow, and prepared to eat. Andrey brought out some small, round, mid-white loaves and rye bread, dried river fish and fresh cucumbers. The train attendant brought a china teapot and mugs, which we hired for a few roubles. Hot water came from a fiendish small boiler set in an alcove at one end of the carriage, which was always kept hot. We feasted on sweet tea, bread and fish. We then slept through the slow rocking journey, and awoke at six in the morning to find ourselves in Rybinsk.

Rybinsk lies at the confluence of the Volga and Sheksna rivers, just below Rybinsk Reservoir, a huge artificial lake, over 120 kilometres long, created in the 1940s by damming the two rivers to produce a hydroelectric power scheme. We travelled a few miles downstream on the Rackyet ('rocket'), an astonishing Russian-built hydrofoil that used to ply up and down the small towns of the Volga, touching the shore on either side at landing stages, until we reached our destination, Tikhvinsk.

The Tikhvinsk site was famous as the source of some beautiful skulls of amphibians, the first of which had been collected by Ivan Efremov in the 1930s. The amphibians of the Permo-Triassic were mainly temnospondyls, a group that probably contains the ancestors of modern frogs and salamanders. As in modern frogs, the temnospondyls had broad low skulls, with a wide curving front to the snout and jaws lined with dozens of modest sharp teeth, ideal for snapping at fishes. The temnospondyls probably lay

38 *The earliest Triassic Rybinskian fauna from Russia, as seen at the Tikhvinsk locality. In the background the basal archosaur* Chasmatosuchus *wanders past the dicynodont* Lystrosaurus. *In the centre are some small reptiles: a therocephalian (left) looks at the procolophonid* Tikhvinskia *and the prolacertiform* Boreapricea, *which has an insect in its mouth. Amphibians include the hefty temnospondyl* Wetlugasaurus *on the bank, and Benthosuchus swimming.*

in wait on the beds of streams and lakes, or in the shallows among plants, and engulfed their prey by suction. Opening a broad mouth rapidly under water creates a region of low pressure inside, and any fish, or other debris, nearby is sucked in. The braincase was tiny and the cranium and the lower jaw were of similar depth, again as in modern frogs and salamanders. They could open their jaw by raising the skull, an ideal arrangement for suction feeding on the bottom.

The sediments at the Tikhvinsk site are classified by Russian stratigraphers as part of the Rybinskian Gorizont, or horizon, a unit of rocks laid down in the Early Triassic, perhaps three or four million years after the great extinction. The Rybinskian Gorizont can be traced over the Moscow region, and equivalent units occur in the southern Volga and Urals regions. The shales contain plant remains and molluscs that point to deposition in rivers and lakes, but perhaps with some marine influence from an Early Triassic sea in the Baltic region.

The striking thing about the Rybinskian fauna (Fig. 38) is how depleted it is in comparison with the Late Permian Vyatskian assemblage represented by the great collections of pareiasaurs, dicynodonts, cynodonts and other reptiles we'd seen in the Moscow museum. The commonest animals are the temnospondyl amphibians *Benthosuchus* and *Thoosuchus*, each 0.7–1.5 metres long. Much rarer is the archosaur *Chasmatosuchus*, which might have preyed on the temnospondyls, or on fishes, and the small, superficially lizard-like procolophonid *Tikhvinskia*. The dominant Permian group, the mammal-like reptiles, are represented in the Rybinskian fauna by just one small form, *Scalopognathus*, which is actually known from only a single jaw. Other reptiles have been reported by Russian authors, but the remains are partial and equivocal.

This is exactly what we saw in the Karoo Basin. The complex latest Permian ecosystems of the *Dicynodon* Zone disappeared suddenly, and the earliest Triassic *Lystrosaurus* Zone faunas were monocultures of the dicynodont *Lystrosaurus*, with a few, extremely rare, small reptiles in association. The Russian situation was rather different – no *Lystrosaurus* at Tikhvinsk, and mainly amphibians. But perhaps this was caused by the environments there, the deposits of black shales suggesting possibly a deep lake. Or perhaps *Lystrosaurus* was already extinct. The Russian *Lystrosaurus georgi* comes from the Vokhmian Gorizont, at the very base of the Triassic, and a few million years older than the Rybinskian Gorizont at Tikhvinsk.

First sight of the Sakmara

Our appetites whetted by this first visit, Glenn Storrs and I were keen to see the classic bone-beds of the Urals. The Permo-Triassic of the Moscow basin was good, and it had produced excellent fossils over the previous centuries, but localities are isolated and small, surrounded by vegetation, farms and towns. Geologists prefer deserts, coasts and ravines, where the rocks are exposed. We arrived in July 1994 on our second visit before heading for the Urals. We went at the invitation of Valentin Tverdokhlebov and Vitaly Ochev in Saratov. Ochev is a distinguished palaeontologist who has worked on the fossil amphibians and reptiles of the Permo-Triassic of the Urals since the late 1950s. Glenn and I knew him by reputation, since he had published widely in both Russian and international scientific journals. Tverdokhlebov was less known to us. He is a field geologist, specializing in sedimentology – the study of ancient sediments and the interpretation of the evidence they provide about environments and climates. He knows every inch of the South Urals area since he has worked for the Geological Survey of the USSR, and now of Russia, for decades, producing geological maps and searching for valuable minerals and ores.

Andrey Sennikov took us to the great Kazan Station in Moscow where we were to board our train to Orenburg. The train left about midnight, as do most long-distance trains in Russia. We hired our teapot and bedding, as usual, and turned in for the night. By morning, we had passed through Lyubertsy, Kolomnal and Ryazan, and were well out on the great plain to the southeast of Moscow. Birch forests and farmland extended for hundreds of kilometres in all directions. By early afternoon, we had passed Syzran, and had our first sight of the Volga. The railway track crosses the Volga at a narrow point, but even so the bridge is nearly 3 kilometres long. The river is so wide that the other bank is not visible. We soon arrived at Samara, on the east bank of the Volga, an industrial city of over a million inhabitants, and a major administrative centre. Factory blocks proclaimed ‘Slava Oktyabr’skoi Revolyutsii’ (‘long live the October Revolution’). This city, like so many others, had recently changed its name: older atlases show it as Kuybyshev.

In Samara, as we waited for people to get off in the afternoon sun, we saw another train pulling in from the south, from Tashkent, the capital of Uzbekistan, one of the smaller Central Asiatic republics. As the doors opened, large numbers of huge melons rolled on to the platform. In fact, the third-class carriages were entirely full of melons, on the floor, stacked on the beds and in the corridors. The small brown-skinned Uzbek farmers in their long dark robes and woollen hats, and their families, rolled out on to the platform, following their fruit. We understood from our companions that it was economically viable for them to buy cheap tickets to make the two-day journey into the heart of Russia to sell their melons. And it was a great deal safer than entrusting their produce to commercial carriers.

Eventually, after a 30-hour journey, and a second night on the sleeper, we drew into Orenburg Station. Valentin Tverdokhlebov and Vitaly Ochev, and their crew, were there to meet us. Tverdokhlebov is a tall, thick-set man with dark hair and a tanned face, clearly more used to field-work than to the office. He spoke only Russian, but I managed to make myself understood. Glenn amazed the Russians by speaking to them fluently in their own tongue: he had learnt Russian at school, largely because of his great passion for Russian choral singing. Ochev hopped about eagerly, speaking a mixture of English and German. Like many older educated Russians, he had learnt one or more western languages as a child. It was hard to remember that Ochev was over 70, but still hugely active. The remainder of the crew were there, and one of them, Leonid Shminke, spoke good English and he was to act as our liaison. We piled into a Waz minibus and set off. The Waz minibus is ubiquitous and always khaki. The name Waz (properly UAZ) stands for Ul'yanovskii Avtomobil'nyi Zavod (Ulyanovsk Car Factory). These buses carry nine people, and they look very like a classic Volkswagen camper of the 1960s.

We headed out of Orenburg, travelling eastwards through the villages of Sakmara, Chorny Otrog to Kolhoz Pravda ('collective farm truth'). After the village of Gavrilovka we passed on to a well-made graded dirt road, or *graida*, and then shot off to the side and through farm fields to the banks of the Sakmara. There, among the trees, Tverdokhlebov and his colleagues had built a magnificent camp site. Five or six substantial tents were ranged round a clearing. The Sakmara at this point is about 20 metres wide, flowing strongly between high banks. The Russians had cut neat steps down the side and had built a small jetty. Then we heard a cheerful cry of 'What on earth are you doing?' from a boat, and an old Russian came ashore with a bucket full of fish. This was Peter Tchudinov, a noted palaeontologist, expert on synapsid reptiles of the Permo-Triassic. But he was retired now, and only here for the fishing, as he made clear in his perfect English.

Getting to grips

We had just over a week on the banks of the Sakmara. Evidently, the Saratov crew came to this spot each year. In the old, Soviet days, Tverdokhlebov had been commissioned to complete a map, or search around the area for some specific minerals. The whole southern Urals region is rich in iron, copper, zinc and uranium, as well as oil and gas, and geologists have been enormously important in developing major industries in the area. We now looked forward to some field work. We were probably the first westerners in the twentieth century to see the classic Russian fossil reptile localities. Some of them had been visited, briefly, by Murchison in 1841 and by Twelvetrees in the 1880s, but free travel in remote parts of Russia by foreigners was virtually impossible during Soviet times.

Our intention was to try to understand the succession of animals in the Permo-Triassic rocks of the Urals. We had read numerous accounts in the Russian journals, some of them, fortunately, translated into English. These listed locality after locality, and they gave great catalogues of species found at each site. But it is very hard to get a real feel for a segment of the history of life just from written accounts. We wondered also how accurate the accounts were. If a Russian were to try to understand the palaeontology of England, or North America, or South Africa, from old written summaries, he or she might receive quite the wrong impression.

On the first field day, we drove back to the main road, back to Orenburg, and then straight south towards the border with Kazakhstan, only a few kilometres away. In heading south, we crossed the Ural River in the middle of Orenburg, and thereby crossed from Europe into Asia. We were heading for the Sol'Iletsk district. Here, in the banks of the Donguz and Berdyanka rivers, abundant remains of fossil amphibians and reptiles were dug up in the 1940s and 1950s by Ivan Efremov and his successors, including Ochev and Tchudinov, who were with us.

At one site we found dozens of bones of the amphibian *Chroniosuchus*, not a temnospondyl, but a member of the other major Permian group, the anthracosaurs. These had narrower skulls than the temnospondyls. Elsewhere, along the Donguz River, Ochev led us to site after site where he had excavated great dicynodont skeletons using bulldozers in the 1950s and 1960s. The bulldozers were clearly an important innovation, and they certainly helped him to excavate huge numbers of skeletons. Indeed, he seems to have been so efficient at excavating that we didn't find any at all.

We went on to visit some of the old copper mines in the Upper Permian, also south of Orenburg; lumps of copper ore could still be picked up and the old workings were still visible, but bones are now extremely rare in the mine tailings.

The following day was a Lower Triassic day. We drove from the camp to the *graida*, and headed westwards to Saraktash, some 15 kilometres away. During the drive, we had passed over a sequence of Upper Permian rocks and into the Triassic. We headed north for the village of Petropavlovka. Here, on the banks of an ancient oxbow lake, formed in a bend of the Sakmara River, were a series of famous fossil sites, numbered clearly by the Russian authors, Petropavlovka I, II, III, IV, V and VI. Here we appreciated the value of visiting the sites. The Russian geological papers give these site numbers, and list the fossils from each, but there was no map. We were lost until that point.

Tverdokhlebov had come equipped with Soviet geological maps of the whole area, and we were impressed and amazed to see them. They were of exceptionally high quality, in some regards better than the equivalent British or American maps. Not only did they show the topography, roads and villages, as well as an overlay of the rock units and geological structures, but they also showed the high and low stands of rivers and river vegetation. Each year, as the ice and snow melt, the Ural, and all the rivers coming off the Ural Mountains, flood. The date of flooding is fairly predictable, as is the extent of the area of flooding and silt. The bushes and reeds along the river banks then have seasonal bursts of growth. All this is shown on the map.

We noticed that each map had a number and that the word 'Sekretna' had been heavily scored off with a thick pen. Might it be possible now for us to buy maps? *Perestroika*, *glasnost*, and so on? No. Although beautifully mapped and engraved, and representing dozens of person-years' work, only 100 copies of each map had ever been produced. Each numbered map was lodged and catalogued in specific government offices. Even Valentin Tverdokhlebov, who had produced a number of the Orenburg region maps himself, did not have his own copies. He had to sign them out and return them. Could we copy them? No.

We visited all the Petropavlovka sites, and saw where the sparse fauna had been collected, some amphibians and procolophonids, as at Tikhvinsk, but also a larger predator, the archosaur *Garjainia*, perhaps 2 metres long and with a deep-sided skull, clearly adapted for feeding on medium-sized amphibians and reptiles. The Petropavlovka beds are somewhat younger than those at Tikhvinsk, representing the Yarenskian Gorizont, and the faunas had diversified. Instead of four or five species, there were more like ten in the Yarenskian. Life was recovering after the end-Permian event, but it still had a long way to go before the ecosystems could rival their richness before the crisis.

The Permo-Triassic boundary

After a rest day, and an extremely drunken evening party with the local farmers, we set out into the field again. Glenn and I, not used to downing vodka by the pint, were rather white-faced as we sat in the back of the van. We drove to a bluff near the camp, called Sambulak, and followed through a sequence of sediments. This was to be one of the most astonishing days of the entire trip.

The rock succession at Sambulak showed a major change half-way up. The basal part consisted of cycles of mudstones and limestones that had been laid down on land, each beginning with an erosional layer and ending with a soil, perhaps representing repeated climatic changes from wet to dry over several thousand years. Here and there we were shown sites where reptile skeletons had been extracted in the past. At the top was a vast thickness of conglomerate, a coarse unit composed of great boulders of exotic rocks – greenish schists, pinkish granites, white quartzites. These boulders had clearly come from the core of the Ural Mountains, some 200–300 kilometres away.

Valentin Tverdokhlebov had mapped out the distribution of the conglomerate beds, and they formed vast alluvial fans, the outwash at the head of major river systems. Clearly, there had been heavy rainfall over the proto-Urals 250 million years ago, and they must have been eroding down rapidly. Great rivers in flood had transported massive boulders down over the plain. At the top was nothing: the softer rocks above had worn away much later to form the modern landscape. Following across country, though, the overlying sandstones and mudstones could be found.

'Kak Gorizont?', I asked.

'Eta granitsa Permo-Trias' replied Tverdokhlebov nonchalantly.

This took a minute or two to sink in.

'The Permo-Triassic boundary?', I repeated to Glenn. 'He says this is the Permo-Triassic boundary. How does he know? What's the evidence?' Glenn questioned Valentin Tverdokhlebov in detail in his more fluent Russian. If this was truly the Permo-Triassic boundary, it was important to understand how accurately it had been dated, what environmental changes had happened in the switch from wet-dry cycles to massive alluvial fans, and how the plant and animal life had been affected.

Back in camp that night, as the mosquitoes feasted on us and the dragonflies soared and swooped after them, Glenn and I had a great deal to talk about. If this really was the Permo-Triassic boundary, we had to find clear-cut evidence to justify that claim. Then we could actually follow inch by inch what had happened all those years ago. But we had to leave the Sakmara the next day to begin the long journey home. We must come back.

Across the boundary

Our 1994 trip had just been a short taster. Glenn and I organized a more serious expedition for July and August 1995, in collaboration with Tverdokhlebov and Ochev. We camped in the same place as before, but with an even larger crew. Two of my students, Andy Newell, now employed by the British Geological Survey, and Patrick Spencer, an expert on small reptiles, went out in advance to join the Russians. Glenn and I arrived two weeks later with David Gower, another student, now working at the Natural History Museum in London, and Darren Partridge, a fossil enthusiast who works as a vet.

When we arrived in camp, we found 13 residential tents and a large kitchen tent containing a full-sized domestic cooker, operated from a gas cylinder, two cooks and two dogs. In all there were 18 Russian geologists and others – most were there, no doubt, for a reasonably-priced holiday by the river. We had agreed to pay for the costs of the camp (the Saratov geologists had paid for everything in 1994), and we handed over the sum of $5000 in roubles. I had taken the dollars to Russia in small denominations, new 1- and 5-dollar notes, and changed them into roubles at the airport. I had felt very vulnerable on the train, travelling with a bundle of 18 million roubles in a wad as thick as a brick in a moneybelt round my neck.

We soon settled into camp life, rising early and eating a huge cooked breakfast, usually consisting of *casha* and gristle. *Casha* is the general name for cereal, and we were served buckwheat, *grechikha*, which is harvested from low plants that grow well in the steppe soils, and tasting a little like rice. The *casha* was always topped with some meat and gravy. This was served for breakfast, but generally also for lunch and dinner. After a hard day in the field, and the weather was usually hot, we were happy to return to camp each evening and swim in the river.

We planned four weeks of fieldwork, and hoped to be able to document the sedimentology in detail to work out exactly how environments and faunas had changed through the Late Permian and Early Triassic. Every day we went out to a new site. Valentin Tverdokhlebov had visited them all before, but we were keen to see everything ourselves. So he indulged our demands to see more and more of the Permo-Triassic rocks. We raced up and down ravines making sedimentary logs, that is, recording bed-by-bed the nature of the sediments and any included fossils or sedimentary structures. Sedimentary structures can give clear guidance on the environments of deposition. For example, different kinds of preserved ripples show the direction of water movement, and may indicate whether the sands were deposited on a beach or in a river. Mudcracks and rain prints of course indicate exposure to the air. Impressions of ancient roots point to a soil. Channels small and large can indicate the type of river, whether meandering or braided, and so on.

In teams of two we were able to cover the ground fast, logging hundreds of metres of sedimentary sequences, right through the Late Permian and the Early Triassic, and frequently including the Permo-Triassic boundary. One person would carry the tape measure and would shout out the thickness, the rock type and any sedimentary structures or fossils. The other wrote it all down.

The beauty of sedimentary logs is that you can convert all the measurements and hasty scribbles about rock types in your field notebook into a simple diagram back in the lab. The diagram (Fig. 39) gives a summary of sometimes hundreds of metres or kilometres of thickness of sediments in a cartoon form that can be read directly. While you are on the ground, fighting through scrub and undergrowth up a winding ravine, the grand pattern may seem obscure. Laid out on a large sheet of paper or on the computer screen, the broader picture may become clear.

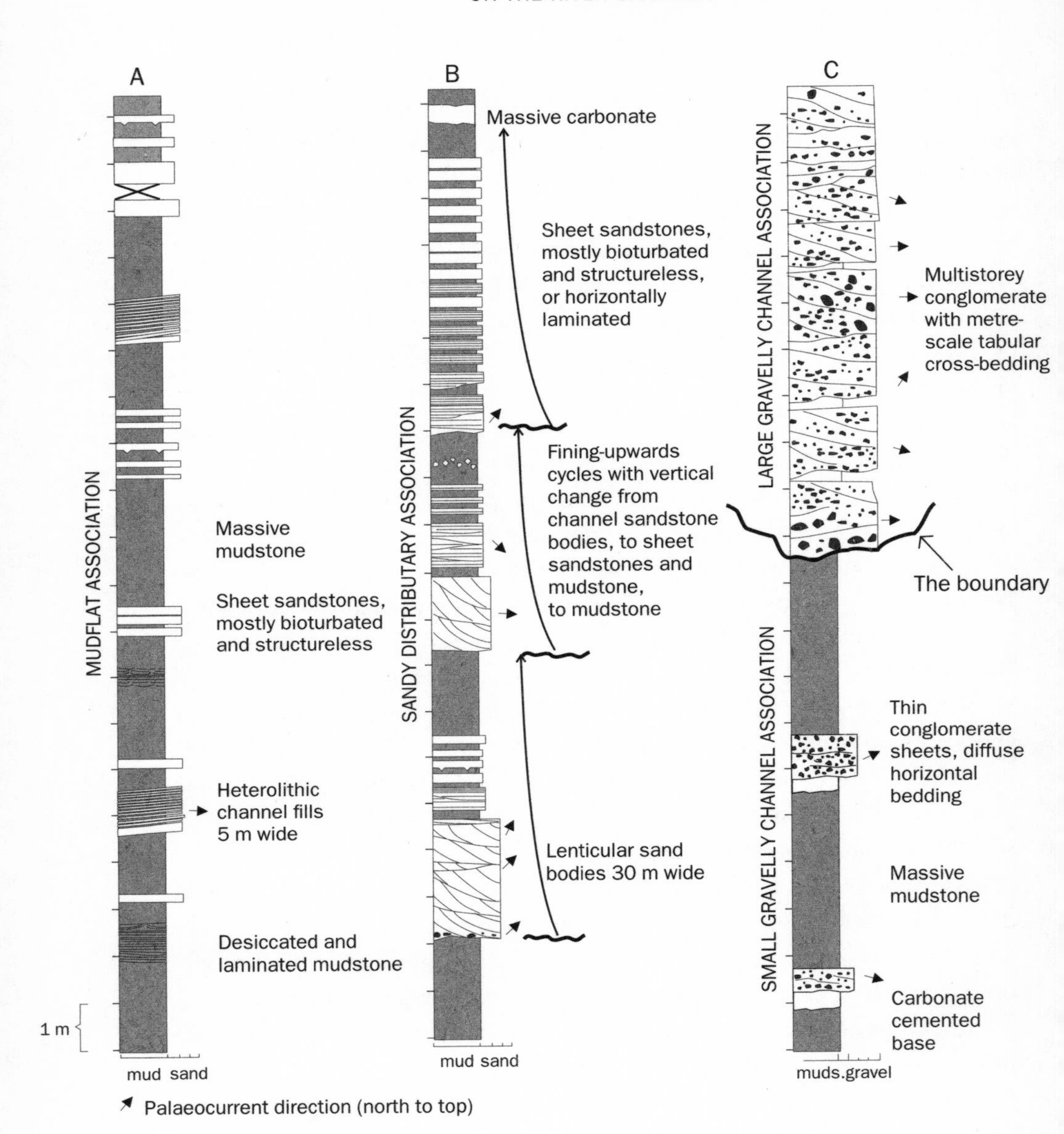

39 *A typical sedimentary log, showing the succession of rocks deposited by Late Permian and earliest Triassic rivers washing down from the Urals. The succession is divided into three columns, beginning at the bottom left, and ending at the top right. The sediments show evidence for mudflats low down (column A), passing up into sandy channels (column B), gravelly channels and then a major gravelly channel (column C).*

Megafans

One thing that was immediately obvious in the field was the apparently sudden appearance of massive conglomerates as we had seen at Sambulak on our previous expedition. This was not a local feature of just that site, but appeared everywhere around the South Urals. Something dramatic had happened – a large-scale change in sedimentation patterns.

Andy Newell, our sedimentologist, worked through all the sedimentary logs and was able to establish the scale of what had happened at the Permo-Triassic boundary in Russia.[1] The rock succession begins with mudflats, homogeneous mudstones with rare, thin sandstones. The muds were deposited in shallow pools that were surrounded by plants, and reptiles and amphibians occasionally wandered across, as shown by some fossilized footprints. Cracks in a number of the mud layers show that the pools dried up from time to time and impressions of salt crystals are evidence that some were at least partially saline – they perhaps flooded and dried up many times, concentrating the salts.

The mudflats are followed by a sequence containing much more sandstone, representing sandy distributary systems. The sandstones occur as channels and as sand sheets, suggesting a variety of types of rivers, from well-defined flows in long-lasting channels, to dramatic floods that presumably swept huge amounts of sand out over a wide area, clothing the whole landscape. This kind of flooding implies a semi-arid climate, with occasional heavy rainfall that erodes the surrounding hills and washes masses of sand over the whole area.

These channel and flood sands are followed by the third sedimentary association, small gravelly channels. The channels are about 100 metres wide, and 1 or 2 metres deep, and they are composed of hand-sized pebbles of quartzite, limestone and marble. These represent wide, shallow rivers travelling at some speed, and carrying in rocks from 100 kilometres away, from the foothills of the Ural Mountains. The pebbles are well rounded, confirming that they have been rolled and tumbled some distance by the rivers.

Then come the massive conglomerates we first saw at Sambulak. These are on quite a different scale from anything that had gone before – here again something dramatic has clearly happened. Individual layers are relatively thin, 1 to 3 metres deep, but they may be 3 to 5 kilometres wide. The included fragments are mainly pebbles, but boulders commonly occur.

As at Sambulak, the pebbles are of quartzite, limestone and marble, as in the thinner conglomerates below, and come from the western edge of the Ural Mountains. Added to this array of transported rocks are also boulders of chert and igneous rocks whose source has been traced to the core of the Ural Mountains, as much as 900 kilometres to the east.

Valentin Tverdokhlebov had been able to track most of the rock types found mixed randomly in the massive conglomerates back to their original sources in the heart of the Ural Mountains. Each of the conglomerate bodies clearly mixes material from a number of source areas, and this proves that there were several huge, fast-flowing torrents which hurtled down the mountain sides, rolling and tumbling great boulders. As the torrents came to the lower slopes, they merged into wider rivers, and the boulders of marble and chert and igneous rock became mingled in enormous rivers that carried millions of tonnes of rocks.

By careful mapping, Tverdokhlebov[2] was able to work out the exact shapes of these huge outwash systems, termed alluvial fans, or, more properly, megafans, since they are several orders of magnitude larger than the gravel fans below. According to his maps (Fig. 40), each megafan had an

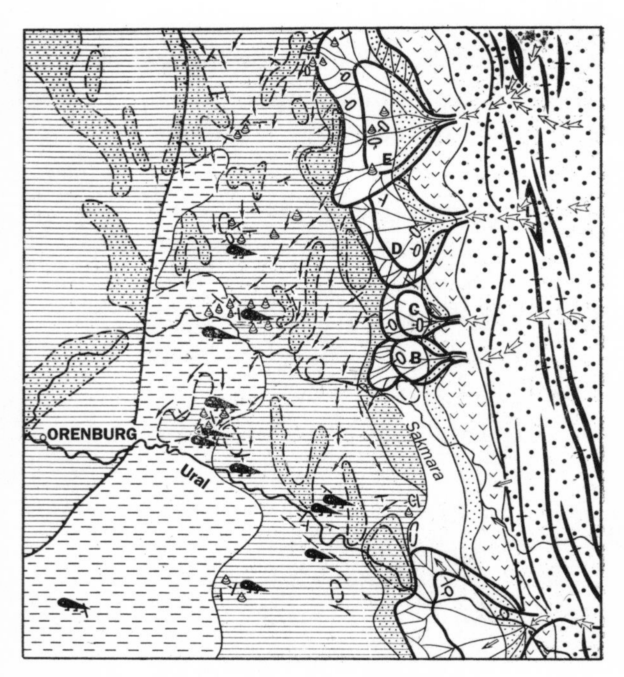

40 *Map of the distribution of earliest Triassic megafans washing off the Ural Mountains to the right. River flows are indicated by arrows which pass into the fans, five of which are shown (A–E). In front of the fans, to the left in the diagram, are finer sediments of rivers and lakes.*

area of some 300 square kilometres, and extended 15 kilometres out from the Ural foothills.

Crisis

So why the huge change in sedimentation pattern? Newell suggests that the mudstone to sandy distributary system to small gravelly fan succession in the latest Permian is a linked series of deposits. All of them imply the same kind of processes, as sediment was washed into the South Urals area from the surrounding hills at different rates, which depended partly on rainfall and partly on the subsidence rates.

Some 60 million years before the end of the Permian, two great tectonic plates representing the precursors of what are now Europe and Asia met, and slowly pushed against each other, raising the Ural Mountains as they were forced together. A similar process is going on today, with India thrusting against the rest of Asia – its stately northwards progression is forcing the Himalaya ever higher (but only by centimetres per century).

In the Late Carboniferous, as the Ural Mountains began to rise, the surrounding landscape sank. At first, seas covered much of that part of Russia, and thick marine sediments accumulated. As these filled up the basins, land was eventually formed in the Late Permian. During the Permian, subsidence rates decreased as the Ural Mountains stabilized, and the old plate movements fizzled out. The change from mudstones to sandstones to small gravelly channels in the latest Permian represents the last phases of filling of the basin, a shift from low-lying areas with ponds to higher levels more in line with the foothills of the Urals.

Then come the massive conglomerates. The easiest explanation might be that there was a sudden increase in rainfall. However, all the evidence of the sediments points to a reduction in rainfall, and a spread of dry conditions. More likely is that there was a huge reduction in vegetation cover at this time, exactly at the Permo-Triassic boundary. The loss of plants, as Roger Smith, Peter Ward and colleagues had suggested in South Africa, leads to massive runoff. First, all the soils that had been held in place by plant roots would go. Any rainfall would then have a devastating effect on the stripped hillsides; the bare rock would rapidly be cut into deep ravines, and fast-flowing torrents carrying boulders would smash further rocks out of place, until the river became a moving liquefied flow of rock debris.

So the two teams, Smith, Ward and others in South Africa, and our group in Russia, had seen exactly the same processes in action, in two widely separated parts of the globe. Evidently something had happened at the Permo-Triassic boundary to strip the world of vegetation, and this is borne out by the subsequent 'coal gap' and the 'fungal spike', as we have seen. In looking for an explanation of the end-Permian crisis, as we shall in the next chapter, it is important to bear these facts in mind.

The reptiles take a dive

Skeletons of amphibians and reptiles are found throughout the Late Permian rock sequence around the South Urals. They disappear abruptly at the top of the mudstone-sandstone-small gravelly fan succession, and then reappear above the megafan deposits. In one sense, this disappearance is a local phenomenon, because you would not expect to find relatively delicate bones in a catastrophic flood deposit. But, above the megafan beds, the amphibians and reptiles that survived the crisis into the earliest Triassic in Russia are a poor assemblage, as we saw earlier in this chapter.

The complex latest Permian communities of small procolophonids, larger herbivorous dicynodonts and pareiasaurs, preyed upon variously by therocephalians, cynodonts and the sabre-toothed gorgonopsians like *Inostrancevia*, collapsed. The earliest Triassic assemblage, the so-called Lower Vetluga (Vokhmian) Community, Efremov's old Zone V, was reduced to *Lystrosaurus* as the sole, reasonably sized herbivore, one species of procolophonid, and some rare therocephalians and diapsids as insectivores and carnivores that preyed mainly on the smaller reptiles.

This information had been reviewed in English by Efremov and colleagues, and by Everett C. Olson, a distinguished American vertebrate palaeontologist who had visited Russia several times in the 1950s and 1960s. However, those publications[3] were now very out of date, and much had been found since. The problem was partly one of language – Glenn could speak Russian well, but I laboured to make myself understood, and found reading Russian a very slow process. So we hatched the plan of publishing a book, in English, that would summarize everything that was known of the Russian Permo-Triassic reptiles. After much anguish, and many heated debates, the book[4] finally appeared in December 2000, nearly 700 pages long and containing 30 articles by various combinations of

34 palaeontologists, Russian and western. That was one thing. But we still did not have a clear picture of how the faunas were changing.

As we raced around the South Urals we saw site after site, each of which had produced some skeletons of amphibians and reptiles. We found the mass of new information hard to take in, and asked if anyone had actually done a census of the rise and fall of the different animal groups through time. Indeed, Valentin told us, he had compiled a huge catalogue of all the sites. When we returned to Saratov with him after the expedition, an interesting journey of 700 kilometres in the front of a huge Gaz 66 truck, Valentin showed us his card index. In it, 400 or so sites are listed, each with a determination of its geological age and a list of the fossils that had been found there.

After this series of trips, we had to do a great deal of writing. First, we successfully translated Valentin's amazing card index into a detailed summary of all the localities, their faunas, ages and environments. This provided materials for a detailed statistical study of the rapid extinction of most amphibians and reptiles at the boundary, and especially showed how the mass extinction had massively perturbed ecosystems for as long as 20 myr. after the event, well into the Triassic. After a gap of ten years, we revisited the Russian sites, and went further afield, exploring sites around Orenburg, Vladimir, Kirov and Kotlas, in 2004, 2006, 2008 and 2009.

The final solution

So what caused this, the biggest of all mass extinctions? The nature of what happened is now much clearer than it had been, say, in 1990. Careful dating has shown that the event took place 252 million years ago, and that species losses were anything from 90 to 95%. This was no local phenomenon, since it has been detected in rocks from China to Spitsbergen, from Greenland to South Africa, from Russia to Australia. In every case, whether looking at events on land or in the sea, the rate of species loss seems to have been similarly huge. There were no safe refuges, nowhere to hide.

There is also no evidence for selectivity, except that the survivors tended to be widespread species. But with such a tiny survival rate – 10% or less of species made it through – it is clear that plants and animals were being wiped out with almost no regard to their adaptations. Certainly on land the large animals all disappeared, but this could be explained as part of a chance process. If there are 100 species, of which 10 are large, a 95% loss of species is likely to kill all the large animals. Add to that the probability that, as today, large animals are rarer than small animals (i.e. smaller population sizes), then it is easy to see why all the rhinos and elephants might disappear, but a few rats and squirrels might survive. It would be wrong, though, to say that this proves that individual rats and squirrels are better adapted to survive crises. It is perhaps only their greater abundance that protects them as a species.

We have some hints of the environmental changes too. In the sea, the rocks show an increase in anoxia. Many of the surviving marine creatures seem to have been peculiarly adapted to living in such conditions of low oxygen. There are hints too of low productivity, meaning a lack of organic matter in food chains, so many of the surviving species were presumably able to survive on very little food.

On land, as the recent studies in South Africa and Russia have shown, the end of the Permian is marked by a sudden change in sedimentation, with megafans composed of huge boulders. In neither case can this be explained by a dramatic increase in rainfall. Indeed, the evidence suggests increased aridity, so the dramatically heightened levels of erosion and runoff can best be explained by a sudden loss of vegetation and soils, perhaps worldwide. Soils show that climates also became warmer.

So what caused the crash? The event must have been sudden. It must have reduced oxygen levels, increased temperatures and reduced rainfall, all on a worldwide scale. It must have had the ability to push all of life virtually to the brink. The tentacles of the killing agent reached into shallow seas and into the deepest oceans. On land, they penetrated lowland basins and mountainous regions, rivers and lakes. What kind of crisis could have been so profound that it killed reptiles on land, and brachiopods and corals on the sea floor? It cleared the Earth of vegetation, even if for a short time. This is more profound than any of the puny threats that humans have devised so far, whether nuclear bombs or mass forest clearance. A global rise in temperature of half a degree in a century as a result of industrial pollution and global warming? That fades into insignificance beside the crisis 252 million years ago.

11
WHAT CAUSED THE BIGGEST CATASTROPHE OF ALL TIME?

I remember as a child reading Arthur Mee's *Children's Encyclopaedia* in which the kings and queens of England were catalogued. Henry VIII was bad because he had six wives. But his daughter, Elizabeth I, was good: she had had an unhappy childhood – her mother was beheaded and she was declared a bastard by Parliament – and as Queen she overcame smallpox, plots to kill her and the Spanish Armada. Henry died unhappy, while Elizabeth had a long and glorious reign.

This sort of account may make history easier for children to understand, and the weaving of moral tales used to be a favourite way to encourage children to become good citizens. But this approach is now much criticized. Historians are taught not to judge the past by present-day values. Charles Darwin said that certain human races were superior to others. By modern standards, he clearly was a racist. But by the standards of England in the 1850s, he was probably considerably less racist than many other people.

This interpretation of history is called 'Whig', a term invented by Herbert Butterfield, in a little book of that title[1] published in 1931. Butterfield was concerned that history was often written as if there were a preordained end to the story, and that moral lessons were to be learned. I had always wondered why Butterfield chose the term 'Whig'. I knew that the Whigs and Tories were the main political parties in Britain in the earlier part of the nineteenth century, and that the Whigs became the Liberals after about 1860. It seems that Butterfield's term derived from the practice of history writing in the 1830s, when Thomas, Lord Macaulay (1800–59), who was politically a Whig, adopted the directional and moralistic approach. Benjamin Disraeli (1804–81), the great Tory prime minister and author, constructed an alternative view of English history and parodied the Whig approach. (As for the strange term 'Whig', it seems to have been derived from a derogatory term applied to a group of Scottish presbyterians during the seventeenth century who opposed the Catholic succession to the throne. In full, the term was originally 'whiggamore', coming perhaps from the old Scots, to 'whig', or urge along, and 'more', a mare.)

As for the history of kings and queens, so too for science. Looking back to the state of understanding of the end-Permian crisis in 1970, or 1980, or even 2000, we can easily point to errors of judgment. Why were palaeontologists and geologists then so blind to the truth? (Can we be sure we have the truth now?) They were looking at the biggest crisis in the history of life and of the Earth and they didn't see it. They must have been deluded, or incompetent, or both. Not so. It is hard for scientists, just as for anyone else, to throw off everything they have been taught. And when the evidence is somewhat intangible, it is understandable if scientists err on the side of caution.

Pervasive conservatism

As we saw earlier, it was commonplace in the 1960s to deny that anything of note had happened at the end of the Permian. Lone voices, like Otto Schindewolf in Germany, who claimed that there had really been an astonishing and catastrophic crisis, were widely ignored. By 1970, however, through the work of Norman Newell and Jim Valentine, who had carefully catalogued the rise and fall of different groups of animals in the sea, the end-Permian mass extinction stood out in ever sharper relief. John Phillips' intuition of the 1840s was borne out, and the hiatus between the life of the Palaeozoic and the Mesozoic could not be explained away by some failure of preservation.

Geologists did not, however, then assume that the great dying-off had been catastrophic. The evidence just wasn't there. Indeed, although we now know that Schindewolf was broadly correct, he had based his judgments on really rather inadequate evidence. Dating of the Permo-Triassic boundary in the 1960s was subject to considerably more error than it is now, and detailed, inch-by-inch studies of high-quality Permo-Triassic boundary sections, such as those in China, Italy and Greenland, had yet to be done. Indeed, a common assumption then was that the boundary was marked by a gap in deposition, and nothing much could be got out of it anyway.

In 1967, Frank Rhodes, a British palaeontologist who had become President of Cornell University, gave the best explanation of the end-Permian event then possible,[2] that it had been caused by 'the multiple interactions of a wide variety of physical and biological factors'. Even in 1993, Doug Erwin, in his masterly review,[3] was unable to go much further: 'I believe that the extinction cannot be traced to a single cause, but rather a multitude of events occurring together'. There had been much to-ing and fro-ing before these comments were written, but no hint of a consensus model that fitted all the evidence.

Continental fusion

Most popular around 1970 was the view that continental fusion had caused a long-term extinction event that lasted perhaps 20 million years. Norman Newell, as early as 1952, had noticed that there had been a major marine regression[4] during the Late Permian. He argued that a general fall in sea level would have led to a reduction in ocean volumes, and hence a fall in available habitats on the sea bed and in the water column above. So, as sea levels fell, he argued, species of marine animals would have died out progressively, until the numbers were much reduced. Of course, at that time the true magnitude of the end-Permian extinction was perhaps not appreciated.

Newell at first could not explain why sea levels had fallen, but plate tectonics provided the answer. Although Alfred Wegener (1880–1930), a German meteorologist, had proposed continental drift in 1915 – and some of his best evidence came from Permo-Triassic plants and animals – many geologists rejected the whole concept. They argued that the Earth's crust was obviously solid and not mobile, and in any case there was no mechanism to drive the plates around.

The discovery of the function of mid-ocean ridges and the existence of symmetrical rock sequences on the seafloor around 1960 proved Wegener had been right. Running in a strip down the middle of the Atlantic is a zone of active upwelling of magma, the Mid-Atlantic Ridge. As magma wells up, it forms fresh rock on either side of the ridge, and slowly the plates beneath the eastern and western halves of the ocean move sideways. This was proved by the discovery of symmetrical, datable successions on the floors of both eastern and western plates, which showed a history of slow opening of the Atlantic Ocean over some 200 million years. As the Atlantic oceanic plates move east and west, the continental plates beneath North America and South America are also moving west, while those beneath Europe and Africa move east. Since the Earth is of fixed size, plates are also being consumed or crushed in other parts of the world.

It is now known that the tectonic plates move at the same rate as the growth of human hair. That is about 5 centimetres per year. This stately pace of the continental and oceanic plates has been established by back-calculations from the present position of the continents to their former locations, confirmed by measurements bounced by laser off the moon. As the opening North Atlantic pushes North America west, it converges on Russia. The Pacific plate, carrying San Francisco and Los Angeles on its eastern edge, is heading northwest, towards Japan. India and Australia continue their northwards migration.

With the acceptance of plate tectonics in the 1960s, Wegener's model for the ancient supercontinent Pangaea was also deemed to be correct. During the Permian period, isolated continents began to fuse together. As we have seen, Europe and Asia had already combined in the Mid Carboniferous with the uplift of the Ural Mountains. Laurasia, the northern continents, fused with Gondwana, and this was largely completed during the Late Permian. So, Jim Valentine and Eldridge Moores argued,[5] as the continents fused, ancient seaways closed off. Worldwide, it was calculated, sea levels fell by 200 metres or more. How could this cause extinctions?

Valentine and Moores argued for two elements to the killing model. First, the global fall in sea level reduced the overall number of habitats on the sea floor and in the water column, as Newell had argued. But there had been other major changes in sea level during the past 600 million years, and many of them were not associated with huge levels of extinction. What was special about the Late Permian episode was the second killing mechanism, the reduction in endemism. Endemism is a biological term referring to the restricted geographic distribution of certain species. So, as the continents fused together, animals and plants on land could clearly move freely from place to place. This would mean that endemic, or local, species would either become cosmopolitan, or they would die out, overwhelmed by invading forms from elsewhere. In the seas too, as isolated seas and basins disappeared, all their endemic species would go too. Imagine, Valentine and Moores were suggesting, that the Mediterranean, the Caribbean and the China seas, and all the seas around Australia and the Southeast Asian islands, were to disappear. All the diversity of their endemic species of corals, molluscs and fishes would disappear into an homogeneous global fauna. There would be a global crash in diversity.

This theory seemed initially appealing, and it did seem to account for extinctions on land and sea. It was also safe: no need for cosmic rays or impacts. However, there are three reasons why it doesn't explain the end-Permian crisis. First, some censusing experiments based on modern distributions of marine organisms showed that the loss of a few inland seas could never account for a loss of 90–95% of species. Second, the assumption of a major regression during the Late Permian has been questioned. Some geologists deny that there was any regression at all, while others[6] argue that it was at most a minor affair, since there is little evidence for the emergence of land or for large-scale erosion. Indeed, sea level was actually rising rapidly at the time of the extinction.

The third reason to reject the Pangaea/regression model is that it is not fast enough. In 1973, when it was proposed, everyone was convinced that the extinction had lasted for 20 or 30 million years, so a slow process lasting for much of the Permian was required. Now that the event is thought to have been much more rapid, regression and continental fusion just will not do.

But how can one go from a postulated long-term Late Permian extinction lasting for 10 million years or more to an instant event? If all the extinctions are now concentrated right at the end of the Permian, what happened to the earlier ones? Have they all been subsumed into the terminal Permian in some kind of large-scale correction of a mega-Signor-Lipps effect (artificial backward smearing of apparent times of extinction). Perhaps there was more than one event?

How many events?

As we saw in Chapters 7 and 8, the end-Permian mass extinction is dated at 252 million years ago, and close studies of the Meishan section in China have shown that 95% of species died out at that time, or within the following 180,000 years. There are still debates about the precise timing of the event, and whether it occurred at the same time everywhere, and at the same time on land and in the sea.

Actually there seem to have been three pulses of extinction across the Permo-Triassic boundary, separated by considerable spans of time: one took place 5–10 million years before the Permo-Triassic boundary, and another 5–10 million years after. First hints of these extra peaks of extinction came from studies of continental vertebrates. At a time when synoptic plots of marine diversity change just showed a gradual decline through the Late Permian, the plot[7] of amphibian and reptile diversification highlighted three extinction peaks, one at the end of the Capitanian Stage, one at the end of Changhsingian Stage, and one during the Olenekian Stage. The Changhsingian Stage is the last time division of the Permian, named after the Chinese rock units of the Meishan section. The end of the Capitanian Stage, lower down in the Late Permian, named from classic sections of the Capitan reef in Texas, also corresponding to the end of the Guadalupian epoch, is dated at 260 million years ago, while the end of the Olenekian Stage, named after the River Olenek in the South Urals, corresponding to the end of the Early Triassic, is dated as 240 to 245 million years ago. So was the extinction a triple whammy?

The end-Capitanian mass extinction, sometimes called the Guadalupian mass extinction, saw the loss of up to 58% of marine genera, according to calculations by Steven Stanley of Johns Hopkins University and X. Yang of Nanjing University: less than at the end-Permian event itself, but more than during the KPg event. Many families and genera certainly disappeared, with major reductions in the diversity of ammonoids, rugose corals, bryozoans, fusuline foraminifera and articulate brachiopods. Of major marine groups, only the blastoids disappeared. The scale and timing of this event have been debated, with some arguing it was a drawn-out affair, and others identifying major, instantaneous losses perhaps just before the end of the Capitanian.

There were also noticeable extinctions on land at about the end of the Capitanian Stage. The dinocephalian reptiles, with nearly 30 genera of plant- and meat-eaters present in the *Tapinocephalus* Zone of the Karoo Basin, disappeared. So too did some basal dicynodont relatives and some flesh-eating synapsid groups, and essentially the same patterns of loss are seen in Russia.

The Olenekian event, just before the end of the Early Triassic, is often forgotten. After all, life was recovering during the Triassic. But there was indeed a third blow, five million years after the end-Permian event. There were high rates of extinction among ammonoids, and perhaps among bivalves. In addition, many groups of amphibians disappeared. But more detailed study of marine sections is required. On land, the Karoo Basin cannot offer much help since it is not clear that the sediments of the Beaufort Group span this interval, but the Russian successions show a major loss of amphibian species.

Despite the interest of these extra extinction events, bracketing the key, end-Permian mass extinction some five million years before and after, they are rather poorly defined at present, and are certainly smaller in magnitude than the big event. If it turns out now that the end-Permian mass extinction was a geologically rapid event, is this a reason to look for extraterrestrial mechanisms? Some people think so.

Cosmic rays and impacts

The impact scenario for the KPg event (see Chapter 5) immediately led researchers to seek evidence for impact at the time of other mass extinctions. Intense efforts have been made to find any hint, however feeble, that the Earth was struck by a giant asteroid or comet 252 million years ago, and a recent strong contender has been put forward, as we saw at the start of the book. Of course, extraterrestrial catastrophe had been posited before.

In 1954, Otto Schindewolf[8] had suggested that a burst of cosmic rays had caused the end-Permian mass extinction. He felt driven to this conclusion since he was convinced from his studies of the Permo-Triassic boundary sections in the Salt Range of Pakistan that the boundary was abrupt, and that life had been wiped out rather rapidly. Schindewolf, as we saw (Chapter 4), was ridiculed for his far-out ideas, and certainly no independent evidence has been found for cosmic rays. What of impact?

There was great excitement in 1984 when an iridium anomaly was announced at the Permo-Triassic boundary in China.[9] This came only four years after Luis Alvarez and his team had published their epochal model for KPg impact, and enhancement in the rare element iridium was then the key evidence for impact. Two teams of Chinese researchers reported 8 parts per billion and 2 parts per billion of iridium in the boundary layer, 10 to 40 times normal background levels. In 1986, however, two American teams reanalysed the Chinese boundary clays and found barely detectable levels of iridium; if anything their values were lower than might be expected in normal sediments.

Some other possible hints of impact have also been rejected. In the Chinese sections, some iron-rich spherules were reported, but these seem to have been more typical of a volcanic origin than an impact.[10] But there was no shocked quartz, such a potent indicator of impact at the KPg boundary.

Then the news broke. Greg Retallack, well-known expert on ancient soils from the University of Oregon, showed some pictures at the 1995 annual meeting of the Geological Society of America that were claimed to be shocked quartz from the Permo-Triassic boundary of Australia and Antarctica. As Richard Kerr, science journalist, reported,[11]

> the hallways buzzed with palaeontologists and geologists exchanging opinions on Retallack's photos, which purportedly showed faint bands of glass-filled fractures within the grains. Retallack thinks the fractures formed in the shock of a massive impact and notes that similar grains have been linked to the Cretaceous-Tertiary extinction. The hallway buzz was more cautious . . .

Retallack's critics, including Philippe Claeys, then at the Humboldt Museum in Berlin, noted that his grains had been overprinted with later stress damage, and that most of them showed only one set of planar deformation features (PDFs). Shocked quartz from KPg sections typically shows three or more sets of such linear features. Kerr goes on,

> Retallack counters that Claeys and others haven't seen all there is to see. Under the microscope, where the full depth of a quartz grain can be viewed by changing the depth of focus, all the grains can be seen to have at least three sets of PDFs, he says; one has seven.

The full account was published three years later, and here Retallack and colleagues showed illustrations of the shocked quartz grains, some apparently showing two sets of PDFs. The rock successions in Australia and Antarctica also showed iridium spikes at the Permo-Triassic boundary. The authors accepted, however, that the iridium levels were lower than in most KPg sections, and that the PDFs in the shocked quartz were not so clear, so the evidence is still equivocal.

Buckyballs and impact

Further excitement broke out in February 2001, as we saw in the Prologue, with the announcement of a meteor impact. The evidence for this came in the form of extraterrestrial noble gases in fullerenes at the Permo-Triassic boundary in China, Japan and Hungary. Fullerenes are large molecules of carbon, composed of 60 to 200 carbon atoms arranged as regular hexagons around a hollow ball. They are often called buckyballs, after Richard Buckminster Fuller (1895–1983), inventor of the geodesic dome, which their natural structure resembles. Fullerenes can form in meteorites, in forest fires, even within the mass spectrometers that are used in studying them.

In the 2001 *Science* paper, Luann Becker of the University of Washington,[12] and colleagues, reported that they had found helium and argon, two of the noble gases, trapped inside the 'cage' formed by the carbon atoms. Their samples came from Hungary, China (from the Meishan section) and Japan, from precisely the Permo-Triassic boundary. The samples from Hungary did not produce any particular signal, but the Chinese and Japanese samples gave strong evidence for impact. When Becker and her team analysed the helium and argon from these sites, they found that it was chemically the same as helium and argon derived from meteorites, and it was present in relatively large quantities. So, they argued, the Permo-Triassic boundary fullerenes must have come from the impact of a meteorite. The results were supported and criticized in equal measure by other geologists and geochemists when the publication came out.

Nothing daunted by such uncertainty, the press went mad. Nigel Hawkes of *The Times* of London was judiciously cautious, and simply reported the results, but his piece was accompanied by a lurid colour illustration showing a massive asteroid hitting the Earth. Michael Hanlon of the *Daily Mail* was less circumspect:

> There was no warning. One minute, the Earth was quiet, just as it had been for countless millennia.... Then it came. Out of the sky, with a roar that must have sounded like the slamming of the doors of Hell itself, an object that would change the course of life for ever on our planet. When it hit, the cataclysm it unleashed was beyond imagining. The ground shook with the force of a million earthquakes, and a crater hundreds of miles across was punched in the crust.

So there we have it. The case is proved. Well, actually not yet.

Laboratory procedures and dating problems

The critics, inevitably, looked highly sceptically at the claims of the Becker team. Several geochemists raised doubts about the laboratory procedures that had been used: after all, the analyses were subtle and involved minute quantities of chemicals. Others urged caution: 'This is tricky stuff, and until it is confirmed there is little reason to get too excited', said Doug Erwin soon after the *Science* paper appeared. Becker responded, 'I have a feeling we're either going to go down in flames, or we're going to be heroes.'

The measured, published criticisms followed in *Science* seven months later, in September 2001. First, K. A. Farley and S. Mukhopadhyay from Caltech, in Pasadena, California, stated that they had reanalysed samples from exactly the same sites and using exactly the same laboratory procedures, and yet they had failed to replicate the results of Becker and her team. The Caltech scientists studied material from the boundary layers in the classic Meishan section, from bed 25, the boundary unit (see Chapter 7). They dried, weighed and crushed the sediment samples, heated them to 1400°C to fuse them into a glass, and heated the sample further to 1800°C to drive off helium. They found only minute quantities of helium, about 100th the amounts reported by Becker, exactly as expected in normal rocks that had not been affected by impact. The conclusion of Farley and Mukhopadhyay was:

> We thus find no evidence for the impact-derived [helium] reported by Becker et al. Our analytical procedure for [helium] is as sensitive and precise ... as that used by Becker et al., so the discrepancy between our results and theirs is probably not analytical in origin. ... Without confirmation of fullerene-hosted [helium] in Bed 25, both the occurrence of an extraterrestrial impact and the cause of the mass extinction ... must remain open questions.

So, Luann Becker and her colleagues had not found an enhancement of helium in their samples from Hungary, and Farley and Mukhopadhyay had failed to confirm high levels in the classic Chinese samples. So what of the third locality, Japan? Becker had obtained samples from the Sasayama section in southwest Japan, but in a second critical comment in *Science*, Yukio Isozaki from the University of Tokyo argued that the Permo-Triassic boundary is missing in this section. The samples analysed by Becker came from at least 80 centimetres below the boundary, and he noted major geological disturbance of the rock succession. So, Isozaki concluded, the Japanese samples are 'significantly older than Meishan Bed 25,' and cannot be related to the end-Permian event, whether it was an impact or not.

These two critiques would appear devastating. However, Luann Becker and her colleague Robert Poreda, responded robustly. The Caltech team had measured the wrong sample, it seems. Becker and Poreda had focused on a carbon-rich layer at the base of Meishan bed 25, and it was in this horizon, which corresponds to the exact time of maximum extinction intensity, that they found the fullerenes and the enhanced levels of extraterrestrial helium. Farley and Mukhopadhyay had sent Becker and Poreda some of the sample *they* had analysed, and it turned out to consist mainly of sand grains, so was obviously from a fractionally higher level within bed 25 from Meishan.

Meanwhile, Becker and Poreda had reanalysed the materials from bed 25 that they had included in their first paper, and they also analysed some replicate samples from the same horizon, but collected at a different time. In all cases, they confirmed high values of helium, 100 times or more the normal, 'background', levels. This aspect of the work will run and run, as yet more samples are checked and rechecked in different laboratories. The work is complex, and requires careful weighing, drying and heating, as well as analytical instruments of astonishing precision. One sneeze or snuffle by the technician at the wrong moment, a poorly calibrated weighing machine or an off day for the mass spectrometer, and the readings will be meaningless.

Isozaki's comments on the dating of the Sasayama section in Japan were rather devastating, and Becker and Poreda had to admit that 'the [Permo-Triassic boundary] cannot be precisely defined in any of the Japanese sections because of poor stratigraphic control'. They maintained, however, that since no one could prove their samples did not come precisely from the boundary bed, they would assume that they had. Soon after, in 2004, Becker and her team published further findings to support their theory, namely the outline of a crater from sediments off the northwest coast of Australia, as well as some shocked quartz and other impact evidence, similar to that which had been identified at the KPg boundary (see Chapter 5). However, this evidence was also roundly and rapidly criticized by large numbers of geologists. Crater experts said that her 'crater' was a trench. Geochemists said the 'shocked quartz' was not really shocked at all - not even mildly surprised. Still lacking the full checklist of impact evidence, the case that the end-Permian mass extinction was caused by a meteorite strike has languished ever since.

For the moment, therefore, I shall adopt here a conservative stance, and assume that the end-Permian crisis was not caused by an impact. The shocked quartz, and the fullerene and helium evidence may be hotly debated, but there is some other evidence concerning the end of the Permian that is much less debated – and it turns out to be just as dramatic and compelling.

The Siberian Traps

At the end of the Permian, 252 million years ago, giant volcanic eruptions took place in Siberia, spewing out some 4 million cubic kilometres of basalt lava, covering 5 million square kilometres of eastern Russia to a depth of 400 to 3000 metres. This region, known as the Siberian Traps, is equivalent to the area of the European Community. Many now think that these massive eruptions, confined to a time span of less than one million years in all, caused the end-Permian crisis.

The suggestion was first made in the 1980s. Russian geologists had explored the Siberian Traps long before then, but were not sure of their age. Over the huge area of their distribution, the volcanic rocks lay variously on top of sediments that could be dated from Silurian to Permian by their included fossils. So that meant they must at least be younger than the Permian rocks they covered.

The Siberian Traps consist of basalt, a dark-coloured igneous rock. Basalt is composed of plagioclase feldspar and pyroxene, with smaller quantities of olivine and magnetite or titaniferous iron. The iron and magnesium in basalt give it its black colour. The name is derived from the Latin *basaltes*, meaning a touchstone – a black stone that was used by chemists to test silver or gold by the colour of the streak they left. Basalts form at temperatures between 1100 and 1200°C – higher than the temperature of formation of granite, for example.

Basalt does not always come out of 'textbook' conical volcanoes. Classic volcanoes do produce basalt, as on Hawaii and in Italy for example, but also the lighter-coloured andesite and rhyolite lavas, which are ejected in rapid and sometimes explosive eruptions. In the case of flood basalts, the lava is ejected, seemingly much more sluggishly, from long fissures in the ground. A major basalt flood eruption occurred in southern Iceland in 1783, when lava was erupted from the Laki fissure, 25 kilometres long. Iceland famously lies athwart the Mid-Atlantic ridge, and fresh ocean floor is created by means of the eruptions; but here the ocean floor forms land. The Laki eruption lasted from June to November in 1783, and during those six months, about 150 cubic kilometres of lava were erupted. The lava flowed widely, covering an area of 600 square kilometres to an average depth of 250 metres. Flood basalts typically form many layers, and may build up over thousands of years to considerable thicknesses. They produce a characteristic kind of landscape called trap scenery, as the different layers, through time, erode back, producing a kind of layered, stepped appearance to the hills. The word 'trap' comes from the old Swedish word *trapp*, meaning a staircase.

Ancient flood basalts cover huge areas of the Earth's crust, on all continents, and many of them have been tied to former mass extinctions. The most notable link has been made between the Deccan Traps in India and the KPg extinction event, as was mentioned in Chapter 5. Early efforts at dating the Siberian Traps produced a huge array of dates, from 280 to 160 million years ago, with a particular cluster of dates between 260 and 230 million years ago. According to these ranges, geologists in 1990 could only say that the basalts might be anything from Early Permian to Late Jurassic in age, but most likely they spanned the Permo-Triassic boundary. A wide age range was not unexpected, since with thicknesses of as much as 3 kilometres in places, it could be postulated that the Siberian Traps had actually been erupted sporadically over a period of tens of millions of years. Similar broad age ranges were associated with the Deccan Traps, and other thick flood basalt successions.

In 1990, then, geologists were divided about the significance of the flood basalts. Some enthusiasts, notably Vincent Courtillot of the University of Paris, argued that the major flood basalts coincided with mass extinctions. The basalts had destroyed life, not so much by flowing over plants and animals and killing them, as by the expulsion of gases which altered the atmosphere and climate in catastrophic ways. Other enthusiasts, of course, were conversely trying to explain all mass extinctions by asteroid impacts.

The vagueness of the radiometric dates did not help Courtillot's case. So long as the mass extinctions seemed to have been geologically rapid but the basalt accumulations spanned tens of millions of years, a link could not be made. In 1988, Courtillot claimed that the Deccan Traps had been erupted in less than half a million years, while other dating specialists disputed his measurements, and claimed that a longer time span was involved.[13]

Dating was all. The geologists could make out as many as 45 separate layers within the thickest sections of the Siberian Traps, in the northwest of their distribution, near Noril'sk. It would clearly be important to date the bottom and top of the pile to assess the entire time span of the eruptions. New methods of more precise radiometric dating became available in the 1990s. In one early effort, Ajoy Baksi and Edward Farrar[14] seemed to discount any role for the Siberian Traps in the end-Permian mass extinction. Their dates, using a new argon-argon dating method, spanned from 238 to 230 million years ago, corresponding to the Mid Triassic, and long after the end-Permian event. This gave Courtillot and his supporters pause for thought.

But not for long. Redating by other research groups, using a variety of radiometric methods, yielded dates exactly on the boundary, and the range from bottom to top of the lava pile was about 600,000 years. In 1993, Doug Erwin, in his overview of the Permo-Triassic crisis, considered all the evidence and identified a role for the Siberian Traps eruptions, but only a relatively minor one, as one of several sources of carbon dioxide. Indeed, he[15] compared the end-Permian event to the plot of Agatha Christie's novel *Murder on the Orient Express*, in which the victim,

> a singularly loathsome man, is found murdered in his compartment on the Orient Express with twelve knife wounds. As Hercule Poirot listens to the stories of each passenger on the coach, he finds they eliminate one another as suspects . . . all eleven passengers plus the porter (making a good English jury) stabbed the man once each in retribution for his kidnapping and murder of a baby girl many years before. . . . In the case of the end-Permian mass extinction a more complicated explanation may be required by the evidence. . . . I believe that the extinction cannot be traced to a single cause, but rather a multitude of events occurring together . . .

Erwin and many other geologists were yet to be convinced. Perhaps the Siberian Traps had a hand in the extinction, but they were not the sole cause. Or were they?

The killing traps

In 1992, Ian Campbell of the Australian National University and his colleagues[16] presented a killing model for the Siberian Traps. They argued that the eruptions led to extinction by the injection of huge amounts of sulphur dioxide and dust into the atmosphere. Analogously with the KPg impact model, the dust blacked out the Sun and caused darkness throughout the time of the eruptions. Global darkness leads to global freezing, and widespread extinction of life on land and in the sea. Campbell and colleagues presented five lines of evidence:

- The Siberian Traps represent the largest volcanic event on land of the past 600 million years.

- The Siberian Traps are associated with huge copper-nickel-sulphide ore bodies, and thus the eruptions were probably rich in sulphur dioxide gas.

- The erupting magmas passed up through sediments containing the evaporitic mineral anhydrite, a rich source of sulphur that would have added more sulphur dioxide gas.

- The basalts are associated with a great deal of tuff, material flung out of the volcanic vent into the air.

- The tuffs contain large amounts of rock from the underlying sediments.

The tuff and rock fragments were unusual for a flood basalt eruption, and Campbell and colleagues argued that these materials proved that the Siberian eruptions had been especially violent, and thus capable of injecting vast amounts of dust and sulphate gases into the atmosphere.

Volcanoes produce two main gases, sulphur dioxide and carbon dioxide. Sulphur dioxide is a 'greenhouse gas', and it initially causes warming. However, it soon reacts with water in the atmosphere to produce sulphate aerosols that absorb the sun's radiation by backscattering, and hence cause cooling. It is well known that carbon dioxide, another greenhouse gas, causes warming of the atmosphere by trapping the heat from the Sun's rays near the Earth's surface. Increases in the amount of carbon dioxide in the atmosphere increase the heat-trapping effect, and temperatures worldwide increase. Carbon dioxide emissions, among others, from industry and from cars are said to be producing measurable increases in temperature today.

In a critique of the model proposed by Campbell and colleagues, Tony Hallam and Paul Wignall noted that there is essentially no geological evidence for dramatic global cooling, the key effect of the hypothesis. All the evidence actually points to global warming. Nor does the timing seem to match. Lasting for nearly one million years, the eruptions took much longer than the extinctions. And finally, Hallam and Wignall suggest that the tuffs and rock fragments may have been misidentified. Field evidence suggests that they are more likely the products of erosion of recently erupted lavas which were transported in rivers during a quiescent phase between eruptions.

Despite these seeming flaws, many geologists have been impressed by the link between the eruption of the Siberian Traps and the mass extinction. Perhaps Campbell and colleagues were broadly right, but they picked the wrong model. It is important to tease apart the sequence of global events in much more detail. What climatic changes were going on in the latest Permian and the earliest Triassic, and can these be explained by massive eruptions?

Global warming

For a number of years, Steven Stanley of Johns Hopkins University argued that most, if not all, mass extinctions were caused by global cooling.[17] His evidence was fourfold: that mainly tropical species disappeared; that the extinctions were gradual; that there was no limestone deposition after the event; and that evidence for ice could be found in the sediments.

But none of these observations is true for the end-Permian event. Temperate and polar species disappeared at a high rate and the extinction was geologically rapid. There is extensive limestone formation in the tropical belt throughout the Early Triassic, showing that climates were warm to hot. Also, there is no very good evidence in the rocks for icy conditions, except near the then north and south poles – which is not particularly surprising.

At one time, it was suggested that the end-Capitanian event might have been associated with global cooling. This was the smaller mass extinction that happened ten million years before the end of the Permian. There was a marine regression at the time which saw the end of the Guadalupian reefs of Texas. It had been proposed that this lowering of sea level was related to glaciation: as ice accumulates around the poles, water is frozen into the ice and withdrawn from the sea. However, more detailed studies have suggested to David Bond and colleagues that the late Capitanian extinction event coincides with the eruption of the Emeishan basalt volcanics in south China, and that the killing model was analogous to that for other mass extinctions, namely global warming, acid rain and oceanic anoxia.

Tony Hallam and Paul Wignall make a strong case[18] for the exact opposite – global warming – at the end of the Permian. In their book, the appropriate section is headed 'Global warning', an interesting typographical error. Evidence for a dramatic increase in global temperatures comes from the oxygen isotopes, from the sediments and from the record of plant life.

As we saw in Chapter 7, oxygen isotopes can be used as a palaeothermometer, and there was a drop in the value of the $\delta^{18}O$ ratio of about six parts per thousand. This corresponds to a global increase in temperature of as much as 15°C. This shift in the oxygen isotope ratio was first detected in a borehole record across the Permo-Triassic boundary at Gartnerkofel in Austria,[19] and it has been confirmed by measurements from Spitsbergen and elsewhere. An increase of 15°C may not sound much, but it would make profound differences to the distributions of plants and animals on land and in the sea. Recall that climatologists have been getting very excited recently about an increase of half a degree in global temperatures during the past century.

The sediments seem to point to increasing temperatures too. Palaeosols, literally 'ancient soils', can be a very good thermometer of past climates. Greg Retallack of the University of Oregon, who suggested that he had found shocked quartz at the Permo-Triassic boundary of Antarctica and Australia, has shown a major shift in soil types across the boundary.[20] For example, Late Permian soils in those regions are peats, suggesting cool conditions and winter freezing, such as seen today in Siberia and in northern Canada (latitudes 50–70°N). These are followed, however, by earliest Triassic warm temperate palaeosols, more comparable to modern soils seen in Ohio and Mississippi and in central Europe (latitudes 35–50°N). Rich, productive swamp soils of the Late Permian disappear entirely. The shift from cool to warm conditions is confirmed by an increase in the weathering of the rocks and increased rainfall. Retallack argues that his studies in Antarctica and Australia confirm the development of a 'high-latitude greenhouse' in the earliest Triassic southern polar regions.

'Absence of evidence is not evidence of absence', as we always teach our students. None the less, there is no evidence of ice anywhere in the world during the Early Triassic; no tillites (fossilized tills, the masses of fine debris and rocks commonly called boulder clay deposited beneath or in front of a glacier), no scratches or U-shaped valleys formed by putative glaciers, and no dropstones. Dropstones are often rather debatable. They are a genuine phenomenon, which occurs when pebbles and boulders, stripped from the land by a glacier, are carried out to sea and fall to the sea floor as the base of the iceberg melts. The difficulty is in recognizing them in the ancient rock record. One person's dropstone is another person's odd or inexplicable pebble. Still, the absence of evidence of ice recorded in Early Triassic sediments can be taken as evidence that there was no ice – until someone finds a counter-example of course.

Stronger evidence for global warming comes from the record of fossil plants. The tree-like seedfern *Glossopteris*, widespread throughout Gondwana, died out at the end of the Permian, as did the other cold-adapted plants associated with it. They were replaced instead by warm temperate floras dominated by the conifers *Voltzia* and *Voltziopsis*. In contrast to the diverse *Glossopteris* floras of the latest Permian, the new floras were depauperate, meaning that there were relatively few species. The few surviving plants were also designed to cope with hostile habitats, being generally low-growing, and having adaptations to withstand difficult growing conditions. Life was tough in the 'post-apocalyptic greenhouse', as Retallack terms it.

Global superanoxia

There is a dramatic change also in marine sediments at the Permo-Triassic boundary, as we saw. In many locations, ranging from the tropics to the poles, the diverse and fossil-rich limestones, mudstones and sandstones of the latest Permian switch to monotonous black mudstones, containing very few fossils. This is a classic shift to anoxia. Since the shift is so large, and so pervasive, it can justifiably be termed 'superanoxia'.

The colour of mudstones is a good guide to their conditions of deposition. Our students learn the mantra, 'red is oxidized, green is reduced, black is anoxic' (this is broadly, but not completely, true). The colours depend on the mineral content, and the mineral content can relate to conditions at the time of deposition. The red/green colour pairing reflects the kind of iron oxide in the mudstone. Red colours come from haematite, produced in the presence of oxygen, just like rust. Green, or grey, colours indicate a different state of iron called goethite, which has lost oxygen, either at the time of deposition, or after burial, by a process termed reduction.

Black mudstones mostly obtain their colour from high proportions of carbon in organic matter. Organic matter can only survive in the absence of oxygen since under normal conditions, with oxygen present, it would be consumed by microbes and animals. Even if the scavengers are not there, the carbon would rapidly combine with oxygen to form carbon dioxide.

Black mudstones commonly follow the Permo-Triassic boundary. This was seen in the succession at Meishan in China, and others nearby, where the earliest Triassic black shales also contain pyrite, more commonly known as fool's gold. Pyrite, or iron sulphide, forms only in anoxic conditions, such as in the stinking black slime at the bottom of stagnant pools, but more commonly in similar masses of rotten organic matter on the sea floor.

Anoxic mudstones are not a peculiarity of China alone. Paul Wignall and Richard Twitchett[21] of the University of Leeds have examined Permo-Triassic boundary sections throughout Asia, Europe and in northern latitudes, and they find the same phenomenon. Spitsbergen, for example, lies to the north of Norway and to the east of Greenland, at 78°N, and it was also in north polar latitudes in the Permo-Triassic. Wignall and Twitchett found abundant evidence for earliest Triassic anoxia there: a switch to black shales, the presence of pyrite, the absence of burrows and the scarcity of bottom-living organisms.

The association of fossils and sediments can be revealing. Under normal conditions in shallow seas, numerous organisms live in and on the sea bed. Corals and other reef-builders are fixed to the sea floor. Gastropods and arthropods creep about looking for food. Worms, bivalves and other animals make burrows into the sea-bed muds and sands, either for protection or in search of food. The latest Permian sediments show evidence of all of this biotic activity. Then it stops. The earliest Triassic black mudstones contain no corals or other fixed sea-floor beasts. Burrowing is absent. The only living things are some fishes that swam high above the sea floor, and the paper pectens such as *Claraia*. As we saw in Chapter 8, *Claraia*, and associated forms, seem to have been specialized in living in low-oxygen conditions.

Global superanoxia was associated with a crash in productivity in the earliest Triassic. Carbon isotope curves show a dramatic negative swing exactly at the Permo-Triassic boundary, as shown in the Meishan section in China (Chapter 7), and in every other Permo-Triassic succession that has been studied, whether marine or continental. The isotope curve swings back to normal values within a few hundreds of thousands of years after the mass extinction event. What was going on?

All kinds of explanations have been proposed for the carbon isotope shift. The negative shift, reflecting an increase in release of carbon-12, could have been caused by erosion of Permian coals. These coals, formed from *Glossopteris* and other forest plants, locked up large amounts of organic carbon-12. After erosion, the carbon-12 might have been washed into the sea, and accumulated fast enough to cause the carbon ratio to shift. But the dramatic reduction in the carbon isotope curve right at the Permo-Triassic boundary could indicate a collapse in productivity.

Productivity is a measure of biological activity, indicating the richness of life and its ability to process carbon and other materials. A productivity collapse occurs when most living things die. The surface of the land would be covered in the trunks of dead trees, and mounds of stinking, rotten plants and animal carcasses. The sea floor would be awash with dead corals, shellfish and fishes. The carbon-12 locked up in these plants and animals is suddenly incorporated into the sediment, and the ratio of carbon-12 to carbon-13, the normal inorganic state, shifts dramatically in the direction of carbon-12. But would the death of all life at the time be enough to explain the negative shift of four to six parts per thousand in the $\delta^{13}C$ ratio?

So how does this all work? Eruption of the Siberian Traps led to global warming, superanoxia and a productivity crash? Or was there more?

The methane burp

It all depends on a rise in carbon dioxide levels. What happens in the atmosphere happens in the oceans as well, as changes in the gas composition or temperature of the air penetrate the upper levels of the ocean. Also, changes in productivity and the cycling of organic carbon on land also affect the situation in the sea, as material is fed in via rivers.

Single-cause models for catastrophe are persuasive. And yet traditionally trained geologists are a cautious breed sometimes. In 1997, in their book on mass extinctions, Tony Hallam and Paul Wignall[22] presented detailed accounts of the Siberian Traps and of the major climatic changes associated with the end-Permian event. Because of problems of matching the timing of events, and because they were uncertain whether these fissure basalt eruptions could have been explosive enough to inject the necessary dust and gases high into the atmosphere, they concluded that the eruptions did not trigger the extinctions. In their flow chart explaining the sequence of events, the Siberian eruptions do not figure. Four years later, Paul Wignall included the eruptions as the major cause, but with some caveats. Why the shift of opinion?

The main concern in 1997 was that the eruptions themselves could not have supplied sufficient carbon dioxide into the atmosphere to cause the global warming of 6°C and the levels of superanoxia that had been detected. Other sources were needed. In 1997, all the postulated carbon dioxide came from oxidation of coals in southern Gondwana. If huge amounts of coal are exposed to the air, much of the carbon in the coal will combine with oxygen in the air to produce carbon dioxide. But this source has always been problematic. Could it really produce enough carbon? And at a fast enough rate?

It's not enough simply to find a new source of carbon dioxide; that source has to be capable of overwhelming the usual atmospheric feedback systems. Under normal conditions, carbon dioxide is produced by animals breathing out, by burning of wood and fossil fuels, and from volcanoes. Negative feedback systems have a tendency to cope with fluctuations and to bring excesses of one input back to a standard level. 'Negative feedback' means that a process is countered by the opposite, or negative, process, so regulating the effects of the process and maintaining a steady state. 'Positive feedback', on the other hand, means that the process is enhanced by more of the same, with further positive processes operating in the same direction.

So with the atmosphere. Excess carbon dioxide is mopped up by plants (during photosynthesis plants absorb carbon dioxide and produce oxygen) and through weathering. Carbon dioxide is stripped out of the atmosphere by rain water, forming weak carbonic acid, which then dissolves limestones on the ground. The carbon combines with the weathering products of the limestone, and the oxygen is given off as carbon dioxide. If you drip an acid on to limestone, the limestone will fizz – this is the carbon dioxide bubbling off.

By 2001, a trendy new carbon source had been identified. And this was one that was fast. Gas hydrates are crystalline solids composed of a cage of water molecules trapping gas inside. The water cages can trap various gas molecules, including carbon dioxide and hydrogen sulphide, but the commonest gas hydrates trap methane, a gas composed of carbon and hydrogen. Gas hydrates form at water depths greater than 500 metres, and particularly in polar regions. Because of the high pressures at such depths, the gas hydrates are amazing gas concentrators; if 1 million litres of methane hydrate is brought to the surface suddenly, 160 million litres of gas can be released.

Since the 1970s, when gas hydrates were discovered, they have been identified deep in sediments around the margins of most continents, and particularly around the poles. The huge frozen masses of ice and compressed gas fill pore spaces within the sea-floor sediments, and occupy vast fields that can be detected by means of geophysical soundings. Worldwide, it is estimated that gas hydrates contain up to 25,000 billion tonnes of carbon, about twice the amount of carbon held in all fossil fuels on earth. If some perturbation hits one of these gas hydrate bodies, and the gas is released, huge volumes of carbon dioxide or methane would bubble up through the ocean and explode from the surface, temporarily displacing the normal atmosphere above.

Could this be a killer? On 3 December 1872, the ship *Marie Celeste* was found adrift off the Azores. There was no sign of life on board, either above or below decks. There were no clues to explain why the crew had disappeared. Indeed everything appeared to be quite normal. In the crew's quarters, clothing lay folded neatly on bunks and washing hung on lines; in the galley, breakfast had been prepared and some of it had been served. Could the crew have been killed by the release of a massive bubble of methane hydrate? Millions of cubic metres of methane or carbon dioxide erupting from below would have a devastating effect, but would then be swallowed up into the atmosphere, leaving no trace. Why were all the crew missing? This will never be known – perhaps the pulse from below and the stagnation of the atmosphere made them all run on deck and jump overboard in search of fresh air.

What if numerous gas hydrate bodies, all round the world, were to have been released at the same time? Evidence has now been found for such a mass gas escape, a so-called methane burp, 55 million years ago, at the end of the Palaeocene. At that time there was a pulse of global warming, by 5 to 7°C over approximately 10,000 years, as shown by oxygen isotopes and the record of fossil plants. It has been suggested that this pulse of warming was caused by the release of 2000 billion tonnes of methane hydrate into the atmosphere.

The pulse of warming was brief, and conditions rapidly returned to normal. Gerry Dickens of the University of Michigan and colleagues suggest[23] that this is good evidence that the warming was caused by gas hydrates. The rapid heating led to the death of many species, the excess organic matter from dead plants and animals was washed into the sea, and carbon dioxide levels in the atmosphere were quickly reduced by the incorporation of the organic matter into oceanic deposits and by increased weathering following the loss of plant cover.

The end-Palaeocene methane burp did not lead to a major extinction event. The effects were worldwide, and many species died out, but the Earth returned to normal soon enough, and most species recovered. Could such a model be enough to kill almost all life?

Explaining the carbon isotope spike

The gas hydrates were probably not the main killer at the end of the Permian. But they may help to explain the massive negative shift in the carbon isotope curve, which dropped by four or five parts per thousand. This perhaps does not sound much of a shift towards the lighter isotope of carbon, carbon-12; but it actually represents a global shift in the entire carbon budget – the introduction of billions of tonnes of light carbon into the oceans.

What are the possibilities? The influx of isotopically light carbon could have come from the collapse of productivity that happened at the Permo-Triassic boundary, and the entry of huge amounts of rotting wood and animal carcasses into the sea. But, Paul Wignall has calculated, this would not be enough. It is estimated that the entire biomass of life on Earth today contains 830 billion tonnes of carbon. If all life is killed instantaneously, that amount of organic carbon-12 could be washed into the sea and buried. But there are already some 50,000 billion tonnes of inorganic, heavier carbon-13 in the ocean-atmosphere system, so the addition of 830 billion tonnes, less than a 50th of the amount, would make very little difference to the ratio.

Even the Siberian Traps eruptions could not have supplied enough isotopically light carbon. If the volume of basalt produced was 2 million cubic kilometres, that would have produced 10,000 billion tonnes of carbon, which would have been a mixture of carbon-12 and carbon-13. So, again, it was not enough to cause the carbon isotope shift. Even with the best figures, the Siberian Traps eruptions could have produced only 20% of the shift that actually happened.

Geologists have embraced gas hydrates with fervour – almost with a sigh of relief. When the calculations are done, nothing else has enough light carbon, nor can act fast enough. The carbon in gas hydrates is isotopically very light, with a $\delta^{13}C$ value of -65 parts per thousand. The release of only 10% of the estimated 10,000 billion tonnes of carbon contained in gas hydrates today would be sufficient to cause the shift in the $\delta^{13}C$ ratio by -5 parts per thousand. The secret is the very light composition of carbon. Although the Siberian Traps may have released the same mass of carbon, its isotopic weight was much heavier, and could not have produced the observed negative spike.

The killing model

Paul Wignall has put everything together into a single flow chart (Fig. 41). The key crisis seems to have been the eruption of the Siberian Traps. Worldwide devastation was caused by the production of different gases during the eruptions, and these gases were presumably pumped into the

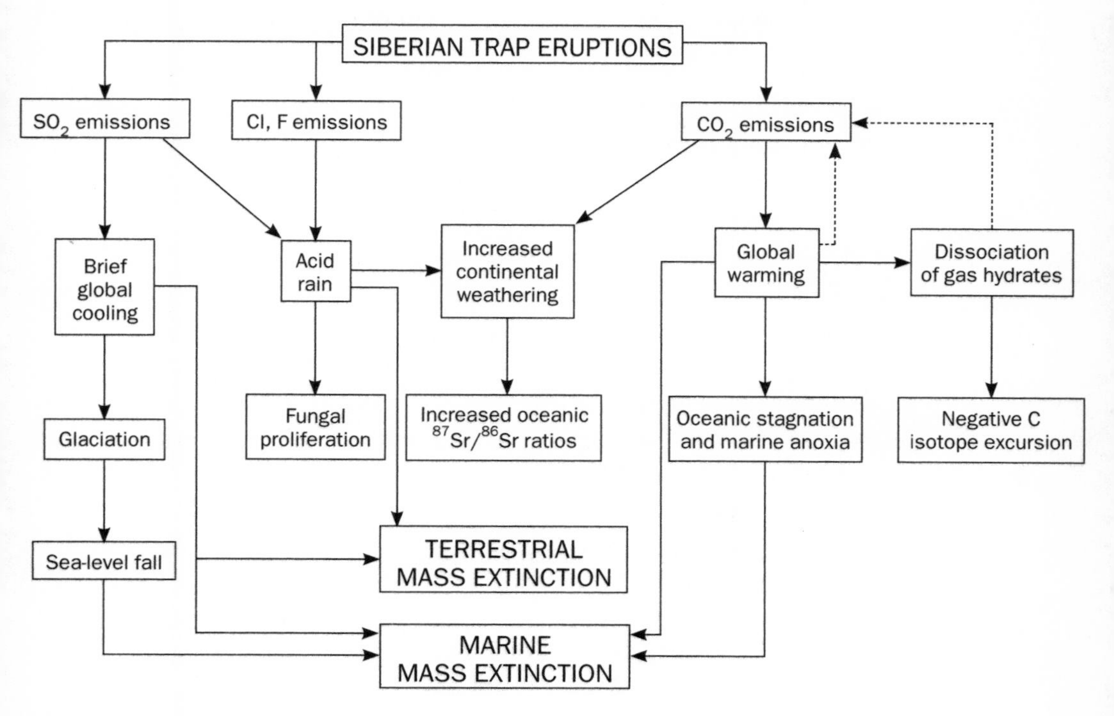

41 *The cycle of death. A summary diagram showing how the eruption of the Siberian Traps led to major atmospheric changes and to the collapse of most of life on Earth 252 million years ago.*

atmosphere sporadically during the entire span of the eruptions. Perhaps a single major eruption could have been absorbed by the Earth, and the short-term disturbance of the atmosphere-ocean system corrected by normal feedback processes. But repeated eruptions may have been too much, and perhaps led inexorably to total collapse of all normal interactions between the physical world and life.

Four gases from the eruptions may have been to blame. Carbon dioxide had the longest-lasting effects, leading immediately to global warming and anoxia, which persisted for hundreds of thousands of years. Each pulse of eruption may have reprimed the effects, and prevented any normal feedback systems from kicking into operation. The release of gas hydrates added to the misery.

Sulphur dioxide was also produced. This gas has a shorter residence time in the atmosphere, but the cooling effects of the sulphates may have caused a snap glaciation in some parts of the world, with associated falls in sea level as marine water was frozen into ice. Whether there was such a glaciation, and how long it lasted, cannot be said at present. And, as we saw above, there is limited geological evidence for freezing, but if it was short-term, as the theory suggests, one would not necessarily *expect* to find evidence preserved in the rocks.

Chlorine gas may also have been produced. In conjunction with the sulphates and the carbon dioxide these would produce acid rain. When combined with water, these gases form hydrochloric acid, sulphuric acid and carbonic acid, and hydrofluoric acid may also have been released. If such a delightful cocktail of acids were to rain out of the sky, normal plant life would have been devastated, just as today acid rain kills forests. With normal plants dramatically reduced, animal life on land would go too. Perhaps this postulated acid burst wiped out much of life on land and led to the fungal spike, the mushrooms and moulds being the first land life to be able to recover.

Acid rain also, of course, increases the rate of normal weathering on land, and the loss of plants would make it worse as soils were stripped off. Retallack and colleagues certainly detected this in the record of soils across the Permo-Triassic boundary, and increased rates of runoff of sediment into the sea are indicated also by a shift in strontium ratios. A dramatic increase in the ratio of strontium-87 to strontium-86 across the Permo-Triassic boundary suggests that huge amounts of terrestrial material were entering the sea via rivers.

The whole end-Permian crisis may have been made even worse by a runaway greenhouse effect. Normally, the atmosphere-ocean system will correct imbalances, and return carbon and oxygen levels to normal. This is a negative feedback process. If carbon dioxide levels increase, burial of organic matter, weathering or proliferation of forests will eat up the excess gas. However, a runaway greenhouse is a positive feedback system. An increase in carbon dioxide, for example, is not countered by processes that mitigate the effect. On the contrary, it triggers processes that add yet more carbon dioxide to the atmosphere.

The end-Permian runaway greenhouse may have been simple. Release of carbon dioxide from the eruption of the Siberian Traps led to a rise in global temperatures of 15°C or so. Cool polar regions became warm and frozen tundra became unfrozen. The melting might have penetrated to the frozen gas hydrate reservoirs located around the polar oceans, and massive volumes of methane may have burst to the surface of the oceans in huge bubbles. This further input of carbon into the atmosphere caused more warming, which could have melted further gas hydrate reservoirs. So the process went on, running faster and faster. The natural systems that normally reduce carbon dioxide levels could not operate, and eventually the system spiralled out of control, with the biggest crash in the history of life.

The view from the burrow

What did all of this look like at the time? Imagine the scene in the Karoo Basin in *Dicynodon* Zone times, which we encountered in Chapter 9. *Dicynodon* itself (Fig. 42), the medium-sized plant-eater that was most abundant at the time, may have been able to make burrows in which it could escape from the normal rigours of the tropical-monsoonal climate in which it lived. As the crisis approached, *Dicynodon* would have scuttled along the river bank and plunged into his cool burrow, expecting that it would all pass in a day or so and he could crawl out again.

The first basalt eruptions began thousands of years before, and far away, in Siberia, and continue, sporadically. None of the noise of the explosions would be heard in Africa, nor would *Dicynodon* have seen any of the erupting lava, ash or gas. But air temperatures might bounce up a little. Locally, around the eruption site, there might be a snap freeze caused by the emission of sulphur dioxide, but that would be a short-term phenomenon, soon overwhelmed by the warming effects of the carbon dioxide emission. The first eruptions pass pretty well unnoticed.

Then, a year or so later, there is a larger eruption. A further snap freeze is replaced by a greater rise in air temperature. This time *Dicynodon* feels it. It is the dry season, the time between the annual monsoonal rains, and it hurts. Life is balanced on a fine margin between survival and death during the dry season in any case, as we saw in Chapter 9. Even a one degree rise in temperature can kill off more plants than normal, and then more herbivorous animals fail to survive through to the rainy season. The blast of heat sends *Dicynodon* into his hole.

This time, the eruption initiates some further processes. The cocktail of gases ejected into the atmosphere rises high into the stratosphere and

encircles the globe. The gases and fine dust distort the normal appearance of the heavens – sunrises and sunsets look weird, with splashes of red, yellow and purple, and this is seen all round the world. A few days later, the perturbation triggers catastrophic acid rain. The chlorine, fluorine, sulphur dioxide and carbon dioxide emitted from the volcano combine with rain water in the high clouds to produce a cocktail of hydrochloric, hydrofluoric, sulphuric and carbonic acids. For millions of square kilometres around the eruption site, the acid rain burns off most of the plants. First to go are the trees and larger plants. Even as far away as South Africa, the effects can be seen. Plants lie dead where once they grew. They rot and decompose. *Dicynodon* creeps about, looking for some palatable morsels, but finds very little – just some mushrooms, mosses, ferns and club-mosses nestling in damp crevices around the river banks.

Then comes a distant rumbling, unheard in South Africa, but the *coup de grâce* none the less. Since the eruptions began, some 10,000 years earlier, the atmosphere and sea surface have warmed by two or three degrees, and the frozen northern polar region begins to melt around the fringes. The polar regions in the Permian were much smaller than they are today, with limited ice caps. But frozen tundra extends hundreds of kilometres away from the poles, and the ocean margins are frozen too. A great icy mass of gas hydrate, locked in the sediments at the margin of the polar sea, is warmed by a degree or two, and it suddenly gives way. First a few bubbles, then many, and finally a huge expansion of gas – 160 times the original volume. What was once frozen and at high pressure, becomes gas at normal temperatures and pressures, and a vast volume of methane and carbon dioxide bursts upwards through the oceans and shoots out into the atmosphere, raising spouts of seawater hundreds of metres into the sky.

There had been a long-term decline in the proportions of oxygen in the atmosphere through much of the Permian, and it reached a low point at the time of the extinction. In the course of a few days, *Dicynodon*, feebly searching for scraps among the stinking decay of plants in southern Africa, and already feeling a rise in temperature of two or three degrees, is now hit by a further devastating blow. Normal levels of oxygen are being driven downwards – he is gasping for air. And, day-by-day, the asphyxiation becomes worse.

42 Dicynodon *explores the devastated post-apocalyptic landscape.*

After a week, heavier rains come. The monsoon has begun. The rain is still acidic, although much of the acid has now been washed out of the system. And the rain carries away all the stinking vegetation down the slopes, into the rapidly filling wadis. Jostling tree trunks, branches and mats of leaves rush down to the sea, where they are dumped a few kilometres offshore, at the end of the estuarine tracts. But most of the soil is washed away too. Without the binding roots of the plants, the soil is vulnerable. After a few days, there is almost no earth left, just bare rock, with pockets of soil clinging on in hollows where mosses, ferns and club-mosses have survived. The countless billions of tonnes of organic carbon locked into the plants and the soil across the whole Karoo have been stripped into the sea. Almost nothing remains. With the soil went the worms, spiders, centipedes, flies, beetles and everything that could not hang on to the rocks and the rushing torrents of water.

Carbon dioxide levels in the atmosphere are still higher than normal. And there are no negative feedback processes. Normally, the carbon dioxide would be removed from the atmosphere by photosynthesis, and the monsoonal rains would have been followed by a dramatic greening of the land. Dried-up trees would spring into life, producing leaves from their gnarled branches. The bare earth would miraculously sprout low ferns and seedferns, as dormant seeds broke into life. But the soil has gone, the land is just naked rock. Nothing like this had happened since the Precambrian, some 300 million years before, when life on land had not yet evolved.

Dicynodon follows the water downhill to the sea, half-starved now, having gone without food for more than a week. Mushrooms, which seem to be all that can flourish, are the only thing available, and they are far from his preferred diet. His instincts suggest there will be food where the water is. But he is wrong. Everything is topsy-turvy in this apocalyptic world. Just as the plants on land were killed by the acid rain, so too the seaweeds that fringed the shores. The increased carbon dioxide levels in the atmosphere have penetrated the top dozens of metres of the sea, and the plankton has been decimated. Within a week, all the shoals of fishes that used to rely on the plankton as their staple diet have starved too; then on up through the food chain: the sharks and larger fishes that fed on the planktivorous smaller fishes die too. The whole sea is poisoned. Rising temperatures have imposed anoxia.

Dicynodon, and all the other animals – the closely related smaller and larger dicynodonts, the bulky, knobbly plant-eating pareiasaurs, the small scuttling procolophonids, therocephalians, millerettids, and the large, sabre-toothed gorgonopsians – are close to death. They wander about on a blank, rocky landscape. They have difficulty in breathing, and their backs are burnt at midday by the hotter-than-normal sun.

Then comes the third eruption. It is not by itself a particularly huge eruption. But it leads to a further cycle of acid rain. Temperatures rise again by a further fraction of a degree. Another vast bubble of methane and carbon dioxide is released in the far north from a frozen gas hydrate deposit. *Dicynodon* is now living through a runaway greenhouse effect. Nothing can turn back the devastating rise both in carbon dioxide and in temperature. He curls up and dies, along with nearly every other living thing on land.

In the sea, the vast influx of organic matter from the land – all the dead plants, animal carcasses and soil – have carpeted the seabed in a stinking, black slime. The decaying organic matter consumes oxygen and gives off hydrogen sulphide. Seafloor life – all the reefs and their denizens, as well as the creeping worms and arthropods, and the burrowing molluscs and shrimps – die. Their carcasses are incorporated into the fetid, black, anoxic bottom slime. Stripped of oxygen from the atmosphere above, and with the black mud below, the oceans go into a spiral of anoxia: oxygen levels fall, step-by-step, until almost nothing survives. Only a few worms, brachiopods and molluscs that can exist at depth in oxygen-poor conditions manage to live through these harsh conditions.

The eruptions continue, at random intervals, to pulse basalt lavas over Siberia. Hundreds of metres of fresh rock accumulate. Sometimes eruptions are separated by days or weeks; at other times, there may be a standstill for a few thousand years, and the rare surviving plants and animals manage to re-establish themselves for a short time, before a further cycle of devastation begins. Some of the eruptions are small and have limited global effects, of course, but others are large enough to lead to the global effects just described. Some day, with ever-more precise study, it may be possible to tease apart some of the detail of the separate phases of the eruption of the Siberian Traps, and whether the killing happened all at the start of the eruption cycle, or whether the process was drawn out over half a million years.

Is this what really happened?

All the pieces of the cataclysm 252 million years ago have been put in place. Research in the past ten years has led to an astonishing, earthbound scenario for almost complete devastation of the Earth and of its inhabitants. The killing model makes sense in terms of what is now understood about eruptions and their effects on the atmosphere, about oxygen and carbon cycling in the earth-life system, about the composition of the oceans, and about gas hydrates. But much of this is very new work, and it might have to be modified in the future.

For many, there is a lingering desire for something more apocalyptic, more instantaneous, some *deus ex machina* such as a huge extraterrestrial impact. Surely, they argue, if the KPg event, when 50% of species disappeared, required a vast meteorite, we need something even more catastrophic to kill off 90% or more of species? But the evidence for impact at the Permo-Triassic boundary is limited at present. That might all change of course any day, if the helium/fullerene story is confirmed, if a crater of suitable age is discovered, or if large amounts of shocked quartz and iridium turn up in rock successions in different parts of the world. However, I'll bet on the Siberian Traps coupled with gas hydrates for the moment.

What was so special about the Siberian Traps, and the whole end-Permian scene, that could have allowed the crisis to develop? The Siberian Traps were the biggest flood basalts of all time. But other Large Igneous Provinces include the Emeishan, Central Atlantic Magmatic Province (CAMP), the Ontong-Java Plateau lavas and the North Atlantic Igneous Province, of 262, 201, 122 and 62–55 million years ago; only the first two were associated with mass extinctions, the Late Capitanian and end-Triassic respectively, so massive outpourings of lava did not always lead to similarly dire environmental consequences.

Maybe it was simply a coincidence of factors. The end-Permian was the only time when the continents were fully assembled into a single supercontinent and when there was a large flood basalt eruption episode. During the later large-scale basalt eruptions, the continents had drifted apart and maybe life had diversified sufficiently in the different continents and oceans to be able to resist a range of severe climatic changes. One almost certainly has to add the coincidence of a massive methane burp, although there is no independent evidence yet for this (apart from the difficulty of otherwise explaining the remarkably large and rapid negative carbon isotope shift). Thorough testing of all the hypotheses by Robert Berner of Yale University has shown[24] that massive methane release has to be the main cause of the dramatic carbon shift, with lesser contributions from volcanic degassing and mass mortality. This isn't proof, but he ran all the possible causes against his well-established climatic models, and very few doubt Berner's pre-eminence in this arcane world of large-scale number-crunching.

Much more has yet to be found out about the end-Permian crisis. Geologists and palaeontologists are far from understanding step-by-step just what happened. But, as we have seen, ideas have sharpened and focused remarkably since 1995, and will doubtless continue to do so. What is new, and equally intriguing, is to try to understand how life might have recovered after such devastation. How does life respond to the loss of all but one in twenty species? These survivors entered a strange world, with ecosystems ravaged and transformed, an alien environment. And yet, remarkably, life did recover.

12
RECOVERY FROM THE BRINK

In 2009, I had a momentous meeting with Chinese researchers in Beijing. We were at Peking University – yes, officially known by the romanized spelling of Beijing – enjoying some beers after a day of conference talks, when I was approached by three professors. Zhong-Qiang Chen and Shixue Hu spoke perfect English, and while Qiyue Zhang spoke only a few words it was clear that all three were hugely excited. Chen, who was then at the University of Western Australia in Perth and is now a Professor at the China University of Geosciences in Wuhan, told me, 'these guys have dug away a whole hill in Yunnan, and it's full of thousands of fishes and marine reptiles. They want to work with a vertebrate palaeontologist.'

I asked about the new finds. Hu and Zhang conferred, and Hu told me, 'we are making geological maps around Luoping in the northeast of Yunnan Province, and we found amazing fossils at different levels. These are Anisian in age, maybe 8 million years after the end-Permian mass extinction, and they show a key step in the recovery of life.' I was invited to join work in the field in 2010, as the guest of Hu and Zhang, funded by the Geological Survey of China in Chengdu. This was completely unexpected: I had been invited to Beijing to take part in the celebrations for the 150th anniversary of publication of Charles Darwin's *Origin of Species*. Darwin is a great hero of science in China, so this was a wonderful and fitting outcome, to be invited to explore one of the largest evolutionary events of all time.

I was especially keen to see some of the marine Permian and Triassic remains in south China, home of the global stratotype of the Permo-Triassic boundary at Meishan (see Chapter 7). I knew that there were great thicknesses of Triassic limestones and mudstones above the level of the Meishan section, and these documented life in the sea in shallow and deep waters. In fact, some localities in Yunnan and Guizhou, two large provinces in the far southwest of China, were world-famous for their exquisite fossil preservation.

Places such as Panxian, Xingyi, and Guanling had already yielded thousands of exquisite fossils of Triassic marine reptiles: the long-necked nothosaurs and pachypleurosaurs, dolphin-shaped ichthyosaurs, shell-crushing placodonts and more. Associated with these were large numbers of the fishes that formed much of the diet for the swimming reptiles, which were themselves hugely important, documenting a re-emergence of fish groups after the crisis. Many of these Triassic fishes were neopterygians, the 'new' fish group that includes most fishes in the oceans and rivers today.

I was also excited to see these Chinese sections because they had begun to yield remarkable new information about the recovery of life after the end-Permian mass extinction. The thickness of the rocks and the enormous geographic area they covered – some 2000 kilometres from east to west – contributed to their importance. While we had been working in the red beds of Russia (see Chapter 10), Chinese and international scientists had been publishing striking new evidence about how life rebuilt itself from near annihilation.

Disaster taxa

After the mass extinction, life in the sea[1] was sparse and monotonous, essentially identical from pole to Equator. All the rich diversity of life in the reefs and sea floors of the Late Permian had gone. The clear differences between the species that lived in the tropical Tethys Ocean and those of the northern Boreal Ocean had disappeared. As we saw, the survivors were a restricted assemblage of rather unusual animals.

What is becoming clear is that all the rules change after a profound environmental crisis. Disaster taxa prove the point (Fig. 43). These are species which, for whatever reason, are able to thrive in conditions that make other species quail. After the end-Permian crisis, the inarticulate brachiopod *Lingula* flourished for a brief spell, before retiring to the wings. *Lingula* is sometimes called a 'living fossil', since it is a genus that has been known for most of the past 500 million years, and it lives today in low-oxygen estuarine muds. It is an inarticulate brachiopod – not because it can't speak (although that is doubtless true), but since it lacks a complex

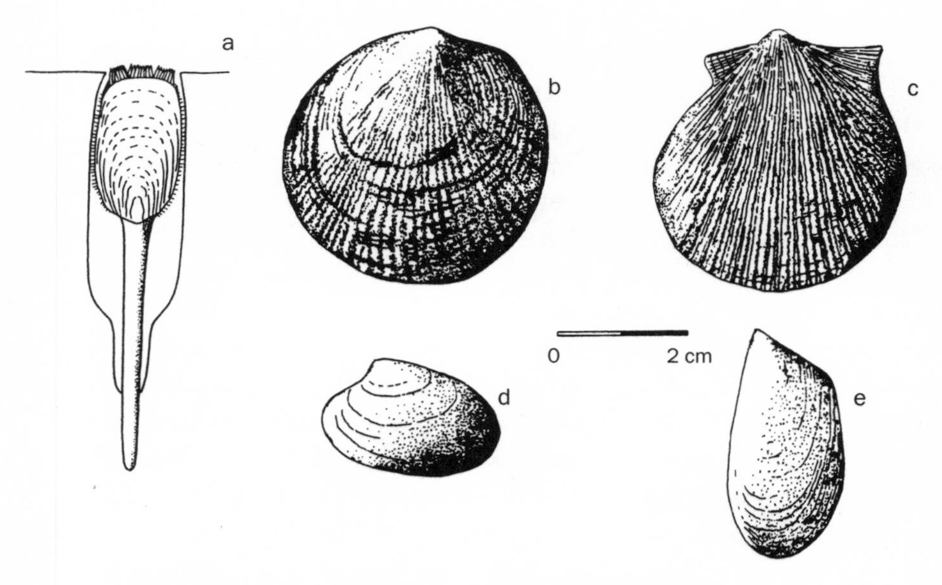

43 *The earliest Triassic 'disaster' forms: the brachiopod* Lingula *(a), and the bivalves* Claraia *(b),* Eumorphotis *(c),* Unionites *(d) and* Promyalina *(e). These five magnificent shelly creatures were virtually all that survived through the end-Permian mass extinction, and they dominated black, anoxic sea floors worldwide.*

hinge mechanism, as seen in the more abundant articulate brachiopods. So, after a brief burst of excitement for *Lingula* in the anoxic Early Triassic oceans, it faded back into the proper obscurity that has characterized the rest of its long history.

Among bivalves, famously, a group of four disaster taxa, *Claraia, Eumorphotis*, *Unionites* and *Promyalina*, are found in the black, anoxic shales everywhere. Of these, the first three radiated during the Early Triassic, producing a cluster of new species. They were presumably responding to the fact that precious little else was around on the sea floor at the time, and they clearly had the adaptations necessary to survive in the low-oxygen, low-productivity post-extinction conditions. Overall, the diversification of bivalves was a slow process, with new forms appearing at a slow rate. Something like the pre-extinction diversity was only recovered by the Mid Triassic.

Other marine groups seemingly recovered faster. The ammonoids, coiled swimming cephalopod molluscs, just squeezed through the end-Permian crisis, with the survival of only two families, the Otoceratаceae and Xenodiscidae. The ammonoids re-radiated in two pulses, first in the early part of the Early Triassic, and then in the later Early Triassic and Mid Triassic, when they exceeded their pre-extinction diversity with over 150 genera. This fitful recovery by ammonoids has been much debated. Some have argued that their rapid rebound in the first 2 million years of the Triassic is proof that life, overall, recovered quickly. Others note that ammonoids were always naturally fast evolving, and so they would speciate rapidly when given a chance, but they would also be likely to succumb to further environmental deteriorations. Their initial rise, and then extinction at about 249 million years ago pointed to at least one sharp Early Triassic extinction event, which has been much explored recently.

The 'reef gap' following the end-Permian extinction was one of the most profound pieces of evidence of major environmental crisis, although its extent has been debated. The generally accepted viewpoint is that the rich tropical reefs of the Late Permian had all gone, and nothing faintly resembling such a reef was seen for 10 million years after the event. This can hardly be a result of poor study or collecting. Not a single coral specimen, a bryozoan or any other reef animal has been found. What were once huge structures, often tens or hundreds of kilometres across, dominating many coastal strips, had disappeared entirely.

An alternative view has been that reefs, like some groups of ammonoids, recovered fast. Low reef mounds, dated at only 1.5 million years after the crisis, have been reported from Nevada and Utah. These reefs were constructed from sponges and serpulid worms (worms that lay down a calcareous tube and can often be seen today meandering as white, chalky tubes over mussel shells, for example). Most would debate whether such structures can be claimed as evidence for the early recovery of reefs and the closing of the 'reef gap'. They can be called reefs because they are mounds on the seabed composed of the skeletons of organisms, but they were modest in scope and did not themselves lead to the re-establishment of the coral reefs of modern aspect that came later.

Even in the Mid Triassic, more extensive evidence for reefs shows they were still unusual in form. They were composed of a motley selection of Permian survivors, a few species of bryozoans, stony algae and sponges.

There are various kinds of reefs. We think of structural reefs as typical – great walls of coral skeletons, often built up over millennia, and sometimes tens of kilometres long. But the Mid Triassic reefs were modest affairs termed 'patch reefs': low amalgamations of reef-like creatures forming a little cluster on the sea bed. The scleractinian corals were there, close relatives of the corals that abound in tropical seas today; they were diverse, but rare. It took another 10 million years before these corals had become relatively common, in the Late Triassic, and before reefs grew in size and complexity again. But they were still much smaller than in their Late Permian heyday.

Delayed recovery and models

Evolutionary biologists are keen on numerical models, and one of the most widely used is the density dependent resource-limited logistic curve. A logistic curve is S-shaped, showing an initial slow rate of increase, then a steeply rising portion, followed by a levelling-off, or asymptote, just like the collector curve we discuss in Chapter 13 (p. 303, Fig. 46).

In ecology, logistic curves were used as models in some seminal work in the 1930s on competition. If a small population of bacteria was seeded on some agar growth medium in a petri dish, the population would begin to expand, slowly at first, then increasingly rapidly, but it would stop as it reached the edges of the dish. The explanation is simple: populations in nature will normally expand in an exponential manner until something stops them. In this case, they run out of space and food, and these are termed limiting factors, evidence that the expansion of the bacterial population is a density-dependent process.

These principles have also been used to understand competition for resources between species, a fundamental concept in ecology. If you set two populations of bacteria growing in the petri dish, starting at opposite ends, they will grow like mad until they contact each other, and then the rate of increase slows down. In the end, the two populations may partition the food equally, or one population may overwhelm the other, out-compete it, and occupy the bulk of the space. Each population or species in such a case has its own logistic curve, and the limiting factors on infinite expansion are both resources (food, space) and competition from the other species.

The numerical ecologist Ricard Solé from the Complex Systems Laboratory in Barcelona, and colleagues, used this kind of thinking to propose a model for recovery from mass extinctions[2]. In the model, they tried to account for three different kinds of recovery processes according to their speed, namely fast, medium, and slow (Fig. 44). In the 'fast' case, diversity bounces back in a more or less unconstrained manner. In the 'medium' case, recovery follows a logistic curve, and the new ecosystems rebuild themselves in a diversity-dependent manner, where species numbers rise quickly, and then the rates of expansion reduce as ecospace is filled up. The third model, for slow recovery, introduces a significant lag time before species numbers build up. This model requires species interactions in order to permit new species to arise, and so new species cannot emerge simply where opportunity permits.

The long lag time in this third model might suggest that this was what was going on after the end-Permian crisis. If life took 8–20 million years to recover, there must have been a set of processes that slowed down the usual propensity of species to multiply rapidly. However, there was probably much more to it than simple biological interactions. Growing evidence is suggesting that the post-extinction world continued to present challenges, and recovering groups, such as ammonoids, suffered setbacks time and again.

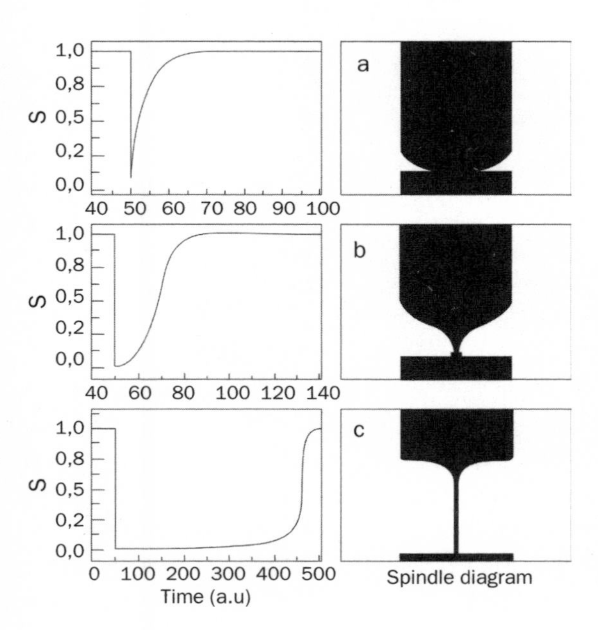

44 Biodiversity recovery could follow one of three trajectories: a, *an immediate linear response;* b, *a logistic recovery; or* c, *a simple positive feedback pattern of species interaction that follows a hyperbolic trajectory. These represent, respectively, quick, medium, and slow recovery.*

Continuing harsh conditions

Evidence suggests that early-recovering ammonoids were hit by small extinction events several times in the Early Triassic, indicating some continuing pattern of poor physical conditions on the Earth. Indeed, several hints of this had been noted before an important paper in 2004, in which Jonathan Payne of Stanford University, and colleagues, showed[3] that major isotopic perturbations continued for 5–6 million years after the end-Permian mass extinction. They counted as many as four light carbon spikes (Fig. 45), each one equivalent in scale to the peak at the Permo-Triassic boundary (see pp. 270–274).

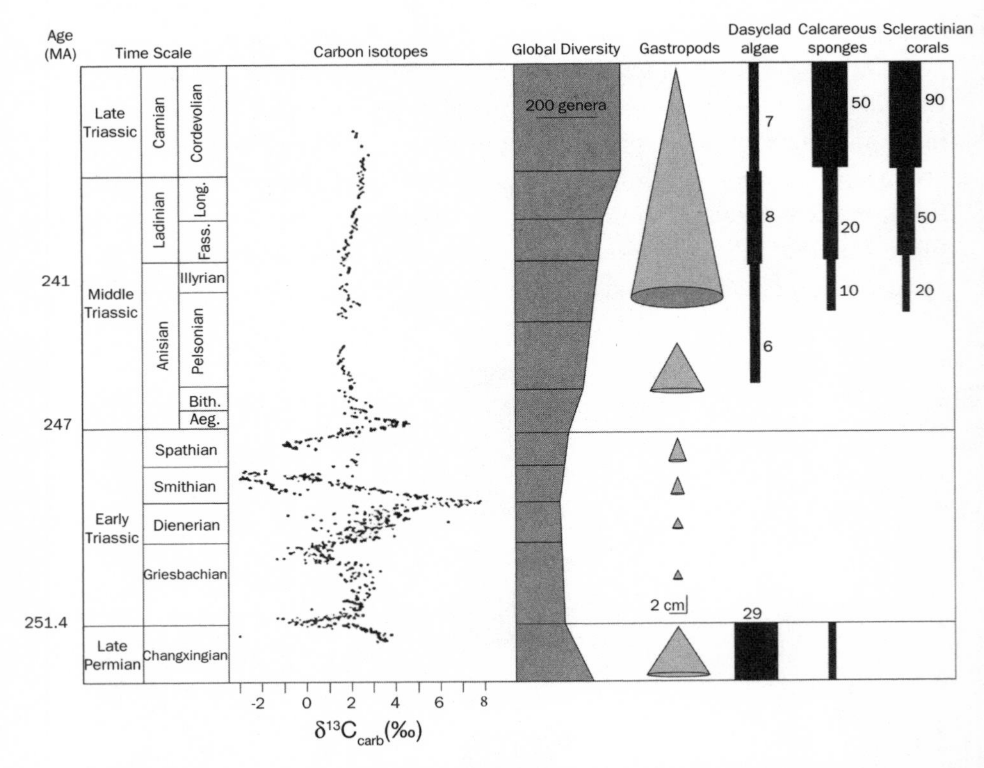

45 Composite carbon isotopic curve for the Changxingian-Carnian (Cordevolian) compared to the pattern of biotic recovery from the end-Permian extinction. Note three sharp, negative spikes in the carbon isotope curve, at the PTB, Smithian-Spathian boundary and late Spathian, each corresponding to an episode of intense warming. Fass., Fassanian; Long., Longobardian; Bith., Bithynian; Aeg., Aegean.

These repeated carbon perturbations have raised two main questions: how were they produced, and what were their effects? Their source is still mysterious. The peak in light carbon isotopes at the Permo-Triassic boundary should have exhausted any supplies of methane hydrates from deep in the oceans. After the frozen gases had been released, it should then have taken many millions of years for the reservoirs to replenish themselves from the steady, but slow, rain of dead plankton onto the seabed. Either the hydrate reserves were renewed faster than had been expected, or light carbon-12 was being released from other sources, such as the passage of volcanic lavas in Siberia through Carboniferous and Permian coal deposits.

Whatever the cause of the repeated light carbon peaks, they happened, and they indicate oceanic and atmospheric conditions just as grim as at the Permo-Triassic boundary – global warming, acid rain, ocean anoxia and the like. There is also geological evidence for anoxia and warming. Immediately after the Permo-Triassic boundary, and at intervals through the Early Triassic, unusual sedimentary structures and fossils appeared. Some geologists have even talked of a reversal to conditions that had existed in the sea before the Cambrian, over 540 million years ago, before typical marine animals became established.

The similarity is eerie for geologists who are used to doing fieldwork in Precambrian rocks; it looks like a return to a time before large animals existed on Earth. These Early Triassic limestones bear layered microbial mats and wrinkle structures on their surfaces. These are evidence that very simple forms of life, the cyanobacteria, sometimes called blue–green algae, had spread over the sea floor as a green slime. Cyanobacteria are closely related to the very first organisms on the Earth, and they form mats and sheets on the sea floor today, but only in very unusual conditions of low oxygen or high salinity. The reason that such simple slimes and oozes do not occur in normal aquatic conditions is that they are generally eaten by more complex animals such as molluscs and worms. These unusual biosedimentary features can only exist in the absence of complex life.

So, in the Early Triassic, at times, all the usual sea creatures were driven out, and weird, early-Earth conditions returned for thousands or hundreds of thousands of years. Such conditions persisted until the oxygen and carbon in the atmosphere had stabilized and the sea floor and the surface of the land became habitable again for plants and animals. No wonder the recovery of ammonoids and other life in the sea and on land was stopped in its tracks. Further, these repeated episodes of environmental crisis reset the pattern of recovery several times, making it difficult to apply theoretical models of ecosystem reconstruction, such as those in Fig. 44.

These earlier studies did not explore the nature of the environmental changes. In 2012, Yadong Sun, then a doctoral student at the China University of Geosciences in Wuhan, provided evidence[4] for the startling observation that temperatures in the atmospheres and oceans had risen to as much as 36°C at the Permo-Triassic boundary, then even higher, to 40°C, 250.7 million years ago in the late Smithian age of the Early Triassic, the time when ammonoids and other groups had been hit a second time. These very high temperatures were calculated using standard methods from records of oxygen isotopes taken from conodont fossils, widely regarded as a reliable technique. They are notably higher than modern sea surface temperatures, typically within the range of 24–30°C in the tropics.

If temperatures really did rise as high as 36 or 40°C, they would not only have been uncomfortable for life in the sea and on land, but fatal. This may seem surprising, as such temperatures are at about human body temperature, the normal level we set for a shower. However, prolonged exposure to such very hot conditions essentially boils a plant or animal and causes certain key life functions to shut down. It is well known that not many living things are comfortable when temperatures are above 35°C for long. For example, plant photosynthesis switches off at this temperature. In terrestrial animals, proteins are damaged at 45°C, but the normal maximum temperature for survival is much lower for marine animals, more like 35°C. They acquire enough oxygen from the water at low temperatures, but as the environmental temperature rises, they become distressed and their demand for oxygen increases until they cannot extract enough to fuel their faster-running metabolism. Indeed, Sun noticed that there are no records of any fishes in the seas or tetrapods on land around the equatorial belt in much of the Early Triassic. Faced with such high temperatures most animals and plants apparently fled north or south into the then-temperate belt.

So, a long spell of massive heating drove plants and animals from the equatorial regions in the middle of the Early Triassic. Extinctions occurred among ammonoids, bivalves and conodonts in the sea, and some tetrapod groups on land. After isotopes and temperatures stabilized again, towards the end of the Early Triassic (Fig. 45), the recovery picked up once more, and this time went further. This is where the Luoping locality in Yunnan comes in.

Luoping and complex ecosystems

I had never been to the south of China, and was looking forward to the trip. Yunnan is thousands of miles from Beijing, deep in the southwest of China and in contact with Thailand, Laos and Burma, well within the tropics. In parts of Yunnan there are elephants, parrots and monkeys, quite different from the cooler, drier climates around Beijing and the north.

At Kunming, the capital of Yunnan, I was taken to the Camellia Hotel. There, after flying for nearly 24 hours (from Bristol to Amsterdam, to Beijing, and finally Kunming) I had dinner with Professors Hu and Zhang and several students; the spicy Sichuan cuisine kept me awake for another few hours.

After a night of profound sleep, we set off towards Luoping, in the northeast of Yunnan Province, a distance of 230 kilometres. Kunming is a busy, chaotic city with modern hotels and roads, but also delightful flash-backs to former, more rural times. As we drove out of the city along a dusty older road, we saw butchers jointing whole cow carcasses, simply squatting on the pavements outside their shops. Children clustered round to watch the interesting spectacle. The journey to Luoping began on a modern motor-way, the G80, with pay booths every few miles, but we then switched to the G324, a highway from an earlier time, mostly in good condition, but with some stretches that required attention. Everywhere, massive, overloaded trucks ground slowly up the long inclines, sprinkling water over their multiple tyres to make sure the brakes worked.

We cut off the main highway a few kilometres short of Luoping town, and meandered through fields of deep brown-coloured earth that smelled highly organic. Neat crops of broad-leaved tobacco stretched out in all directions, tended by men and women with broad-bladed hoes. Patient buffalo, with their long, backwards-arcing horns, stood by their wooden carts at the edges of the small roads. One looked a little cross, his cart filled with his own dung, ready to be stacked around the precious tobacco plants. The country roads are all in good order, built high above the field levels with cemented stone supporting walls on each side, so water runs off into the fields to the advantage of the crops.

The villages were neat, comprising forty or fifty compounds, each with a house, often bearing gaudy red good luck messages and symbols. The yard housed the farm equipment, hay and straw, neat stacks of firewood, herbs and clothing drying on lines. Children walked to school in neat uniforms, eager for their education. In one village, a funeral procession was underway, the coffin borne in front and villagers and friends behind, carrying large, circular, colourful funerary banners, chatting loudly, and setting off firecrackers. The bodies are buried in earth barrows, one per person, fronted with elaborately carved limestone façades, built into hillsides overlooking the fields.

We reached our target, the village Da'aozi, or Da'wazi, which both mean 'Large depression', an odd name for a village on a hillside. Passing through the village and up into the limestone hills, we entered the southern Chinese countryside, recalling the classic willow pattern typical of fine Chinese art, with tall, sugarloaf mountains rising straight out of the fertile fields. Such scenery only occurs in these tropical latitudes, where millennia of monsoon rain have eroded the limestone into tall spines, and where loess, fine desert dust, blown from the Gobi Desert, has filled up the hollows to form flat, fertile fields. Swinging round a corner of the rough track we saw our target.

Luoping 1, as the first quarry is called, is one of many such limestone hills, which the Chengdu geologists targeted in 2008 and 2009 for a major excavation[5]. Even from a distance, the top of the hill was clearly denuded of vegetation and opened as a quarry; hundreds of tons of rock had been tipped down the hillside in front. We meandered up a long stony path between small ploughed strips growing sweetcorn and leaf crops, fields belonging to the farmers of Da'wazi. Past a small concrete field station hut, we reached the first excavated benches of the quarry.

Professor Zhang and his students, alongside teams of local farmers who were paid $10 a day, had cleared their way up to the top of the hill, exposing 183 beds, removing overlying soil and vegetation, and sweeping the surfaces clear. They had cleaned a section 16 metres thick, and each of the 183 beds bore a number painted with whitewash to guide the field geologists as they documented the rock section. They found fossils throughout the section, identifying at least 11 fossiliferous beds. Some contained bivalves and other seabed shellfish, with rare echinoderms (sea lilies) and conodonts here and there. These showed that the sediments had been laid down in water, under the sea, and that the water was relatively shallow, perhaps less than 100 metres deep.

One or two organisms derived from land were also found – twigs, plant leaves and a millipede. These terrestrial fossils were rare, but they were in good enough condition to show that they had not been transported far, and so the site must have been close to land, maybe within 10 kilometres, in the Middle Triassic.

This succession of rocks is part of the Guanling Formation, a rock unit that covers hundreds of square kilometres of eastern Yunnan and western Guizhou provinces. The conodonts prove an age of Pelsonian, a part of the Anisian Stage, and dated at about 242 million years ago, 10 million years after the end-Permian mass extinction.

Most importantly, Professor Zhang and his crew recovered thousands of specimens of lobsters, fishes and marine reptiles. These three groups are the most important, and they show two things. First, some of the lobsters and fishes belong to rather advanced groups that were best known from the succeeding Jurassic period. So, the new finds at Luoping extend their ranges back, deep into the Triassic. Palaeontologists always like to find the 'oldest X', in part because there is a competitive element to their personalities, but also because such finds may help to clarify the importance of some major evolutionary events. If such events occur earlier than expected, then some of the assumptions about triggers and consequences have to be revised.

The second reason that the Luoping fossils are important is that they document a complex ecosystem, with established 'top predators'. The top predators were some long-snouted fishes called *Saurichthys*, very common at Luoping, and known hunters that lurked, like elongate pike, behind rocks and waterweeds, and shot out after their prey. More importantly, the marine reptiles from Luoping, the ichthyosaur *Mixosaurus*, the nothosaur *Nothosaurus*, and the long-necked *Dinocephalosaurus* fed on fishes and on smaller reptiles, providing a new layer at the top of the food chain, something that had not been seen in the Permian. The Luoping biota, or assemblage of plants and animals, we argued, documented the final stage in the recovery of life, 8 million years after the devastation of the end-Permian mass extinction.

Crawling over the exposed benches, I could see dozens of small fishes at some levels, an ichthyosaur preserved *in situ* under a flagstone, and many burrows and other fossils indicating that life was rich and diverse. We retreated to the field station, a small concrete box in which a custodian sleeps, to have a cup of tea and to escape the hot sun. Then, miraculously, a TV crew emerged, having carried their cameras, batteries and other equipment up the winding path. The presenter was a very pretty young woman, who hopped about the site in leather trousers and sparkling high-heeled white shoes. It is a mystery how she preserved her poise along the farmers' track and on the dusty site. I was asked to say, in English, what I thought, and my Chinese colleagues explained the importance of the Luoping site for a news programme to be aired that evening.

We then walked out and drove on to the town of Luoping. Here we lodged in the Hengsheng Hotel. I later discovered this means 'Better forever' or 'Onward and upward', and it is owned by the local tobacco-growers trading company for their officials when visiting. The posters outside more or less encourage the citizens of Luoping to smoke as their patriotic duty. Throughout the hotel, of course, there are 'no smoking' signs.

Bouncing back on land

Life in the sea recovered, then, after 8 million years or so of grim environmental conditions and repeated setbacks. The marine Triassic rocks of south China, spanning from the Permo-Triassic boundary at Meishan to even younger units at Luoping, showed how some groups had recovered, died out, recovered, died out, and finally established a stable recovery in the Anisian, when environmental conditions settled back to the pre-extinction norm.

The same story seems to be found on land. We saw earlier how the latest Permian reptiles of southern Africa and Russia were organized in complex communities, with several scales of herbivores, presumably feeding on different grades and sizes of plant.[6] Preying on them in turn were, likewise, several scales of carnivore, from small to medium to large to very large. The top carnivores were the sabre-toothed gorgonopsians, which fed on hippopotamus-sized herbivores such as pareiasaurs and larger dicynodonts.

Then, after the crisis, the world was filled with a single reptile, the dicynodont *Lystrosaurus*. Never before had one vertebrate existed in such numbers and on every continent. Recall too that *Lystrosaurus* made up as much as 95% of the fauna, all the other smaller amphibians and reptiles being incredibly rare. Clearly, the post-apocalyptic ecosystem on land was a caricature of a natural system. Not only was it horribly unbalanced, but it seemed to be almost identical worldwide.

The story of reptilian evolution during the Triassic is one of expansion and increasing complexity. The surviving forms such as *Lystrosaurus* branched out and gave rise to new reptilian dynasties. There was some differentiation of faunas from continent to continent. Larger, and smaller, forms appeared. Ecosystems became more complex.

Typical herbivores were still the dicynodonts, descendants of the group that had been important in the Late Permian, but which had been almost wiped out, except for the survival of *Lystrosaurus* and its smaller relative *Myosaurus*. Another synapsid group, the cynodonts, gave rise to some herbivorous side-branches, such as the chiniquodonts and diademodonts. The rhynchosaurs, a diapsid group, distant relatives of the archosaurs, were also important. Rhynchosaurs were 1 or 2 metres long, with broad triangular-shaped skulls. The front of the skull was formed into a pair of curving tusk-like structures, which may have been used in raking together plant food. Their jaws were lined with multiple rows of teeth arranged in a kind of cobbled pavement, and with a groove and blade system that provided precise interlocking of upper and lower jaws. Evidently their plant diet consisted of tough stems and leaves that required firm chopping.

These diverse herbivores were preyed on by a new array of flesh-eaters. Among the synapsids, the cynodonts diversified into several lines of mainly small- and medium-sized flesh-eaters. Dominating the upper end of the scale were members of a new group, the archosaurs. The archosaurs, 'ruling reptiles', had arisen in the Late Permian, but were then rare. In the Early Triassic, they began their inexorable rise, diversifying slowly and expanding their range of niches. By the Mid Triassic the group had split two ways, one line evolving towards the crocodilians, the other to the birds. On the crocodilian line were the rauisuchians, a group of terrifying flesh-eaters that existed right to the end of the Triassic. Rauisuchians ranged in size up to 5 metres long, and their deep, long skulls were lined with fearsome teeth. Here were some effective top predators, not sabre-tooths like the Late Permian gorgonopsians, but able to deal with the largest and slipperiest herbivores of the day.

The dinosaur/bird line of archosaurs has recently proved even more fascinating. Until recently, it was easy to say that dinosaurs evolved more than 25 million years after the end-Permian mass extinction. However, now there are several hints that dinosaurs actually arose much earlier, at about the same time as the Luoping deposits. In Tanzania, palaeontologists have identified some incomplete skeletons as belonging to the nearest relatives of dinosaurs, a small group called silesaurids. If silesaurids were present around 242 million years ago, then dinosaurs, their immediate cousins and thus a so-called 'sister group', must have been as well. This is because any pair of nearest relatives must have shared a common ancestor, and the older of the two sets a minimum estimate of the date of origin. There is also a very incomplete skeleton, called *Nyasasaurus*, from the same rock unit, in 2013 very tentatively pronounced the world's oldest dinosaur; the remains are too incomplete to be sure. In any case these finds show that dinosaurs originated much earlier than expected, right in the maelstrom of the recovery of life from the end-Permian mass extinction. But they took some time to become really dominant in Triassic ecosystems.

So a measure of ecosystem complexity had been recovered by the Middle and Late Triassic. There were small, medium and large plant-eaters, and small, medium and large flesh-eaters. Twenty million years seems relatively rapid for biological recovery after such a devastating mass extinction. But in a sense, ecological recovery was not yet complete. There were no truly large plant-eaters, and the global diversity of species was not yet back to the latest Permian levels. Indeed, no Late Triassic reptilian fauna anywhere in the world was as diverse as any of the Late Permian faunas of Russia or southern Africa.

Within the Late Triassic, at the end of the Carnian stage, about 225 million years ago, life on land was hit by a further extinction event. The dicynodonts, chiniquodonts and rhynchosaurs died out. With all three of the dominant herbivore groups gone, their predators were also affected. The causes of this end-Carnian event are uncertain, but they may relate to a drying event and a major floral change – the replacement worldwide of Gondwanan seed ferns by conifers.

This marks the true beginning of the age of the dinosaurs, when they really became established. With the dominant herbivore groups gone, something was able to evolve to fill their place. The first complete specimens of dinosaurs from the Carnian were small insect-eaters and predators that nipped about trying to avoid the beady gaze of the rauisuchians. Indeed, Carnian dinosaurs were rare; only two or three individuals for every fifty rhynchosaurs or dicynodonts. Then, after the end-Carnian extinction event, dinosaurs appeared all over the place. The group evidently took its chance and radiated to populate the empty ecospace.

The few modestly sized herbivorous dinosaurs that existed in the Carnian evolved into giants like *Plateosaurus*, a biped-cum-quadruped, with a long neck and tail, and jaws lined with sharp-edged, plant-cutting teeth. *Plateosaurus* had a huge sickle-like thumb that may have been used to rake plants together, or for defence. Either way, some specimens of *Plateosaurus* became very large, around 5 or 7 metres long. Some of its relatives achieved lengths of 10 metres or more before the end of the Triassic. At the time, flesh-eating dinosaurs were abundant, but most were only human-sized or smaller. The top carnivores were still the rauisuchians, but even they probably could not prey on full-grown *Plateosaurus*, which were just too big.

After the end of the Triassic, and the extinction of the rauisuchians and some other basal archosaur groups, the dinosaurs diversified further. Large predatory forms appeared, progenitors of *Allosaurus* and *Tyrannosaurus*, which were big enough to tackle *Plateosaurus* and its relatives. It is intriguing that we can now trace the origin of dinosaurs, among many other groups, to the time of recovery after the end-Permian mass extinction. What had been a crisis for the vast majority of Permian life actually had remarkably creative consequences for evolution in the Triassic.

Recovery

The post-crisis rebound can be defined also in purely numerical terms. An easy way to do this is to find out when global diversity hit pre-extinction levels. At family and generic level, it took until the Early Cretaceous, at least 100 million years, for marine life to reach Late Permian levels again. The global pattern of recovery at species level cannot readily be determined, but if there is a trend of increasing time to recovery at lower taxonomic levels, it might have been as long as 150 million years.

After the other 'big five' mass extinctions, the recovery phase was faster. The best-studied has been, of course, the KPg event. For most groups of plants and animals, it seems that the recovery phase lasted for 5–10 million years in total. There was indeed a delay before recovery processes kicked in, but the niches vacated by the dinosaurs were largely filled during the Paleogene. Birds and mammals which survived the KPg event famously radiated in the Paleogene. Indeed, as in the post-Permian world, there was a time when the outcomes were not clear, with new groups popping up and dying out rather more quickly than normal.

The post-KPg world witnessed giant terror birds that preyed on horses. Had evolution proceeded a little differently, such might be the scene today. Mammals had been effective, but small, insect-eaters, gnawers and tree climbers in the dinosaurian world. Birds had been fish- and flesh-eaters, and some had become flightless. So it is perhaps not surprising that some giant flightless birds became top predators in different parts of the world, hunting the smaller mammals. They had deep, massive beaks, and ran at speed, something like an ostrich, snatching their prey from the ground, and, with a flip of the neck, swallowing the struggling mammal just about whole. At that time, the distant ancestor of horses, *Hyracotherium*, was about the size of a terrier dog, so it would have been within the dietary range of some of the predatory birds.

This sort of short-lived experiment is typical of recoveries. The first species to become established may have a good evolutionary run for their money. New species can proliferate at a faster-than-normal rate as vacated niches fill up, and as ecosystems reconstruct themselves. But some of the first-comers may not be able to cling on to their positions, if, for instance, they are not as well adapted to their roles as other species that come in and take over. In the end, the giant terror birds gave way in most places to mammalian predators, the ancestors of cats and dogs. So, during a post-extinction rebound there can be a phase of rapid niche-filling, and then comes the inevitable sorting and stabilization phase, when many extinctions happen and ecosystems readjust to a pattern that may then hold sway for 50 million years or more of relative stability.

13

THE SIXTH MASS EXTINCTION?

As long ago as 1992, Al Gore, then the Vice-President of the United States of America, wrote,

> . . . it occurred to me that . . . we are causing 100 extinctions each day – and many scientists believe we are . . .[1]

This is a startling figure, and the prediction resulting from the calculations quoted by Gore is that all of life will be extinct in 400 to 800 years. Do we believe this? What is the scientific basis for such dramatic predictions? Or should we settle back comfortably with the extreme Bible-belt Americans who think that everything on Earth was created by God for the benefit of humanity, and therefore that anything done by human beings is by definition good?

Probably both positions are gross caricatures, and it would be sensible to be cautious. Al Gore based his statement on the reputable calculations of Paul and Anne Ehrlich in 1990 that perhaps 70–150 species are becoming extinct each day.[2] Scaling this up to current estimates of total global diversity leads to the above alarming prediction of how long life will last on the Earth. Perhaps the daily extinction estimate is too high (see below), but such calculations always lead to startling conclusions about the total time to extinction of all life.

If human activities truly are causing such devastation, then we are witnessing the sixth mass extinction (the other five being the geologically documented 'big five'). In this case, a close understanding of the events of the past will shed light on what is happening now, and what may happen in the future. In particular, people often ask, how long does it take for life to recover after the devastation of a mass extinction? As we saw in Chapter 12, this can vary. Recovery after the KPg event that eliminated the dinosaurs took perhaps 5–10 million years, but after the end-Permian event may have taken as long as 10–20 million years.

But what about Al Gore's alarming suggestions? Are such estimates reasonable?

Modern diversity: 2 million species?

Current estimates of biodiversity, the number of species on Earth, range from 2 million to 100 million. It seems incredible that we have so little idea about how diverse life is today.

The low-end figure of 2 million, quoted still – though rarely – by some conservatives, comes from the most solid evidence of all, a count of every named species. By the year 2015, some 1.9 million species had been named in scientific literature, the majority of them, about 1 million as it happens, being insects. It is hard to give an exact total, since new species of plants and animals may be described in any of the thousands of scientific journals of the world, including publications of museums and institutions small and large. There is, as yet, no central repository of such information, although there are several schemes underway on the internet to provide a reliable, scientifically vetted listing of species names.

Do we stick to 1.9 million as our estimate of current biodiversity? Yes and no. The number is obviously far too low for one reason, and rather too high for another. Too high because of synonymy. Synonyms in biology, as in common speech, are different terms for the same entity. Despite all their best efforts, biologists are only human and they make mistakes. Their enthusiasm for finding new species means that they may mistakenly invent a name for something that has already been named by someone else. This may happen even when they know of all previous work in their field. It may be because an earlier name was given in some particularly obscure journal, or the original description might have been incomplete, poorly illustrated or misleading. Be that as it may, this leads to a constant rate of synonymy, the renaming of species that have already been named. Fortunately, scientists are critical folk, and they delight in identifying the errors of others. Sooner or later, synonyms are rooted out and exposed. That is a foundation of science: all published work is open to scrutiny and checking.

More important in many ways than merely naming new species is the work of systematists, the biologists who study the relationships and diversity of life, constantly revising and re-examining whole groups. They might decide to take in hand a family of orchids or water beetles. In doing so, they travel the world, look at type specimens (the individual examples selected to represent a new species when it was named, sometimes called the name-bearer, and preserved in a museum for later study), and produce comprehensive schemes of characters and relationships. In doing so they may identify synonyms, which they formally subsume under the names originally given. The global rate of synonymy is about 20%, and it always has been. In other words, about one-fifth of new species announced to the world by their proud parent will inevitably be deleted on later revision. Our figure of 1.9 million named living species then falls to about 1.5 million.

This, though, is almost certainly a gross underestimate. New species are being found all the time, and even though one-fifth of them may be spurious, four-fifths are not. How many new species? If you read the serious newspapers, new species of birds or mammals make headlines. A new mammal or bird is discovered only quite rarely, maybe one or two a year, and usually after diligent exploration in remote parts of the world. This might suggest that we really know pretty well all birds and mammals alive today, some 10,000 and 5500 species respectively. The story is different for insects though: entomologists come up with a regular 7000–8000

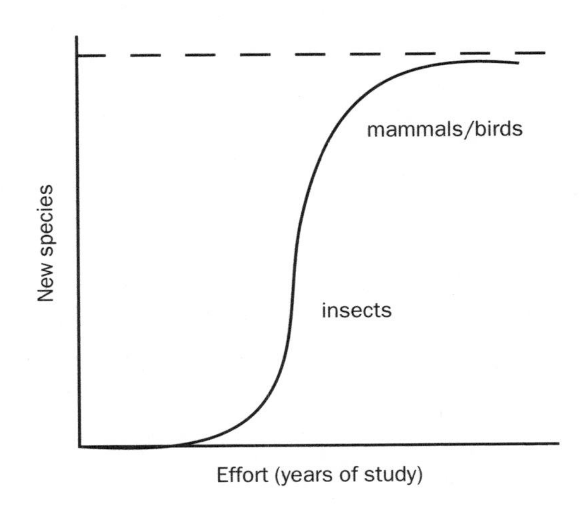

46 *A collector curve for the naming of species. Systematists of birds and mammals have just about found all the species that exist. But entomological systematists are trailing badly. This is not their fault, since insects are thousands of time more diverse than birds and mammals. At full steam, entomologists are describing 8000 or so new species of insects each year, and yet they seem to be no nearer finishing than they were 100 years ago.*

new species each year, and the only limitation on this rate of discovery seems to be the number of entomologists available and their rate of work. Entomologists I know feel that their situation is desperate: if they could only work another hour per day, they could identify another dozen new species a year. Human frailty holds them back, and they lose sleep over it. They've named a million species so far – but there is no sign of an end to their necessary labours.

These contrasts can be appreciated from a collector curve. The collector curve is a well-established way of answering the question: 'When should I stop collecting?' It is simply a graph of the number of new species discovered plotted against effort. In the classic case, a biologist is sent to some remote place and is asked to compile an inventory of species. How long should she take? If she collects for a day, every plant or bug she picks up will be something new to her. After a week, she knows all the common species, and only rarely finds a new form. The law of diminishing returns sets in. Plotted as a graph (Fig. 46), this shows that the rate of discovery is high at first, and then rapidly tails off. The right-hand part of the line is tending towards a final figure, which it may never reach (this is technically called an asymptote), but it is clear roughly what that total is. If this collector curve then represents the worldwide efforts of systematists in identifying new species, mammalogists and ornithologists are well up on the asymptote, and they can predict with confidence that they know pretty well all there is to know. Entomologists, on the other hand, are still struggling up the steep part of the discovery curve, and it is impossible to predict when the rate of discovery will slow down. A rough estimate from the collector curve of all organisms might suggest that the total number of species already described and to be described will reach 5 million.

Insects are all very well. We see them around us, and they are easy to collect. But what about microbes – all the bacteria, viruses, algae, protozoans and other micro-organisms we can't see? Microbiologists just smile when entomologists announce their problems. They haven't even begun to scratch the surface of current microbial diversity. Similarly amused are the systematists who specialize in deep-sea organisms. They can only study the greyish mangled remains hauled up from the depths in trawls and sediment punches. Their task is like that of an extraterrestrial who swoops high over the Earth, drops a sample grab down from time to time, and hauls back into his spaceship whatever has been encountered.

What about the meiofauna too, I hear you ask? Indeed. The nearly microscopic, though not quite transparent, soft-bodied bugs and worm-like creatures that live between sand grains and among the soil particles, have barely been studied. No one really knew they even existed until the 1960s, nor really cared, and only a small number of systematists study them today. Perhaps a global inventory of named species is not a valid estimate of total global biodiversity. Another approach might be to adopt a sampling strategy.

Bug guns: 100 million species?

A sampling strategy is just what Terry Erwin applied in the 1970s. In two well-known papers[3] he outlined his method. He chose to look at tropical rainforest beetles, to estimate their biodiversity in a small part of the world, and then to extrapolate from that to the whole world. He sampled the entire arthropod fauna – all the insects, centipedes, millipedes, spiders and their relatives – from the canopy of the tree *Luehea seemannii* from Central and South America. This was done by setting 'bug bombs' under the selected trees, devices that pump powerful insecticide upwards in clouds. When these had gone off, Erwin collected all the dead arthropods that fell to the ground, and he sorted them into species, many of them, as it happened, never seen before.

Erwin estimated that there are 163 species of beetles living exclusively in the canopies of *Luehea seemannii.* There are about 50,000 tropical tree species around the world, and if the number of endemic beetle species in *Luehea seemannii* is typical, this implies a total of 8.15 million canopy-dwelling tropical beetle species in all. This figure excludes forms that live in several tree species. Beetles typically represent about 40% of all arthropod species, and this leads to an estimate of about 20 million tropical canopy-living arthropod species. In tropical areas, there are typically twice as many arthropods in the canopy as on the ground, giving an estimate of 30 million species of tropical arthropods worldwide.

Thirty million species of tropical arthropods? If this were true then that would imply a total diversity of life somewhere between 50 and 100 million species. Some wild-eyed biologists even talked of figures of more than 100 million! Work published in 2002, has, however, challenged some of Erwin's assumptions.

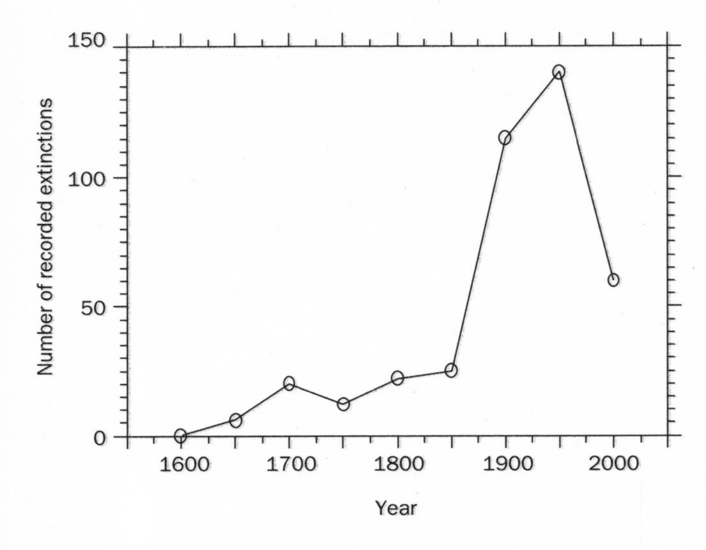

47 *The rate of historical extinctions of species for which information exists. Rates of loss mount in the twentieth century. The apparent drop from 1950 to 2000 is not an indication of an improvement: the figures were recorded in 1990, and the full tally of species lost up to 2000 has not yet been calculated.*

Mature reflection[4] has suggested that Erwin's estimate of total global biodiversity was too high and should be revised down to, say, 10–20 million species, a widely accepted figure. Similar extrapolation exercises have been performed for deep-sea organisms, microbes, fungi and parasites, and they all point to total global biodiversities of 10–20 million. Such estimates are astounding, and they have profound consequences. Should systematists give up the endeavour to describe and name all species since they will never finish the task? Should governments employ many more systematists in order to do the work properly? How can conservationists and planners begin to estimate the effects of pollution and other human activities on biodiversity, since no one has the faintest idea how many species exist today, nor what they are, and where they are?

Modern extinction rates: 70 species extinct per day?

The present rate of extinction can be calculated for some groups from historical records. For birds and mammals, groups that have always been heavily studied, the exact date of extinction of many species is known from such records – many of them, shamefully, not so historical. The last dodo was seen on Mauritius in 1681. By 1693, it was gone, prey to passing sailors who valued its flesh, despite the fact that it was 'hard and greasie'. The last two great auks were collected in the North Atlantic in 1844 – ironically, this pair was bludgeoned to death on Eldey Island off Iceland by natural history collectors. Some sightings were reported in 1852, but these were not confirmed.

Human activity has not caused the extinction only of rare or isolated birds. The last Passenger pigeon, named Martha, died at Cincinnati Zoo in 1914. Only one hundred years earlier, the great ornithologist John James Audubon, reported a flock of Passenger pigeons in Kentucky that took three days to go by. He estimated that the birds passed him at the rate of 1000 million in three hours. The sky was black with them in all directions. They were wiped out by a programme of systematic shooting, which, at its height, covered the landscape with their carcasses as far as the eye could see.

These datable extinctions can be plotted (Fig. 47) to show the rates of extinction of birds, mammals and some other groups in historical time.[5] Data on historical extinctions can give some guidance on overall figures of species loss today, although care has to be taken in extrapolating from a few, well studied groups to all of life. For example, some calculations published in 2014 confirm the wide range of ignorance we have about modern biodiversity, and modern extinction rates. With current estimates of diversity ranging from 2–50 million species, and estimates of extinction rates from 0.01–0.7% per year, we could be experiencing losses of between 500 and 36,000 species per year (which translate as 2–100 per day). But how does this compare to what we would expect under normal conditions? A simple calculation might be that, if there are, say, 20 million species on Earth, and a typical species lasts for 1 million years, we would expect two species to become extinct each year. So, the calculated extinction rates would suggest we are losing species much faster than expected, anything from 250 to 18,000 times the background rate without human intervention. Sobering indeed.

Even if all of life does not go extinct, and it seems likely that some species would be tenacious, or lucky, enough, to escape whatever depredations and environmental catastrophes may be meted out to them by human activities, can we foresee a recovery phase? We know that life bounced back after the previous mass extinctions, but, as we saw in Chapter 12, that recovery phase can last for many millions of years – far longer than the potential survival of any human beings. So we would not see the renaissance of life.

Panic or complacency?

Lessons from the past can be read in two ways. The ecological activist would emphasize how human activities are destroying biodiversity and how this could turn into a cascade of death, as species after species becomes extinct. As tropical forests are cleared and reefs are poisoned, we are losing not only species, but whole habitats. The palaeontological record of mass extinctions then makes grim comparisons. We know that after a mass extinction, life takes a long time to recover. The geologist may say that 10 million years is a short time, but measured in human lifetimes, it is effectively infinity.

Low levels of extinction can turn into high levels. Destroying species and habitats piecemeal might lead to a runaway crisis, as seems to have happened in the past. Once the world becomes locked into a spiral of downward decline, it is impossible to see how any intervention by humans could turn it back. It could be, for example, that removing one or two species from an ecosystem does little damage. The remaining species can adapt and plug the gaps. But if another few species are picked off, then another few, and then a few more, a point may be reached when that ecosystem will collapse. Better to stop destroying the environment before we become locked into such a catastrophic sequence of events. The natural world is complex, and consequences are often unpredictable.

Destruction of forests can kill ocean fish, for example. Plants take up carbon dioxide during photosynthesis and pump out oxygen. Animals require oxygen, but produce carbon dioxide as a waste product. There is a balance here, and that balance could be perturbed by destroying too much of the world's forests. Cycling of carbon is important too, as dead plants and animals are incorporated into the soil, as organic carbon builds up in the bottoms of lakes or is washed into the sea. Nutrients from these sources then circulate in the oceans, providing sustenance for fishes.

A political conservative could also claim justification from the fossil record. Such a person would note that life has always bounced back, even from a mass extinction as profound as the end-Permian event. Evidently, each species locks up a huge evolutionary potential in its genes and, given the chance to explore the full extent of their capabilities, most species seem to be able to proliferate and expand into new niches. Indeed, the conservative, warming to the theme, might suggest that species that have been killed by human intervention were obviously rather feeble, and a bit of extinction is good for the moral fibre. Who needs dodos and great auks anyway?

This conservative viewpoint has gained ground in some political circles. Bjorn Lomborg,[6] a Danish statistician and one-time green campaigner, argues that world resources are not running out, that forest cover across the world has increased and that the world's species are not disappearing at an alarming rate. His views have inevitably been greeted with outrage by many, who claim that he has selected narrow definitions of natural phenomena in order to make his case: farmed, temperate-climate tree nurseries differ from ancient, complex tropical forests. It is startling none the less that it is still difficult to make definitive and universally convincing statements about the state of the natural world today.

Coming back to reality

Much as one might wish to accept such reassuring claims, they are too complacent. Of course some life will survive human depredations. It may be cockroaches or rats, but to claim that humans cannot drive all life to extinction is hardly cause for congratulation.

There are lessons to be learnt from the past. Human activities have done more than eliminate just one or two species here and there. The Maoris in New Zealand killed all the moas, some 13 or more species of impressive, large, flightless birds, thus eliminating an entire family of birds, the Dinornithidae, some time before 1775. The Maoris also killed off other, smaller families of native birds. Similarly, after Europeans arrived on the Hawaiian islands in 1788, 18 bird species disappeared, and another 12 may also be extinct. Of 980 species of native Hawaiian plants, 84 have already been eliminated, and a further 133 have wild populations numbering fewer than 100 individuals.

The famous 'red books' of the International Union for the Conservation of Nature classify different levels of threat to present species. Thousands of species are listed as in danger of extinction, and the lists become longer and longer each time they are revised. Species under threat include much-loved forms such as pandas, tigers and blue whales. Millions of dollars are spent by governments and charities in order to try to conserve such species. Special breeding programmes in zoos help the effort, and sometimes – rarely – these huge efforts allow a species to come off the endangered list. But at what cost? We can eliminate a species in a moment, but conservation is expensive. And of course, while people will pay for the rescue of the panda, the California condor, even the Kerry slug, who will pay for the protection of the countless uninteresting beetles, bugs, scorpions, frogs, snakes and tropical plants that are just as close to extinction? And what of the threatened species we don't even know about?

The extinction estimates quoted by Al Gore have a firm basis in fact. The only substantial reason to question them is that there may be levels of extinction resistance among species. What we have mentioned so far is the disappearance, to a large extent, of extinction-prone forms, endemics restricted to single islands for example. If there is such a sliding scale, and we are busily eliminating the more precarious species, perhaps rates of extinction will decline as humans tackle the more recalcitrant species, the ones that refuse just to give in and die.

On the other hand, human populations are increasing exponentially. The time it takes for the human population to double keeps diminishing. Global populations rose from 100 million to 200 million between the time of Christ and 1500. The 400 million mark was achieved by 1700, 800 million by 1800, 1600 million by 1900, 3200 million by 1980, and it reached 7000 million (7 billion) in 2012, so the doubling time is down to 25 years. At today's levels of human population, some 40% of global productivity has been sequestered for our benefit – including all humans and their domesticated plants and animals. This means that all other species have somehow to get by on only 60% of the oxygen and carbon (though in reality it is more than 60% since human and domestic waste goes into the 'wild' systems) that they had available to them in the days of Julius Caesar.

And even though the exponential rise in global human population is damped, or slowed down, by famines and wars, the rate continues to go up. This brings a pressure that might cancel out any tendency to reduction of current extinction rates. So, argue many, the debate about extinction-prone and extinction-resistant species is irrelevant. They are all going to go anyway, as wealthy nations pump pollutants into the atmosphere, and poorer peoples replace natural habitats with poor-quality farm land. There is no room for complacency.

Unanswered questions

Ironically, extinctions in the distant past are better understood than the current crisis. Reversing that well-worn maxim, it may be that 'the past is the key to the present'. Normally, geologists and palaeontologists bow humbly before scientists who work on modern phenomena. To understand how rivers worked in the Permo-Triassic of the Karoo Basin, the geologist seeks advice from geographers and geomorphologists who study modern river systems. To understand what *Archaeopteryx* looked like, the palaeontologist consults an ornithologist. In extinction studies, palaeontology has (some of) the answers.

As we saw earlier, biologists have failed to gauge current biodiversity, with estimates ranging from 2 million to 100 million species. Biologists have also failed to approximate current extinction rates, and there are robust debates around the figures quoted at the beginning of the chapter by Al Gore. Palaeontologists, on the other hand, can give good estimates of extinction rates, certainly at family and generic levels, and they have relatively reliable ways of turning those into estimates of species extinction rates. So we know how profound the end-Permian crisis was and the scale of the KPg event. We know also how long the recovery took, since timescales were lengthy.

Palaeontologists, of course, become more hesitant when pressed about how long any particular extinction crisis lasted. Their weakness is short timescales, where the error bars on age estimates may exceed the time intervals in question. So, no one can say whether the end-Permian crisis lasted for one day or a few thousand years.

Recently, conservation biologists have been learning some practical methods from palaeontologists. They have realized that it is futile to try to make a complete inventory of all species on the Earth. If it has taken since 1758 to document the 1.9 million named species, and if there are currently 10–20 million species, it will clearly take another 2000 years of hard work to describe and name the remainder. If there are 100 million species on Earth, the task will take 14,000 years. Of course, the political conservative might argue that if we drive more and more species to extinction, these estimates come down, and governments can save money on the salaries of systematists. In view of the futility of attempting to make accurate species counts of world diversity, conservation biologists are experimenting with counts of genera or families, and then using rarefaction to extrapolate what these figures mean at the level of species, just as David Raup did when he explored the scale of the end-Permian crisis.

We have focused here on the worst crisis of all, and it is one of a number of profoundly unpleasant times when life has come to the brink. Not explored here have been those times of apparent crisis when not much happened to life. There have, for example, been numerous vast flood basalt eruptions when nothing really became extinct, except, presumably, those unfortunate organisms that were caught up in the flow of the lava. There have also been many impacts of meteorites and comets on the Earth that have not led to extinction. Some of these impacts were indeed nearly as large as that which caused the Chicxulub crater, and yet the palaeontological record passes them by with not a hint of any elevated extinction rates. These are mysteries, yet to be resolved.

Palaeontologists and geologists are beginning to identify common aspects of mass extinctions. They are looking for species that are more prone to extinction than others. They are now taking more interest in the post-extinction recovery period. The scope and timing of the rebound seems to depend on the magnitude of the preceding extinction. But, perhaps unexpectedly, there is often a long lag period when evolution seems to have been suspended, and this was particularly true for the post-Permian recovery. It is important to identify disaster species, the forms that are able to radiate soon after a crisis. What special features, if any, do they have that allow them to survive, and to re-radiate before anything else can? Or are they just lucky? What of the species that finally rebuild normal ecosystems, and replace the disaster species?

I have tried to show in this book, by weaving history and science together, how arguments are often re-run generation after generation. Scientists are evidently human too. They can be prejudiced, they can be scared or constrained; the concept of catastrophic extinctions in the geological past is a beautiful example of this. Presented in the 1820s, but firmly crushed by Lyell in the 1830s, the idea could only raise its dangerous head again in the 1980s. It took 150 years for geologists to dare to accept the obvious, that there truly have been mass extinctions in the past and that structures on the Earth's surface that look like impact craters actually are impact craters.

To have lived through the tail end of this switch-over has been fascinating. I was taught by anti-catastrophists, but I now preach asteroids and mass extinctions to my students. Even more rapid has been the accumulation of knowledge about the end-Permian event. Everything changed between 1992 and 2015. In that time, the event window focused down from 10 million years to a few thousand years. The Siberian Traps flowed into view as the main culprit. Gas hydrates, undreamt of before the 1970s, suddenly became the answer to abrupt climate changes in the past, and perhaps crucial as part of a runaway greenhouse model for the end-Permian environmental crash. Long drawn-out post-apocalyptic anoxia is everywhere in the Early Triassic.

As one grows older, one realizes how little one knows: 'the more you learn, the more ignorant you become'. The joy of being a scientist is to discover this. When I was beginning my career, I felt that scientific research was a line of work that led to ever greater complexity. As one accumulated information about how the Earth works, all the simple questions would be answered. Then the questions would have to become more intricate and harder to solve.

But the unanswered questions are as big and as simple as you could wish for (although the answers may be so intricate as to be unattainable). How diverse is life? How does the world react to human intervention? What will happen in the next 100 years? Where did life come from? How resilient is life to crisis?

As Georges Cuvier wrote in 1812, at the very birth of geology, in his *Revolutions de la globe*:[7]

> The ancient history of the earth, the ultimate goal towards which all this research is leading, is in itself one of the most fascinating subjects on which the attention of enlightened persons can be fixed. If they take an interest in following, in the infancy of our own species, the almost erased traces of so many extinct nations, they will doubtless find it also in gathering, in the darkness of the earth's infancy, the traces of revolutions previous to the existence of every nation.

GLOSSARY

Ammonites: extinct *cephalopods* with coiled shells which lived abundantly as free-swimming carnivores in Jurassic and Cretaceous seas.

Anapsid: a reptile with no temporal openings in the skull; includes turtles and several extinct groups.

Angiosperm: a flowering plant.

Anoxia: lack of oxygen. The term usually refers to conditions on the sea bed when oxygen is virtually, or completely, absent.

Belemnites: extinct *cephalopods* with straight, bullet-shaped internal guards.

Biodiversity: the *diversity* of life.

Biostratigraphy: the use of fossils to establish the relative sequence of rocks, and to match rocks of similar age in different parts of the world.

Brachiopod: a 'lamp shell'; member of the Phylum Brachiopoda, a group of two-shelled animals that filter feed and generally live attached to the sea floor. Formerly a dominant group, now rare.

Cephalopod: a member of the Cephalopoda, the group of molluscs that includes octopus, squid and cuttlefish, as well as the pearly nautilus and the extinct *ammonites* and *belemnites*. All have big goggly eyes and they are reputed to be more intelligent than other invertebrates.

Clade: a taxonomic group that had a single ancestor and which includes all descendants of that ancestor. Clades can be species, genera, families, orders, phyla or any other formally named group.

Cladistics: the method of establishing phylogenies by identifying clades on the basis of unique shared characters.

Conodont: a toothed phosphatic fossil, probably the jaw parts of an extinct group of vertebrates.

Correlation: matching rock ages from place to place within a local basin of deposition, or globally.

Diapsid: a reptile with two temporal openings in the skull; includes lizards, snakes, crocodiles, birds and several extinct groups.

Dicynodont: 'two-canine-teeth'; a member of the group of mammal-like reptiles that were dominant plant-eaters in the Late Permian, and survived, in the form of *Lystrosaurus*, through the end-Permian mass extinction, to flourish again in the Triassic.

Dinocephalian: member of an early group of mammal-like reptiles that included plant- and flesh-eaters, and that dominated in the earlier parts of the Late Permian.

Diversity: the richness of life, usually measured as the number of species, genera or families, in a restricted area, or worldwide. Other definitions can include measures of abundance, ecological range or genetic range.

Echinoderm: 'leathery skin', member of the Phylum Echinodermata, which includes all the invertebrates with calcite plates and a five-rayed anatomy – the starfish, sea urchins and sea cucumbers.

Ecosystem: the physical and biological components that make up a life environment; the environment plus the organisms that live in it.

Endemic: local in distribution.

Eocene: the epoch lasting from 55–34 myr.

Fauna: the sum total of animals in a defined place and time.

Flora: the sum total of plants in a defined place and time.

Foraminifera: the group of microscopic, single-celled animals with calcareous shells that lived on the seabed and in the plankton, and which are useful in dating rocks.

Fusilinids: a group of *radiolarians*, with a delicate skeleton made from silica; microscopic animals that mainly float in the plankton.

Gastropod: a snail, conch or limpet; member of the group of molluscs that are equipped with a single valve ('shell'), and the shell is often coiled or spired.

Goniatite: member of a *cephalopod* group that dominated oceans through the Devonian to Permian, often with a coiled shell.

Gymnosperm: a member of the plant group that includes conifers and smaller groups such as cycads and ginkgos.

Igneous rocks: rocks that formed from molten magma, either deep within the Earth's crust, or on the surface. Surface igneous rocks typically form from volcanic lavas.

Magnetostratigraphy: the division of geologic time using measurements of the Earth's magnetization as preserved in the rocks. Numerous phases of 'normal' (as at present) and 'reversed' (poles reversed) magnetization may be distinguished, and used for correlation.

Metamorphic rocks: *sedimentary* or *igneous* rocks that have been modified by heat or pressure after deep burial in the Earth's crust.

Miocene: the epoch lasting from 24–5 myr.

Niche: the role and ecological attributes of a species, its diet, its interactions with other species and its range of environmental requirements.

Ostracod: a small arthropod, a relative of crabs and shrimps, that lives inside a two-valved shell, and swims on its back, filter feeding using its legs.

Pecten: a bivalve mollusc with a triangular-shaped shell, often bearing distinct radiating ridges; the symbol of the Shell Petroleum Company.

Pelycosaurs: basal mammal-like reptiles that were present in the Late Carboniferous and Early Permian; includes famously *Dimetrodon* with the sail on its back.

Phylogeny: the shape of evolutionary trees, or an individual evolutionary tree.

Radiation: expansion or diversification; a time when a group speciates and splits rapidly, perhaps as a result of some new opportunity (e.g. after a mass extinction) or as a result of some new adaptation (e.g. the explosion of mice and their relatives in the past 15 myr.).

Radiolaria: silica-shelled microscopic planktonic animals; includes the *fusilinids*.

Radiometric dating: the establishment of exact ages by study of unstable radioactive elements and a comparison of proportions of parent and daughter elements where the half-life of breakdown is known.

Sedimentary rocks: rocks that have formed from sediments, such as mud, sand or lime mud, forming sandstone, mudstone or limestone, respectively.

Stratigraphy: study of the sequence and dating of rocks.

Synapsid: a reptile with only a lower temporal opening; includes mammal-like reptiles and mammals.

Taphonomy: the processes that affect fossils between the death of an organism and the discovery of the fossil. Taphonomic processes include predation and scavenging damage to a carcass, decay processes, compression and distortion during burial, and physical and chemical alterations during and after burial.

Taxonomy: the science of classification; the division of the diversity of life into manageable subgroupings, from species upwards.

Tetrapod: 'four-footer'; Tetrapoda collectively includes all the land vertebrates – amphibians, reptiles, birds and mammals.

Therapsids: the more derived mammal-like reptiles, following after the pelycosaurs; includes all the Late Permian and Triassic synapsids.

Therocephalians: a group of small, insect- and flesh-eating mammal-like reptiles, particularly from the Late Permian and Early Triassic.

Trilobites: extinct marine arthropods, with three-lobed bodies and many pairs of limbs, which dominated early Palaeozoic faunas as carnivores and detritivores.

NOTES

Prologue (pp. 7–17)

1. The quotations are from the website http://www.space.com/scienceastronomy/planetearth/impact_extinction_010222.html. This website, like so many others, has been superseded.

1. Antediluvian Sauria (pp. 18–35)

1. The quotation is from Owen (1845b, p. 638).
2. There are many accounts of the evolution of amphibians and reptiles, including R. L. Carroll (1988) and Benton (2015).
3. The original descriptions of the Yorkshire crocodile are to be found in Chapman (1758) and Wooler (1758). The tale is told evocatively by Osborne (1998).
4. There is a great deal of documentary evidence about the American mastodon debate, on both sides of the Atlantic. The story is told in considerable detail by Greene (1961, chapter 4), and in more compact form by Rudwick (1976), Durant and Rolfe (1984), and Buffetaut (1987).
5. There are many biographies of Cuvier. One of the best is by Outram (1984), who explores Cuvier's life, and the mix of scientific endeavour and administrative power throughout his career. Rudwick (1997) focuses on his scientific work, and he presents authoritative translations of many key works, with incisive commentary.
6. The early history of discoveries of Permian and Triassic fossil amphibians and reptiles in Russia is told by Ochev and Surkov (2000).
7. Excellent accounts of the early discoveries of dinosaurs in England may be found in Colbert (1968) and Cadbury (2000).
8. Biographical details about Richard Owen and his work on dinosaurs may be found in books by Rupke (1994) and Cadbury (2000).
9. The published report by Owen on the terrestrial reptiles is Owen (1842); he established the Dinosauria on page 103.
10. The history of studies of the Permo-Triassic amphibians and reptiles of Russia is summarized by Ochev and Surkov (2000), and they give fuller references to the works of Kutorga, von Qualen, von Waldheim and von Eichwald.

2. Murchison Names the Permian (pp. 36–55)

1. There is no single, comprehensive biography of Roderick Impey Murchison. The *Life* by Geikie (1875) and the more recent biography by Stafford (1989) cover most aspects of his life, but the first is perhaps over-generous to its subject, the second under-generous. Full accounts of Murchison's work and his character are offered in a series of recent books that explore some of the particular controversies in which he was engaged, for example, Rudwick (1985) on the great Devonian controversy, Secord (1986) on the Cambrian-Silurian dispute, Oldroyd (1990) on the Highlands controversy, and Collie and Diemer (1995) on Murchison's work in northeast Scotland.
2. Murchison named the Permian system in the paper published in 1841, and he discussed this further in his Presidential Address to the Geological Society of London in February 1842 (Murchison 1841a, b, 1842a, b). Murchison's entire narrative of his Russian tours in 1840 and 1841 has been transcribed and published by Collie and Diemer (2007). I present some additional observations on his first meeting the Permian redbeds in Benton et al. (2010).
3. Murchison established the Silurian System in his book *The Silurian System* (Murchison 1839).

4. Murchison and Sedgwick established the Devonian System formally in a short paper in 1839 (Sedgwick and Murchison 1839).

5. Murchison's quoted comments on the Permian system, and on Herr Ermann's supposed slur on his originality, come from Murchison (1842b, pp. 665–666).

6. More detail of the Geological Society of London, its purpose and structure, are given by Morrell and Thackray (1981) and Rudwick (1985).

7. The quotation by Murchison about his views on the division of the geological timescale is in his Presidential Address of 1842 (Murchison 1842b, pp. 648–649). Italics are as in the original.

8. The history of the division of the geological timescale, including William Smith's contribution, has been told many times by, for example, Zittel (1901), Hallam (1983) and Rudwick (1976, 1985). William Smith's story is told delightfully by Winchester (2001).

9. The work by John Phillips is outlined by Zittel (1901), Rudwick (1985) and by D. H. Erwin (1993). Phillips published his stratigraphic ideas in many places, including Phillips (1838, 1840a, b, 1841), and his famous diagram of the diversity of life through time is from Phillips (1860).

10. Sedgwick and Murchison described their geologizing trip to Belgium and Germany in Sedgwick and Murchison (1840).

11. Published accounts of the first trip to Russia by Murchison are given in Murchison and Verneuil (1841a, b).

12. The earliest account of Russian geology written in English was by Strangways (1822).

13. The second trip to Russia was described by Murchison (1841a, b, 1842a, b), by Murchison and Verneuil (1842), and by Murchison et al. (1842) on the Ural Mountains.

14. The justification for the biostratigraphic distinctiveness of the Permian is given by Murchison (1841a, p. 419).

15. British Museum (Natural History), Richard Owen correspondence, 20 November 1841.

16. Full details of the background to the publication of the *Geology of Russia* (Murchison et al., 1845), the geological problems, printing costs and negotiations with the Russian government, are given by Thackray (1978).

3. The Death of Catastrophism (pp. 56–70)

1. Lyell's textbook is Lyell (1830–33). The quotation is from volume 1, p. 123.

2. The catastrophism vs. uniformitarianism debate is outlined by Hallam (1983, chapter 2) from a modern viewpoint.

3. The 1810 report by Cuvier is quoted in full in a new translation by Rudwick (1997, pp. 115–126). The two quotations are from pages 124 and 126 respectively.

4. The *Ossemens fossiles* was published as Cuvier (1812), as a much revised edition in 1821–24, and reissued as a third, only slightly revised edition in 1825. The quotations are from Rudwick's (1997, chapter 15) translation, from pages 188–189 and 206–207 (Cuvier's sections 5 and 27 respectively, in his 1812 edition). This preliminary discourse was also issued as a separate publication in four editions in English from 1817 to 1827, in German, and in a separate French (third) edition in 1826.

5. By 1842 Murchison would have hated to have been classed as a Cuvierian, and yet he clearly was. In his Presidential Address, he argued for the primacy of vertebrates over invertebrates in stratigraphy in his dispute with John Phillips. That seems very odd to a modern geologist, who would generally hold quite the opposite opinion. But Cuvier, himself a vertebrate palaeontologist, supplied lengthy arguments in favour of the importance of fossil vertebrates, and the ways in which they would be more subject to catastrophes

than marine invertebrates. Of course Murchison, as a young man in the mid-1820s, was seeking to inculcate himself in everything geological, and he would surely have turned first to Cuvier for inspiration: his *Recherches sur les ossemens fossiles* had just been issued in a second (1821–24) and third (1825) edition, which Murchison might well have read in the original, or in one of the new English translations which had also just appeared. This was before Lyell had begun to publish his *Principles* in 1830, essentially a counterblast to Cuvier. Murchison might have sought to deny it later, but the early training via Monsieur Cuvier had stuck in the back of his mind, despite all efforts to expunge it. And, indeed, why expunge it?

6. The quotation is from Cuvier (1825, vol. 1, pp. 8–9), as translated by Gillispie (1951).

7. The comment on Lyell's advocacy is from Sedgwick (1831). The strategy of Lyell's *Principles* is discussed in detail by Rudwick (1969), Gould (1987, chapter 4) and Hallam (1983, chapter 2).

8. Fitton's comments on advocacy are in a review by Fitton (1839).

9. Lyell's mixing of four meanings of 'uniformity' has been exposed in detail by several historians of science, including Rudwick (1969, 1976, chapter 4), Hallam (1983, chapter 2) and Gould (1987, chapter 4). Astonishingly, most historians had gone along with Lyell's confusions up to that time, and this had cemented Lyell's own presentation of the history of geology as a tension between catastrophists, who were in all regards misguided, and uniformitarians, who were entirely sensible and correct.

10. See chapters in Rudwick (1997) for full translations, and commentaries, on Cuvier's views about geology.

11. Lyell's non-progressionism is discussed by Bowler (1976). Benton (1982) shows how late Lyell continued his campaign. In the 1850s, he was eagerly seeking field evidence for his views, and documented supposed Devonian lizards from Scotland and supposed Silurian tetrapod tracks from North America in his *Elements of Geology* (1838).

12. Professor Ichthyosaurus was reinterpreted as an attack on Lyell by Rudwick (1975), and the case is further discussed by Gould (1987, chapter 4).

13. The quotations from unpublished notebooks by J. D. Forbes are given by Rudwick (1985, p. 74).

14. The letters are quoted by Geikie (1875, vol. 2, pp. 119–121). The comments on 'Lyell's quietude' were to Mr J. P. Martin. Murchison offered strong opposition, in letters, to Lyell's championing of the Silurian vertebrate tracks from North America, and the tracks and lizard skeleton from the supposed Old Red Sandstone of Elgin (see note 11). To Lyell he asserted, 'Hitherto, we know absolutely nothing of land animals in the Silurian world'. On the supposed Devonian lizard from Elgin, Murchison wrote to Sedgwick in 1851, 'And he is to wag his tail next meeting, to the infinite delight of Lyell, who is inebriate with joy . . . I hold . . . that such proofs of the inhabitants of the land and fresh water of those early days can have no influence in changing our general argument founded on marine succession.'

4. The Concept That Dared Not Speak its Name (pp. 71–95)

1. Bob Carroll's statement about the apparent absence of an end-Permian mass extinction of vertebrates is in R. L. Carroll (1988, p. 589).

2. He later accepted the reality of the end-Permian mass extinction among vertebrates (R. L. Carroll, 1997, p. 383).

3. Owen's speculations about Mesozoic conditions and about the extinction of the dinosaurs are in Owen (1842, p. 203).

4. The quotations are from the third edition of the *Origin of species* (Darwin, 1861, p. 348).

5. Huxley's account of dinosaurian diversity is Huxley (1870).

6. Marsh's reviews of dinosaurian diversity are Marsh (1882) and (1895).

7. Buckland's *Reliquiae Diluvianae* ('Relics of the Flood'; 1822) gives a full account of the findings in Kirkdale Cavern, and of his view that the Pleistocene extinction of mammals had been caused by the biblical Flood. Victorian views of Pleistocene extinctions are discussed in detail by Grayson (1984).

8. The quotation is from Agassiz's most famous work on the ice age (Agassiz 1840, p. 314; translation by the author).

9. Boucher de Perthes published numerous accounts of human remains associated with giant Pleistocene mammals from France (Boucher de Perthes 1847, 1857, 1864). Full details are given in the article by Grayson (1984).

10. Orthogenesis ('straight generation') and finalism were versions of a viewpoint that saw evolution as directed along pre-ordained paths. So, much of evolution could be said to have been directed towards the origin of human beings, as the crowning achievement. Bowler (1983) gives a detailed account of the history of such thought in late Victorian times and in the early twentieth century.

11. The quotations are from Woodward (1898, pp. 213 and 418).

12. Arthur Smith Woodward's address to the British Association is Woodward (1910).

13. The quotation is from Loomis (1905, p. 842).

14. The quotations are from Nopcsa (1911, p. 148) and Nopcsa (1917, p. 345).

15. Matthew's ideas about dinosaurian extinction are given in Matthew (1921).

16. The other suggestions about dinosaurian extinction, from the 1920s, were climatic cooling (Jakovlev, 1922), high disease levels (Moodie, 1923), mammals eating dinosaur eggs (Wieland, 1925) and volcanic eruptions (Müller, 1928).

17. The first major review of the extinction of the dinosaurs was presented by Audova (1929).

18. The two surveys of theories for the extinction of the dinosaurs are Jepsen (1964) and Benton (1990).

19 Wilkinson et al. (2012) calculated that dinosaurs collectively produced more than 500 million tonnes of methane per year, equivalent to current methane emission levels from all sources.

20. Will Cuppy, renowned American humourist, gave his view of racial senility in his book *How to Become Extinct* in 1964.

21. The quotation comes from p. 35 of the English translation of Schindewolf's classic book. *Grundfragen der Paläontologie*, published in 1950. The translation, *Basic Questions in Paleontology*, was published in 1993, long after Schindewolf's death, and more as an historical curiosity than as a serious text. The fact that Schindewolf published exclusively in German, and in the German journals, and that his book was not earlier translated into English, confirms how divided the English-speaking and German-speaking palaeontologists were during much of the twentieth century. This has changed now, with German palaeontologists regularly writing in English, as well as in German, and placing their articles in international journals. Schindewolf's contributions are reviewed by Laublicher and Niklas (2009).

22. Camp (1952) and Watson (1957) doubted the reality of the end-Permian extinction event among amphibians and reptiles. Schindewolf (1958) argued strongly against their viewpoints, and gave a fuller riposte in his 1963 'Neokatastrophismus?' paper.

5. Impact! (pp. 96–122)

1. The epochal paper, which presented the first fully worked-out model for dinosaurian extinction by impact was L. W. Alvarez et al. (1980).

2. Numerous accounts of the KPg debates since 1980 have been published. W. Alvarez (1997) is the best insider's view of the work by Luis Alvarez and his team. Other books about the KPg impact and the debates include Raup (1986), which presents a strong palaeontological case for impact, Archibald (1996), which presents the opposite case – for caution about accepting a simple, instantaneous impact model – and Glen (1994), which contains chapters by a number of authors on the scientific disputes.
3. Ken Hsü (1994) gives a personal memoir of the shift from dogmatic anti-catastrophism in the United States to the acceptance of the possibility of major impacts on the Earth during the twentieth century.
4. The story of the Ries crater, including Shoemaker's involvement is told in many places, including guidebooks by Kavasch (1986) and Chao et al. (1978). Stöffler and Ostertag (1983) provide a more technical account of the crater and the evidence for impact. The critical first demonstration that the Ries structure was an impact crater is Shoemaker and Chao (1961).
5. The gradualist ecological succession model for the extinction of dinosaurs, and other animals and plants, at the KPg boundary, was summarized by Van Valen (1984), Sloan et al. (1986), Officer et al. (1987) and Hallam (1987).
6. The distinction between 'hard science' impact supporters and 'soft science' gradualists was noted by Raup (1986, p. 212).
7. The quotation is from L. W. Alvarez (1983, p. 632).
8. The quotation is from Jastrow 1983, p. 152).
9. The comment is from Van Valen (1984, p. 122).
10. The quotation is from L. W. Alvarez (1983, p. 67).
11. These further quotations are from L. W. Alvarez (1983), pp. 638, 640 and 629 respectively.
12. News reports about some of the more ugly aspects of the KPg impact debate include Browne (1985, 1988).
13. The quotation by Robert Bakker is given by Raup (1986, pp. 104–105).
14. A sociological review of styles of argumentation and clashes of different branches of science in the KPg debate is given by Clemens (1986, 1994). See also other papers in Glen (1994) for commentary on how the KPg debates evolved.
15. The supernova theory for the demise of the dinosaurs was championed in particular by Dale Russell, then of the Royal Ontario Museum in Toronto, Canada, and his colleagues (Terry and Tucker, 1968; Russell and Tucker, 1971; Béland et al., 1977).
16. Later papers that further explain the consequences of impact are W. Alvarez et al. (1984a, b). The quotation is from W. Alvarez et al. (1984a, p. 1135).
17. The other articles which appeared at the same time were Smit and Hertogen (1980), Hsu (1980) and Ganapathy (1980).
18. Informed, contemporary news coverage of the KPg debate was given by Lewin (1983, 1985a, b), Maddox (1985), Hoffman (1985) and Browne (1985, 1988).
19. The discovery of the Chicxulub crater was reported by Hildebrand et al. (1991). Current position papers are Schulte et al. (2010) on the simple impact model for the KPg mass extinction, and Archibald (2010) on a multiple causes explanation.

6. Diversity, Extinction and Mass Extinction (pp. 123–155)

1. The quotation is from Cuvier (1812), as translated by Rudwick (1997, 190).
2. D. H. Erwin's (1993, 2006) books are detailed accounts of the end-Permian extinction event.

3. The word 'biodiversity' was introduced in the report by Wilson and Peter (1988) on threats to the global diversity of life.

4. The two fundamental works in the classification of plants and animals are Linnaeus (1753) and Linnaeus (1758) respectively.

5. Darwin (1859) is, of course, his famous *Origin of species*, the starting point of so many fields of biology and palaeobiology. The basic characters of life are reviewed in all general biology textbooks. Recent reviews of the beginnings of life, and later additions of complexity, are Maynard Smith and Szathmáry (1995), Knoll and Bambach (2000) and S. B. Carroll (2001).

6. Plots of diversification of life in the sea were first produced in the 1960s and 1970s. The most famous examples stem from the work by Jack Sepkoski, Jr. (e.g. Sepkoski, 1984). I have produced plots of diversification of life in the sea and on land (Benton, 1995) based on a comprehensive data compilation produced by 100 experts (Benton, 1993).

7. The different mass extinction events are not all explored in detail here. There are many books about mass extinctions, and the best current one is by Hallam and Wignall (1997). It gives a full account of each event, what happened and why, or at least a selection of the hypotheses for each event.

8. The proposed statistical test for mass extinctions was presented by Raup and Sepkoski (1982).

9. The statistical criticism of Raup and Sepkoski's method was that a regression line can only be drawn if it is assumed that the data are a normal population of independent points. Techniques like regression analysis had been developed by biologists and psychologists who used the methods to sort out cause and effect in populations of plants, animals or humans. It was stretching the validity of the approach to use it on fixed measures through time. First, the extinction rate measures are not independent of each other: clearly, they occur in a sequence, and they affect a single evolving system, so there is strong time-linkage between adjacent rate measures. Secondly, it is not possible to have a complete distribution, as with a population of plants, animals and humans. In the extinction rate case, part of the distribution of values, for negative extinction rates, cannot be sampled.

10. The sampling studies of dinosaur extinction in the Hell Creek Formation of Montana, USA, are described by Archibald and Bryant (1990) and Sheehan et al. (1991), as well as in later papers. Archibald (1996) is an excellent book about the fossil record around the KPg mass extinction.

11. Our study of perceptions of the fossil record over the past 100 years is presented in Maxwell and Benton (1990). The comparison of changes in knowledge of the marine fossil record over 10 years is given by Sepkoski (1993).

12. Peter Ward's comparisons of his understanding of ammonite extinctions in the KPg boundary sections of northern Spain are summarized in Ward (1990).

13. Cladistics was enunciated first in English by Hennig (1966), and there are now many accounts of the methods. Forey et al. (1998) is a good primer of the basics.

14. The nucleic acids are deoxyribose nucleic acid, abbreviated as DNA, and ribose nucleic acid, abbreviated as RNA. DNA is the key coding material within the chromosomes within the nucleus of cells, and the various forms of RNA transfer information during cell division and during protein synthesis. Any recent basic biology book will give fuller details. The methods of molecular phylogeny reconstruction are outlined in Hall (2011) and Rosenfeld (2012).

15. The first comparisons of phylogenies and stratigraphy were made by Norell

and Novacek (1992). They looked at 25 phylogenies of mammals and found that three-quarters of them showed good congruence.

16. Benton et al. (2000a) considered 1000 phylogenies, cladistic and molecular, of a wide range of organisms, from algae to mammals, and found good agreement. In addition, when the phylogenies were sorted into those with branching occurring at different points in the geological past, they showed constant levels of agreement through geological time.
17. Jablonski and Raup (1995) sampled a huge database on marine bivalves and gastropods across the KPg boundary, comparing victims with survivors. There was no selectivity on the basis of habitat, size, diet or breeding habits. But geographically widespread genera were more likely to survive than those with restricted geographic ranges.
18. The role of geographic distribution on survival through the KPg event was assessed by Raup and Jablonski (1993). Comparisons of bivalves with temperate and tropical distributions did *not* confirm that tropical forms were more likely to go extinct than temperate, which had been the expected result.

7. Homing in on the Event (pp. 156–179)

1. The quotation is from Teichert (1990, p. 231).
2. D. H. Erwin (1993, p. 226) gives the estimate of 3–8 million years for the end-Permian event.
3. The reduction in timing of the end-Permian event to less than a million years is given by Bowring et al. (1998).
4. The further revision downwards, to less than 200,000 years is given by Shen et al. (2011).
5. Murchison's 'reduced' succession of the Permian is discussed in his big book describing his adventures in Russia (Murchison et al., 1845). Detailed evidence in support of this interpretation is given by Dunbar (1940) and Harland et al. (1982, p. 23).
6. Karpinsky named the Artinskian stage in 1874, and gave a fuller account of it in 1889, with full descriptions of the fossils, and evidence for their Permian, not Carboniferous, affinities. It is named after the town Arti (sometimes spelled Artinsk), on the western side of the Ural Mountains.
7. Ruzhentsev (1950) subdivided the Sakmarian into a lower substage, which he called the Asselian, and an upper Sakmarian substage. The two were later raised to full stage status. The Sakmarian was named after the Sakmara, a tributary of the Ural River in the southern Ural Mountains. The Asselian is named after the Assel River. The other Permian stages are also named after places in Russia: Kungurian (Kungur, a town near Perm), Ufimian (Ufa, a city in the Urals), Kazanian (Kazan, a city on the banks of the Volga), Tatarian (after the Tatars of Tatarstan, a region of Asiatic Russia).
8. The 1937 conference is described by Williams (1938) and a detailed overview of the field trip and the new view of Russian Permian stratigraphy by Dunbar (1940).
9. Dunbar's remark comes in the acknowledgments section of his paper (Dunbar, 1940, p. 280). The formalized principles of stratigraphy are given in detail in Salvador (1994). Good accounts of stratigraphy may be found in textbooks such as Stanley (2008) and Nichols (2009).
10. The Permo-Triassic sections in northern Italy have been described by Wignall and Hallam (1992), D. H. Erwin (1993, pp. 52–57) and Hallam and Wignall (1997, pp. 118–119).
11. Kummel and Teichert (1966, 1973) describe their visits to Permo-Triassic boundary sections around the world.

12. Restudy of the conodonts from the Salt Range is reported by Wignall et al. (1996), where conodonts in the Kathwai Member, previously said to be indicative of Early Triassic age, are reassigned to the latest Permian.
13. Doug Erwin commented on the correspondence of important Permo-Triassic boundary sections in Asia with current political turmoil in his book (D. H. Erwin, 1993, p. 63).
14. The establishment of the base of the Triassic was announced by the International Union of Geological Sciences in 2000. Further details may be found in Yang et al. (1995).
15. The sedimentology of the Meishan Permo-Triassic boundary section is described by Wignall and Hallam (1993), and summarized by Hallam and Wignall (1997, pp. 120–122). Further details are given by Yang et al. (1995), Jin et al. (2000), and Chen et al. (2014).
16. The first radiometric dates for the Meishan boundary clays were done by Claoue-Long et al. (1991) using the $^{206}Pb/^{235}U$ isotopic decay series.
17. The first argon-argon date for the Permo-Triassic boundary was reported by Renne et al. (1995).
18. Detailed dating studies on the Meishan section were done by Bowring et al. (1998), with later corrections by Mundil et al. (2001). The latest work on dating was Shen et al. (2011).
19. Details of the distribution of fossils across the Permo-Triassic boundary in the Meishan section are given by Jin et al. (2000). The data are revised and updated by Song et al. (2013).
20. The Signor-Lipps effect is named after the account of necessary incompleteness of the fossil record by Signor and Lipps (1982).
21. Wignall and Hallam (1993, p. 231) comment on the similarity of the Hushan section to Meishan.
22. The Greenland Permo-Triassic succession is described by Teichert and Kummel (1976), where they present their 'armoured mudballs' hypothesis. Hallam and Wignall (1997, p. 117) said 'armoured mudballs' to that theory. Dating evidence is given by Wignall et al. (1996).
23. Twitchett et al. (2001) and Looy et al. (2001) describe the Greenland Permo-Triassic sections in some detail.

8. Life's Biggest Challenge (pp. 180–203)

1. D. H. Erwin (1993) used the term 'mother of all mass extinctions' as a chapter title. The quotation is from D. H. Erwin (1994, p. 231).
2. The statistical analysis of fusuline extinctions is Rampino and Adler (1998).
3. The quotation is from Darwin's 'big species book', his unpublished longer version of the *Origin* (Stauffer 1975, p. 208).
4. The reanalysis of the fates of brachiopods and bivalves is in Gould and Calloway (1980). Other information on the fate of different groups of shellfish during the end-Permian crisis is in D. H. Erwin (1993) and Hallam and Wignall (1997).
5. A superb account of trilobites is given by Fortey (2000). Extinctions among fossil arthropods are discussed by Briggs et al. (1988).
6. The detailed overview of evolution of bony fishes through the Permian and Triassic is by Romano et al. (2014).
7. Hallam and Wignall (1997, pp. 98–110) offer a detailed overview of the end-Permian extinctions, group by group.
8. Raup (1979) presented the first simple rarefaction analysis that a loss of 50% of families is equivalent to the loss of 90% of species. Criticisms were made by Hoffman (1986) and McKinney (1995), who claimed that this was an overestimate, and that true species-level extinctions were much lower.

9. A Tale of Two Continents (pp. 204–228)

1. The story of Andrew Bain and his fossil collecting is told in his own personal memoir, published posthumously (Bain 1896), as well as by Desmond (1982, pp. 195–198) and by Buffetaut (1987, pp. 176–179).
2. The first publications on the Karoo Permo-Triassic reptiles of the Karoo Basin are by Bain (1845) on the geology, and by Owen (1845a) on the reptiles.
3. Robert Broom's story is told by Watson (1952), and more briefly by Buffetaut (1987, p. 179).
4. Western accounts of the Russian Copper Sandstone reptiles were given by Meyer (1866) and Owen (1876).
5. British palaeontologists who visited the Russian localities and wrote about the fossils in Victorian times, include Twelvetrees (1880, 1882) and Seeley (1894).
6. Ivan Efremov's contribution to Russian palaeontology is assessed by Ochev and Surkov (2000), who also trace in considerable detail the work of his students since the 1950s. Efremov's key presentations of the Permo-Triassic stratigraphic scheme are Efremov (1937, 1941, 1952). His science fiction novels still sell widely. I was able to buy a reprint of his *Tais Afinskaya* (the name of a mythical Athenian goddess) in 1995 in a small shop in a South Urals' village that had little else on offer (some twiggy raisins, oily tins of oily fish, as well as beautiful Western-style tins of fruit imported from Hungary, at prices no one could afford).
7. The current stratigraphic scheme for the Karoo Permo-Triassic is summarized by Rubidge (1995). James Kitching's work on localizing Karoo vertebrate specimens is summarized in Kitching (1977). These works summarize the faunas of the *Dicynodon* Zone, and this is supplemented by synoptic data in Nicolas and Rubidge (2010).
8. Estimates of familial losses among terrestrial vertebrates are taken from Maxwell (1992) and Benton (1993, 1997).
9. The idea that erosion rates increased at the end of the Permian as a result of loss of plant cover is presented by MacLeod et al. (2000) and Ward et al. (2000). Additional evidence for mass runoff comes from the discovery of a dramatic increase in silica (sand) washed into oceans worldwide at the boundary (Algeo and Twitchett 2010). Distributions of the reptiles across the Permo-Triassic boundary in the Karoo are given by Smith and Ward (2001), Smith and Botha (2005) and Smith and Botha-Brink (2014).
10. Retallack (1999) gives a clear and detailed account of changes in soils across the Permo-Triassic boundary in Gondwana.
11. Eshet et al. (1995) describe evidence for a sharp fungal spike just at the Permo-Triassic boundary. Fungi (moulds and mushrooms) reproduce by sending out hyphae, tube-like structures that can give rise to new individuals, and form a huge and complex underground mat. Fungi also produce spores which are thrown out into the air or water, and may spread widely.
12. Looy et al. (2001) and Twitchett et al. (2001) present their evidence about the marine and terrestrial extinction events from the Greenland sections.
13. Broom (1903) described and named *Lystrosaurus* and argued that it lived a semi-aquatic lifestyle. This was questioned seriously by King (1991b) and King and Cluver (1991), but refuted, for the Russian species at least, by Surkov (2002).

10. On the River Sakmara (pp. 229–250)

1. Andrew Newell's account of the sedimentology of the Russian Permo-Triassic sections is Newell et al. (1999). The footprints in the mudflats association were described by Tverdokhlebov et al. (1997).

2. Valentin Tverdokhlebov has published many short papers on the sediments of the Permo-Triassic of the Urals. His account of the alluvial megafans and their source areas is Tverdokhlebov (1971).

3. Early accounts of the Russian Permo-Triassic amphibians and reptiles are Efremov (1940), in German, and Olson (1955, 1957), in English. Olson (1990) wrote an extremely informative account of his visits to Moscow, and his meetings with Ivan Efremov, in the 1950s. Sennikov (1996) gave a brief account of the evolution of Permo-Triassic ecosystems in Russia, as part of the joint research programme between Bristol and Moscow.

4. The major outcome of the collaboration between Bristol and Moscow so far is Benton et al. (2000b), an edited book that contains chapters on the rock successions and the major groups of amphibians and reptiles from the Permo-Triassic, as well as younger dinosaur-bearing units of Mongolia and the Former Soviet Union.

5. Our summary of Valentin Tverdokhlebov's detailed records appeared as Tverdokhlebov et al. (2003, 2005), and the statistical study as Benton et al. (2004). Our subsequent expeditions are described in Benton (2008).

11. What Caused the Biggest Catastrophe of All Time? (pp. 251–283)

1. Herbert Butterfield's book, *The Whig Interpretation of History*, first published in 1931 and reprinted many times since, is a standard source for discussions about historical method. It has been attacked and defended equally vigorously in many works later in the twentieth century.

2. Frank Rhodes made his comment in a review of the Permo-Triassic extinction event (Rhodes, 1967).

3. Quoted from D. H. Erwin (1993, p. 256).

4. N. D. Newell (1952) first noted the Late Permian marine regression and defended the idea in later papers, such as Newell (1967).

5. Valentine and Moores (1973) developed the plate tectonic hypothesis for sea-level fall and mass extinction.

6. The evidence for and against a Late Permian marine regression is presented by D. H. Erwin (1993, pp. 145–155). Hallam and Wignall (1997, pp. 132–133) cast severe doubts on the idea of regression at this time.

7. The three extinction peaks across the Permo-Triassic boundary are shown in my plot of diversification of continental tetrapods (Benton, 1985), and the end-Olenekian peak is shown for both tetrapods and for ammonoids in Benton (1986), based on global family-level and generic-level counts. Two peaks of extinctions among Late Permian tetrapods were confirmed by King (1991a) and Rubidge (1995), based on species counts from the Karoo Basin. Stanley and Yang (1994) also pointed out the end-Capitanian extinction among marine species, especially as seen in the Chinese sections; this event might have occurred earlier, before the end of the Capitanian (Bond et al. 2010). Wang and Sugiyama (2000) made detailed studies of coral extinctions through the Late Permian in China. Sepkoski (1996) also noted the Olenekian extinction event, sometimes called the end-Scythian event, using an alternative name for the time unit.

8. Schindewolf (1958) suggested cosmic rays as a cause of the end-Permian event.

9. The enhancement in iridium at the Permo-Triassic boundary was announced from the Meishan section by Sun et al. (1984) and from another Chinese section by Xu et al. (1985). The re-analyses were done by Clark et al. (1986) and Zhou and Kyte (1988), who found no unusual quantities of iridium.

10. The ferruginous spherules were reported by Yin et al. (1992).

11. Kerr (1995) was first to report the shocked quartz grains from Australia and Antarctica, and Retallack et al. (1998) gave a fuller account.
12. The report of extraterrestrial noble gases in fullerenes is Becker et al. (2001). Quoted newspaper reports are Hawkes (2001) and Hanlon (2001). Erwin and Becker were quoted by Simpson (2001). Published critiques are Farley and Mukhopadhyay (2001) and Isozaki (2001), with a response by Becker and Poreda (2001). The Australian crater was reported by Becker et al. (2004), and critiques include Renne et al. (2004).
13. Courtillot (1990) postulated a short time span for eruption of the Deccan Traps, exactly at the KPg boundary, while Baksi and Farrar (1991) argued that the precision of his dates was illusory. Later work tended to confirm Courtillot's data.
14. Baksi and Farrar (1991) gave radiometric dates for the Siberian Traps that placed their eruption in the Mid Triassic. The dates were revised to the Permo-Triassic boundary by Renne and Basu (1991), using the argon-argon method, and Campbell et al. (1992), using the uranium-lead method. Later redatings, using wider arrays of specimens, confirmed that the Siberian basalts coincided with the Permo-Triassic boundary, and were erupted in less than one million years (Basu et al., 1995; Renne et al., 1995). Self et al. (2014) give an update on the Siberian Traps.
15. The quotation is from D. H. Erwin (1993, pp. 255–256).
16. The volcanic winter hypothesis was presented by Campbell et al. (1992). A critique is given by Hallam and Wignall (1997, pp. 135–137).
17. Stanley (1984, 1988) argued strongly that mass extinctions, including that at the end of the Permian, were caused by global cooling. Hallam and Wignall (1997, pp. 133–134) dispute these claims, but present evidence that the end-Capitanian event might have been the result of global cooling.
18. Hallam and Wignall (1997, pp. 137–138) summarize the evidence for global warming at the end of the Permian.
19. The isotope geochemistry of the Gartnerkofel core was presented by Holser et al. (1989), and of the Spitsbergen section by Gruszczynski et al. (1989).
20. The palaeosol evidence for climate warming is given by Retallack (1999), who also discusses changes in the floras.
21. The Spitsbergen sections are described by Wignall and Twitchett (1996). Anoxic black mudstones are reported from Italy and the United States by Wignall and Hallam (1992), from Pakistan and China by Wignall and Hallam (1993), and from Greenland by Twitchett et al. (2001).
22. Hallam and Wignall (1997, chapter 5) present a detailed account of the end-Permian crisis and their model based on oxidation of coals. Wignall (2001) brings in the Siberian Traps, and other sources of carbon dioxide. Wignall and Twitchett (2002) review the whole superanoxia hypothesis, and Bond and Wignall (2014) bring the volcanic hypothesis up to date.
23. The methane burp hypothesis for the global warming pulse 55 million years ago has been presented by Dickens et al. (1997), and enlarged on by Bains et al. (1999, 2000a, b).
24. Robert Berner outlines his climatic models and the end-Permian perturbations in Berner (2002).

12. Recovery from the Brink (pp. 284–300)

1. The Early Triassic recovery phase for life in the sea has been summarized by Hallam and Wignall (1997), Erwin (2006)

and Chen and Benton (2012). Detailed accounts have been made by McRoberts (2001) on bivalves, Nützel (2005) on gastropods, Twitchett (1999) on the changing environments, Rodland and Bottjer (2001) on *Lingula*, Chen et al. (2005) on brachiopods in general, Brayard et al. (2009) on ammonoids, and Twitchett et al. (2001) on marine recovery in the Greenland sections.

2. The competition/density-dependent models for the recovery of life after mass extinctions are given in detail by Solé et al. (2010), and outlined briefly by Chen and Benton (2012).

3. The summary of carbon isotope data through the Early and Middle Triassic is from Payne et al. (2004).

4. The evidence for truly very hot conditions at the Permo-Triassic boundary and several times during the Early Triassic is provided by Sun et al. (2012). He provides geochemical evidence for the high temperatures, as well as palaeogeographic maps that show the absence of many groups of animals in equatorial regions on land and in the sea.

5. We describe the Luoping biota, its geology and palaeontology, in Hu et al. (2011), Chen and Benton (2012), and Benton et al. (2013).

6. The history of tetrapods on land during the Triassic, and the origin of the dinosaurs, has been reviewed many times, by Benton (2015), for example. Evidence for the very oldest hints of dinosaurs and their early diversification are given by Benton et al. (2014).

13. The Sixth Mass Extinction? (pp. 301–313)

1. Gore (1992, p. 28).

2. Ehrlich and Ehrlich (1990) provide an estimate of the current rate of species loss.

3. T. Erwin (1982, 1983) presented his sampling estimates of tropical arthropod diversity.

4. Total global diversity has been reviewed by many authors, including Wilson (1992) and May (1990, 1992). Read more in Wilson (2002), his latest book. New work by Novotny et al. (2002) suggests that Erwin overestimated the level of endemicity of his beetles; more of them are actually shared between tropical trees than he thought. If correct, this could bring his estimate down to 4–6 million species of arthropods.

5. Current extinction rates have been estimated by Smith et al. (1993) and Pimm et al. (1995). Read more in Wilson (2002), his latest book. The latest and best estimates are in Monastersky (2014), and more extended information can be found in Kolbert (2014).

6. Bjorn Lomborg's views are presented in Lomborg (2001). They have been criticized in numerous reviews, including three articles from the *Guardian*, at http://www.guardian.co.uk/globalwarming/story/0,7369,539558,00.html

7. Quoted from Cuvier (1812), as translated by Rudwick (1997, p. 185).

BIBLIOGRAPHY

Agassiz, L. 1840. *Études sur les glaciers*. Neuchâtel: Jent and Gassman.

Algeo, T. J. and Twitchett, R. J. 2010. Anomalous Early Triassic sediment fluxes due to elevated weathering rates and their biological consequences. *Geology* 38, pp. 1023–1026.

Alvarez, L. W. 1983. Experimental evidence that an asteroid impact led to the extinction of many species 65 million years ago. *Proceedings of the National Academy of Sciences, USA* 80, pp. 627–642.

——, Alvarez, W., Asaro, F. and Michel, H. V. 1980. Extraterrestrial cause for the Cretaceous-Tertiary extinction – Experimental results and theoretical implications. *Science* 208, pp. 1095–1108.

Alvarez, W. 1997. *T. Rex and the Crater of Doom*. Princeton, New Jersey: Princeton University Press; London: Penguin.

——, Kauffman, E. G., Surlyk, F., Alvarez, L. W., Asaro, F. and Michel, H. V. 1984a. Impact theory of mass extinctions and the invertebrate fossil record. *Science* 223, pp. 1135–1141.

——, Alvarez, L. W., Asaro, F. and Michel, H. V. 1984b. The end of the Cretaceous: Sharp boundary or gradual transition? *Science* 223, pp. 1183–1186.

Archibald, J. D. 1996. *Dinosaur Extinction and the End of an Era: What the Fossils Say*. New York: Columbia University Press.

—— and Bryant, L. J. 1990. Differential Cretaceous/Tertiary extinctions of nonmarine vertebrates: evidence from northeastern Montana. *Special Paper of the Geological Society of America*, 247, pp. 549–562.

—— and 28 others. 2010. Cretaceous extinctions: multiple causes. *Science* 328, p. 973.

Audova, A. 1929. Aussterben der Mesozoischen Reptilien. *Palaeobiologica* 2, pp. 222–245, pp. 365–401.

Bain, A. G. 1845. On the discovery of the fossil remains of bidental and other reptiles in South Africa. *Quarterly Journal of the Geological Society of London* 1, pp. 317–318, and *Transactions of the Geological Society of London, Series* 27, pp. 53–58.

—— 1896. Reminiscences and anecdotes concerned with the history of geology in South Africa, or the pursuit of knowledge under difficulties. *Transactions of the Geological Society of South Africa* 2, pp. 59–75.

Bains, S., Corfield, R. and Norris, R. 1999. Mechanisms of climate warming at the end of the Paleocene. *Science* 285, pp. 724–727.

——, —— and —— 2000a. Structure of the late Palaeocene carbon isotope excursion. *GFF* 122, pp. 19–20.

——, Norris, R. D., Corfield, R. and Faul, K. L. 2000. Termination of global warmth at the Palaeocene/Eocene boundary through productivity feedback. *Nature* 407, pp. 171–174.

Baksi, A. K. and Farrar, E. 1991. Ar^{40}/Ar^{39} dating of the Siberian Traps, USSR – evaluation of the ages of the 2 major extinction events relative to episodes of flood-basalt volcanism in the USSR and the Deccan Traps, India. *Geology* 19, pp. 461–464.

Basu, A. R., Poreda, R. J., Renne, P. R., Teichmann, F., Vasiliev, Y. R., Sobolev, N. V. and Turrin, B. D. 1995. High-He^3 plume origin and temporal-spatial evolution of the Siberian flood basalts. *Science* 269, pp. 822–825.

Becker, L. and Poreda, R. J. 2001. An extraterrestrial impact at the Permian-Triassic boundary? *Science* 293, 2343a (U2–U3).

——, ——, Hunt, A. G, Bunch, T. E. and Rampino, M. 2001. Impact event at the Permian-Triassic boundary: evidence from extraterrestrial noble gases in fullerenes. *Science* 291, pp. 1530–1533.

Béland, P., Feldman, P., Foster, J., Jarzen, D., Norris, G., Pirozynski, K., Reid, G., Roy, J. R., Russell, D. and Tucker, W. 1977. Cretaceous-Tertiary extinctions and possible terrestrial and extraterrestrial causes. *Syllogeus* 1977(12), pp. 1–162.

Benton, M. J. 1982. Progressionism in the 1850s: Lyell, Owen, Mantell, and the Elgin fossil reptile *Leptopleuron* (*Telerpeton*). *Archives of Natural History*, 11, pp. 123–136.

—— 1985. Mass extinction among non-marine tetrapods. *Nature* 316, pp. 811–814.

—— 1986. More than one event in the late Triassic mass extinction. *Nature* 321, pp. 857–861.

—— 1990. Scientific methodologies in collision; the history of the study of the extinction of the dinosaurs. *Evolutionary Biology* 24, pp. 371–400. Also available at http://palaeo.gly.bris.ac.uk/essays/dino90.html

—— (ed.) 1993. *The Fossil Record 2*. London: Chapman & Hall.

—— 1995. Diversification and extinction in the history of life. *Science*, 268, pp. 52–58.

—— and Harper, D. A. T. 2009. *Introduction to Paleobiology and the Fossil Record*. New York: Wiley Blackwell.

——, Wills, M. A. and Hitchin, R. 2000a. Quality of the fossil record through time. *Nature* 403, pp. 534–538.

——, Shishkin, M. A., Unwin, D. M. and Kurochkin, E. N. (eds) 2000b. *The Age of Dinosaurs in Russia and Mongolia*. Cambridge: Cambridge University Press.

——, Tverdokhlebov, V. P. and Surkov, M. V. 2004. Ecosystem remodelling among vertebrates at the Permian–Triassic boundary in Russia. *Nature* 432, pp. 97–100.

—— 2008. The end-Permian mass extinction – events on land in Russia. *Proceedings of the Geologists' Association* 119, pp. 119–136.

——, Sennikov, A. G. and Newell, A. J. 2010. Murchison's first sighting of the Permian, at Vyazniki in 1841. *Proceedings of the Geologists' Association* 121, pp. 313–318.

——, Zhang, Q. Y., Hu, S. X., Chen, Z.-Q., Wen, W., Liu, J., Huang, Y. J., Zhou, C. Y., Xie, T., Tong, J. N. and Choo, B. 2013. Exceptional vertebrate biotas from the Triassic of China, and the expansion of marine ecosystems after the Permo-Triassic mass extinction. *Earth-Science Reviews* 125, pp. 199–243.

——, Forth, J. and Langer, M. 2014. Models for the rise of the dinosaurs. *Current Biology* 24, R87–R95.

—— 2015. *Vertebrate Palaeontology*. New York: Wiley Blackwell.

Berner, R. A. 2002. Examination of hypotheses for the Permo-Triassic boundary extinction by carbon cycle modeling. *Proceedings of the National Academy of Sciences*, 99, pp. 4172–4177.

Bond, D. P. G., Hilton, J., Wignall, P. B., Ali, J. R., Stevens, L. G., Sun, Y. D. and Lai, X. L. 2010. The Middle Permian (Capitanian) mass extinction on land and in the oceans. Earth-Science Reviews 102, pp. 100–116.

—— and Wignall, P. B. 2014. Large igneous provinces and mass extinctions: an update. *Geological Society of America, Special Papers* 505, doi: 10.1130/2014.2505(02).

Boucher de Perthes, J. 1847. *Antiquités celtiques et antédiluviennes. Mémoire sur l'industrie primitive et les arts à leur origine* (Vol. 1). Paris: Treuttel and Wertz.

—— 1857. *Antiquités celtiques et antédiluviennes. Mémoire sur l'industrie primitive et les arts à leur origine* (Vol. 2). Paris: Treuttel and Wertz.

—— 1864. *Antiquités celtiques et antédiluviennes. Mémoire sur l'industrie primitive et les arts à leur origine* (Vol. 3). Paris: Treuttel and Wertz.

Bowler, P. J. 1976. *Fossils and Progress. Palaeontology and the Idea of Progressive Evolution in the Nineteenth Century*. New York: Science History Publications.

—— 1983, *The Eclipse of Darwinism: Anti-Darwinian Evolution Theories in the Decades Around 1900*. Baltimore, Md.: Johns Hopkins University Press.

Bowring, S. A., Erwin, D. H., Jin, Y. G., Martin, M. W., Davidek, K. and Wang, W. 1998. U/Pb zircon geochronology of the end-Permian mass extinction. *Science* 280, pp. 1039–1045.

Brayard, A., Escarguel, G., Bucher, H., Monnet, C., Brühwiler, T., Goudemand, N., Galfetti, T. and Guex, J. 2009. Good genes and good luck: ammonoid diversity and the end-Permian mass extinction. *Science* 325, pp. 1118–1121.

Briggs, D. E. G., Fortey, R. A. and Clarkson, E. N. K. 1988. Extinction and the fossil record of arthropods. In Larwood, G. P. (ed.) *Extinction and Survival in the Fossil Record*, pp. 171–209. Systematics Association Special Volume No. 44.

Broom, R. 1903. On the remains of *Lystrosaurus* in the Albany Museum. *Records of the Albany Museum* 1, pp. 3–8.

Browne, M. W. 1985. Dinosaur experts resist meteor extinction idea. Paleontologists say dissenters risk harm to their careers. *New York Times* 1985 (29 October), pp. 21–22.

—— 1988. Debate over dinosaur extinction takes an unusually rancorous turn. *New York Times* 1988 (19 January), pp. 19, 23.

Buckland, W. 1822. *Reliquiae diluvianae; or, observations on the organic remains contained in caves, fissures, and diluvian gravel, and on other geological phenomena, attesting the action of an universal deluge.* London: John Murray.

Buffetaut, E. 1987. *A Short History of Vertebrate Palaeontology.* London: Croom Helm.

Butterfield, H. 1931. *The Whig Interpretation of History.* London: Bell. (Reprinted 1965, New York: Norton.)

Bystrov, A. P. 1957. [The pareiasaur skull.] *Trudy Paleontologicheskogo Instituta AN SSSR* 68, pp. 3–18.

Cadbury, D. H. 2000. *The Dinosaur Hunters.* London: Fourth Estate; New York: Holt.

Camp, C. L. 1952. Geological boundaries in relation to faunal changes and diastrophism. *Journal of Paleontology* 26, pp. 353–358.

Campbell, I. H., Czamanske, G. K., Fedorenko, V. A., Hill, R. I. and Stepanov, V. 1992. Synchronism of the Siberian Traps and the Permian-Triassic boundary. *Science* 258, pp. 1760–1763.

Carroll, R. L. 1988. *Vertebrate Palaeontology and Evolution.* New York: W. H. Freeman.

—— 1997. *Patterns and Processes of Vertebrate Evolution.* Cambridge: Cambridge University Press.

Carroll, S. B. 2001. Chance and necessity: the evolution of morphological complexity and diversity. *Nature* 409, pp. 1102–1109.

Chao, E. C. Y., Hüttner, R. and Schmidt-Kaler, H. 1978. *Principal Exposures of the Ries Meteorite Crater in Southern Germany.* München: Bayerisches Geologisches Landesamt.

Chapman, W. 1758. An account of the fossile bones of an alligator, found on the sea-shore, near Whitby in Yorkshire. *Philosophical Transactions of the Royal Society of London* 50, pp. 688–691.

Chen, Z. Q. and Benton, M. J. 2012. The timing and pattern of biotic recovery following the end-Permian mass extinction. *Nature Geoscience* 5, pp. 375–383.

——, Kaiho, K., and George, A. D. 2005. Early Triassic recovery of the brachiopod faunas from the end-Permian mass extinction: a global review. *Palaeogeography, Palaeoclimatology, Palaeoecology* 224, pp. 270–290.

——, and 16 others. 2014. Complete biotic and sedimentary records of the Permian-Triassic transition from Meishan section, South China: Ecologically assessing mass extinction and its aftermath. *Earth-Science Reviews*, online ahead of print (doi::10.1016/j.earscirev.2014.10.005).

Claoue-Long, J. C., Zhang, Z. C., Ma, G. G. and Du, S. H. 1991. The age of the Permian-Triassic boundary. *Earth and Planetary Science Letters* 105, pp. 182–190.

Clark, D. L., Cheng-Yuan, W., Orth, C. S. and Gilmore, J. S. 1986. Conodont survival and low iridium abundances across the Permian-Triassic boundary. *Science* 233, pp. 984–986.

Clemens, E. S. 1986. Of asteroids and dinosaurs: The role of the press in the shaping of scientific debate. *Social Studies of Science* 16, pp. 421–456.

—— 1994. The impact hypotheses and popular science: conditions and consequences of interdisciplinary debate. In Glen, W. (ed.), *Mass Extinction Debates: How Science Works in a Crisis*, pp. 92–120. Stanford, Ca.: Stanford University Press.

Colbert, E. H. 1968. *Men and Dinosaurs*. New York: Dutton.

Collie, M. and Diemer, J. 1995. Murchison in Moray: a geologist on home ground, with the correspondence of Roderick Impey Murchison and the Rev. Dr. George Gordon of Birnie. *Transactions of the American Philosophical Society* 85(3), pp. 1–263.

—— and ——. 2004. *Murchison's Wanderings in Russia: His Geological Exploration of Russia in Europe and the Ural Mountains, 1840 and 1841*. Keyworth: British Geological Survey.

Courtillot, V. E. 1990. What caused the mass extinction? A volcanic eruption. *Scientific American* 263(10), pp. 85–92.

Cuppy, W. 1964. *How to Become Extinct*. New York: Dover Publications.

Cuvier, G. 1812. *Recherches sur les ossemens fossiles de quadrupèdes, où l'on rétablit les caractères de plusieurs espèces d'animaux que les révolutions du globe paroissent avoir détruites*. 4 vols. Paris: G. Dufour et d'Ocagne. 2nd edition: 1821–24; slightly revised 3rd edition, 1825: 4th edition, 1834–36.

—— 1825. *Recherches sur les ossemens fossiles de quadrupèdes, où l'on rétablit les caractères de plusieurs animaux dont les révolutions du globe ont détruit les espèces*. 3rd edition. 7 vols. Paris: G. Dufour et d'Ocagne.

Darwin, C. 1859. *On the origin of species by means of natural selection, or the preservation of favoured races in the struggle for life*. London: John Murray.

—— 1861. *On the origin of species by means of natural selection, or the preservation of favoured races in the struggle for life*. 3rd edition. London: John Murray.

DeSalle, R. and Rosenfeld, J. 2012. *Phylogenomics: a Primer*. New York: Garland Science.

Desmond, A. J. 1982. *Archetypes and Ancestors: Palaeontology in Victorian London, 1850–1875*. Chicago: University of Chicago Press.

Dickens, G. R., Paull, C. K. and Wallace, P. 1997. Direct measurement of in situ methane quantities in a large gas-hydrate reservoir. *Nature* 385, pp. 426–428.

Dunbar, C. O. 1940. The type Permian: its classification and correlation. *Bulletin of the American Association of Petroleum Geologists* 24, pp. 237–281.

Durant, G. P. and Rolfe, W. D. I. 1984. William Hunter (1718–1783) as natural historian: his 'geological' interests. *Earth Sciences History* 3, pp. 9–24.

Efremov, I. A. 1937. [On the stratigraphic divisions of the continental Permian and Triassic of the USSR, based on tetrapod faunas.] *Doklady Akademii Nauk SSSR* 16, pp. 125–132.

—— 1940. Kürze Übersicht über die Formen der Perm- und Trias Tetrapoden-Fauna der UdSSR. *Centralblatt für Mineralogie, Geologie und Paläontologie, Abtheilung B*, pp. 372–383.

—— 1941. [Short survey of faunas of Permian and Triassic Tetrapoda of the USSR.] *Sovetskaya Geologiya* 5, pp. 96–103.

—— 1952. [On the stratigraphy of the Permian red beds of the USSR based on terrestrial vertebrates.] *Izvestiya AN SSSR, Seriya Geologicheskaya* 6, pp. 49–75.

Ehrlich, P. R. and Ehrlich, A. H. 1990. *The Population Explosion*. New York: Simon & Schuster.

Erwin, D. H. 1993. *The Great Paleozoic Crisis: Life and Death in the Permian*. New York: Columbia University Press.

—— 1994. The Permo-Triassic extinction. *Nature* 367, pp. 231–236.

—— 2006. *Extinction! How Life Nearly Ended 250 Million Years Ago*. New Jersey: Princeton University Press.

Erwin, T. 1982. Tropical forests: their richness in Coleoptera and other arthropod species. *Coleopterists' Bulletin* 36, pp. 74–75.

—— 1983. Beetles and other insects of tropical forest canopies at Manaus, Brazil, sampled by insecticidal fogging. In Sutton, S. L., Whitmore, T. C. and Chadwick, A. C. (eds), *Tropical Rain Forest: Ecology and Management*, pp. 59–75. London: Blackwell.

Eshet, Y., Rampino, M. R. and Visscher, H. 1995. Fungal event and palynological record of ecological crisis and recovery across the Permian-Triassic boundary. *Geology* 23, pp. 967–970.

Farley, K.A. and Mukhopadhyay, S. 2001. An extraterrestrial impact at the Permian-Triassic boundary? *Science* 293, 2343a (U1-2).

Fitton, W. H. 1839. Elements of geology, by Charles Lyell, Esq., F.R.S. *Edinburgh Review* 69, pp. 406–466.

Forey, P. L., Humphries, C. J., Kitching, I. J., Scotland, R. W., Siebert, D. J. and Williams, D. M. 1998. *Cladistics: A Practical Course in Systematics.* 2nd edition, Oxford: Clarendon Press.

Fortey, R. 2000. *Trilobite! Eyewitness to Evolution.* London: Harper Collins; New York: Knopf.

Ganapathy, R. 1980. A major meteorite impact on the earth 65 million years ago: evidence from the Cretaceous-Tertiary boundary clay. *Science* 209, pp. 921–923.

Geikie, A. 1875. *Life of Sir Roderick Murchison . . . based on his journals and letters with notices of his scientific contemporaries and a sketch of the rise and growth of Palaeozoic geology in Britain.* 2 vols. London: John Murray.

Gillispie, C. C. 1951. *Genesis and Geology.* Cambridge, Mass.: Harvard University Press.

Glen, W. (ed.) 1994. *Mass Extinction Debates: How Science Works in a Crisis.* Stanford, Calif.: Stanford University Press.

Gore, A. 1992. *Earth in the Balance.* New ed. 2000. London: Earthscan.

Gould, S. J. 1987. *Time's Arrow, Time's Cycle.* Cambridge, Mass.: Harvard University Press.

—— and Calloway, C. B. 1980. Clams and brachiopods – ships that pass in the night. *Paleobiology* 6, pp. 383–396.

Grayson, D. K. 1984. Nineteenth-century explanations of Pleistocene extinctions: a review and analysis. In Martin, P. S. and Klein, R. G. (eds), *Quaternary Extinctions, A Prehistoric Revolution*, pp. 5–39. Tucson: University of Arizona Press.

Greene, J. C. 1961. *The Death of Adam. Evolution and its Impact on Western Thought.* New York: Mentor.

Gruszczynski, M., Halas, S., Hoffman, A. and Malkowksi, K. 1989. A brachiopod calcite record of the oceanic carbon and oxygen isotope shifts at the Permian/Triassic transition. *Nature* 337, pp. 64–68.

Hall, B. G. 2011. *Phylogenetic Trees Made Easy.* 4th edition. New York: Sinauer.

Hallam, A. 1983. *Great Geological Controversies.* Oxford: Oxford University Press.

—— 1987. End-Cretaceous extinction event: argument for terrestrial causation. *Science* 238, pp. 1237–1242.

—— and Wignall, P. 1997. *Mass Extinctions and their Aftermath.* Oxford: Oxford University Press.

Hanlon, M. 2001. The great dying. *Daily Mail* (23 February), p. 13.

Harland, W. B., Cox, A. V., Llewellyn, P. G., Pickton, C. A. G., Smith, A. G. and Walters, R. 1982. *A Geologic Time Scale.* Cambridge: Cambridge University Press.

Hawkes, N. 2001. Crash 250 million years ago nearly wiped out life. *The Times* (23 February), p. 13.

Hennig, W. 1966. *Phylogenetic Systematics.* Bloomington: University of Indiana Press.

Hildebrand, A. R., Penfield, G. T., Kring, D. A., Pilkington, M., Camargo, Z. A., Jacobsen, S. B. and Boynton, W. V. 1991. Chicxulub crater: a possible Cretaceous/Tertiary boundary impact crater on the Yucatán Peninsula, Mexico. *Geology* 19, pp. 867–871.

Hoffman, A. 1985. Patterns of family extinction depend on definitions and geological timescale. *Nature* 315, pp. 659–662.

—— 1986. Neutral model of Phanerozoic diversification: implications for macroevolution. *Neues Jahrbuch für Geologie und Paläontologie, Abhandlungen* 172, pp. 219–244.

Holser, W. T., Schönlaub, H.-P., Attrep, M., Jr., Boeckelmann, K., Klein, P., Magaritz, M., Pak, E., Schramm, J.-M., Stattgegger, K. and Scmöller, R. 1989. A unique geochemical record at the Permian/ Triassic boundary. *Nature* 337, pp. 39–44.

Hsu, K. T. 1980. Terrestrial catastrophe caused by a cometary impact at the end of the Cretaceous. *Nature* 285, pp. 201–203.

—— 1994. Uniformitarianism vs. catastrophism in the extinction debate. In Glen, W. (ed.), *Mass Extinction Debates: How Science Works in a Crisis*, pp. 217–229. Stanford, Calif.: Stanford University Press.

Hu, S. X., Zhang, Q. Y., Chen, Z. Q., Zhou, C. Y., Lü, T., Xie, T., Wen, W., Huang, J. Y. and Benton, M. J. 2011. The Luoping biota: exceptional preservation, and new evidence on the Triassic recovery from end-Permian mass extinction. *Proceedings of the Royal Society*, Series B 278, pp. 2274–2282.

Huxley, T. H. 1870. On the classification of the Dinosauria, with observations on the dinosaurs of the Trias. *Quarterly Journal of the Geological Society of London* 26, pp. 32–51.

Isozaki, Y. 2001. An extraterrestrial impact at the Permian-Triassic boundary? *Science* 293, 2343a (U2).

Jablonski, D. and Raup, D. M. 1995. Selectivity of end-Cretaceous marine bivalve extinctions. *Science* 268, pp. 389–391.

Jakovlev, N. N. 1922. [Extinction and its causes as a principal question in biology]. *Mysl* 2, pp. 1–36.

Jastrow, R. 1983. The dinosaur massacre: a double-barrelled mystery. *Science Digest* 1983 (September), pp. 151–153.

Jefferson, T. 1799. A memoir on the discovery of certain bones of a quadruped of the clawed kind in the western parts of Virginia. *Transactions of the American Philosophical Society*, 4, pp. 246–259.

Jepsen, G. L. 1964. Riddles of the terrible lizards. *American Scientist* 52, pp. 227–246.

Karpinsky, A. P. 1874. Geologische Untersuchungen im Gouvernment Orenburg. *Verhandlungen der Kaiserlichen Gesellschaft für die Gesammte Mineralogie* 9, pp. 210–212.

—— 1889. Ueber die Ammoneen der Artinsk-Stufe. *Mémoires de l'Académie Impériale des Sciences de St Pétersbourg*, 7ème. Série, 37 (2), pp. 1–104.

Kavasch, J. 1986. *The Ries Meteorite Crater. A Geological Guide.* Donauwörth: Auer.

Keller, G. and Barrera, E. 1990. The Cretaceous/Tertiary boundary impact hypothesis and the paleontological record. *Geological Society of America Special Paper* 247, pp. 563–575.

Kerr, R. A. 1995. A volcanic crisis for ancient life. *Science* 270, pp. 27–28.

King, G. M. 1991a. The aquatic *Lystrosaurus*: a palaeontological myth. *Historical Biology* 4, pp. 285–321.

—— 1991b. Terrestrial tetrapods and the end Permian mass extinction event. *Historical Biology* 5, pp. 239–255.

—— and Cluver, M. A. 1991. The aquatic *Lystrosaurus*: an alternative lifestyle. *Historical Biology*, 4, pp. 323–341.

Kitching, J. W. 1977. The distribution of the Karoo vertebrate fauna; with special reference to certain genera and the bearing of this distribution on the zoning of the Beaufort Beds. *Bernard Price Institute for Palaeontological Research, Memoir 1*, pp. 1–131.

Knoll, A. H. and Bambach, R. K. 2000. Directionality in the history of life: diffusion from the left wall or repeated scaling of the right? *Paleobiology* 26 (Suppl.), pp. 1–14.

Kummel, B. and Teichert, C. 1966. Relations between the Permian and Triassic formations in the Salt and Trans-Indus ranges, West Pakistan. *Neues Jahrbuch für Geologie und Paläontologie, Abhandlungen* 125, pp. 297–333.

—— and —— 1973. The Permian-Triassic boundary in Central Tethys. In Logan, A. and Hills, L. V. (eds), *The Permian and Triassic Systems and their Mutual Boundary.* Memoir 2, Canadian Society of Petroleum Geologists, Calgary, pp. 17–34.

Kolbert, M. 2014. *The Sixth Extinction: An Unnatural History.* London: Bloomsbury.

Laubichler, M. D. and Niklas, K. J. 2009. The morphological tradition in German paleontology: Otto Jaekel, Walter Zimmermann, and Otto Schindewolf. In Sepkoski, D. and Ruse, M. (eds),

The Paleobiological Revolution: Essays on the Growth of Modern Paleontology, pp. 279–300. Chicago: University of Chicago Press.

Lewin, R. 1983. Extinctions and the history of life. *Science* 221, pp. 935–937.

—— 1985a. Catastrophism not yet dead. *Science* 229, p. 640.

—— 1985b. Catastrophism not yet dead. *Science* 230, p. 8.

Linnaeus, C. 1753. *Species plantarum*, 2 volumes. Stockholm: L. Salvii.

—— 1758. *Systema Naturae per Regna Tria Naturae, Secundum Classes, Ordines, Genera, Species, cum Characteribus, Differentiis, Synonymis, Locis*. 10th edition. Stockholm: L. Salvii.

Lomborg, B. 2001. *The Skeptical Environmentalist*. Cambridge: Cambridge University Press.

Loomis, F. B. 1905, Momentum in variation. *American Naturalist* 39, pp. 839–843.

Looy, C. V., Twitchett, R. J., Dilcher, D. L., Van Konijnenburg-Van Cittert, H. A. and Visscher, H. 2001. Life in the end-Permian dead zone. *Proceedings of the National Academy of Science* 98, pp. 7879–7883.

Lyell, C. 1830–1833. *Principles of geology, being an attempt to explain the former changes of the Earth's surface, by reference to causes now in operation*. 3 vols. London: John Murray.

—— 1838. *Elements of Geology*. London: John Murray.

McKinney, M. L. 1995. Extinction selectivity among lower taxa – gradational patterns and rarefaction error in extinction estimates. *Paleobiology* 21, pp. 300–313.

MacLeod, K. G., Smith, R. M. H., Koch, P. L. and Ward, P. D. 2000. Timing of mammal-like reptile extinctions across the Permian-Triassic boundary in South Africa. *Geology* 28, pp. 227–230.

McRoberts, C. A. 2001. Triassic bivalves and the initial marine Mesozoic revolution: a role for predators? *Geology* 29, pp. 359–362.

Maddox, J. 1985. Periodic extinctions undermined. *Nature* 315, p. 627.

Marsh, O. C. 1882. Classification of the Dinosauria. *American Journal of Science, Series 3* 23, pp. 81–86.

—— 1895. On the affinities and classification of the dinosaurian reptiles. *American Journal of Science, Series 3* 50, pp. 483–498.

Matthew, W. D. 1921, Fossil vertebrates and the Cretaceous-Tertiary problem, *American Journal of Science, Series 5* 2, pp. 209–227.

Maxwell, W. D. 1992. Permian and Early Triassic extinction of nonmarine tetrapods. *Palaeontology* 35, pp. 571–583.

—— and Benton, M. J. 1990. Historical tests of the absolute completeness of the fossil record of tetrapods. *Paleobiology* 16, pp. 322–335.

May, R. M. 1990. How many species? *Philosophical Transactions of the Royal Society, Series B* 330, pp. 293–304.

—— 1992. How many species inhabit the Earth? *Scientific American* 267(4), pp. 18–24.

Maynard Smith, J. and Szathmáry, E. 1995. *The Major Transitions in Evolution*. Oxford: W. H. Freeman Spektrum.

Meyer, H. von 1866. Reptilien aus dem Kupfer-Sandstein des West-Uralischen Gouvernements Orenburg. *Palaeontographica* 15, pp. 97–130.

Monastersky, R. 2014. Life – a status report. *Nature* 516, pp. 158–161.

Moodie, R. L., 1923. *Paleopathology*. Urbana, Illinois: University of Illinois Press.

Morrell, J. B. and Thackray, A. 1981. *Gentlemen of Science. Early Years of the British Association for the Advancement of Science*. Oxford: Clarendon Press.

Müller, L. 1928. Sind die Dinosaurier durch Vulkanausbruche ausgeratet worden? *Unsere Welt* 20, pp. 144–146.

Mundil, R., Metcalfe, I., Ludwig, K. R., Renne, P. R., Oberli, F. and Nicoll, R. S. 2001. Timing of the Permian-Triassic biotic crisis: implications from new zircon U/Pb data (and their limitations). *Earth and Planetary Science Letters* 187, pp. 131–145.

Murchison, R. I. 1839. *The Silurian System, founded on geological researches in the counties of Salop, Hereford, Radnor, Montgomery, Caermarthen, Brecon, Pembroke, Monmouth, Gloucester, Worcester and Stafford; with descriptions of the coal-fields and overlying formations*. 2 vols. London: John Murray.

—— 1841a. First sketch of some of the principal results of a second geological survey of Russia, in a letter to M. Fischer. *Philosophical Magazine and Journal of Science*, Series 3, 19, pp. 417–422.

—— 1841b. Geologicheskaya naogyudeniya v Rossii; pis'mo G. Murchisona k'G. Fishera fon Vald'heimu. *Gorny Zhurnal, Moskva*, 1, pp. 160–169.

—— 1842a. Letter to M. Fischer de Waldheim . . . containing some of the results of his second geological survey of Russia. *Edinburgh New Philosophical Journal*, 32, pp. 99–103.

—— 1842b. Anniversary address of the President. *Proceedings of the Geological Society of London*, 3, pp. 637–687.

—— and Verneuil, E. de. 1841a. On the stratified deposits which occupy the northern and central regions of Russia. *Report of the British Association for the Advancement of Science*, 1840, pp. 105–110.

—— and —— 1841b. On the geological structure of the northern and central regions of Russia. *Proceedings of the Geological Society of London*, 3, pp. 398–408.

—— and —— 1842. A second geological survey of Russia in Europe. *Proceedings of the Geological Society of London*, 3, pp. 717–730.

——, —— and Keyserling, A. von 1842. On the geological structure of the Ural Mountains. *Proceedings of the Geological Society of London*, 3, pp. 742–753.

——, —— and —— 1845. *The geology of Russia in Europe and the Ural Mountains.* 2 vols. Volume 1, London: John Murray. Volume 2, Paris: Bertrand.

Newell, A. J., Tverdokhlebov, V. P. and Benton, M. J. 1999. Interplay of tectonics and climate on a transverse fluvial system, Upper Permian, southern Uralian foreland basin, Russia. *Sedimentary Geology* 127, pp. 11–29.

Newell, N. D. 1952. Periodicity in invertebrate evolution. *Journal of Paleontology* 26, pp. 371–385.

—— 1967. Revolutions in the history of life. *Scientific American* 208, pp. 76–92.

Nichols, G. 2009. *Sedimentology and Stratigraphy.* 2nd edition. New York: Wiley-Blackwell.

Nicolas, M. and Rubidge, B. S. 2010. Changes in Permo-Triassic terrestrial tetrapod ecological representation in the Beaufort Group (Karoo Supergroup) of South Africa. *Lethaia* 43, pp. 45–59.

Nopcsa, F. 1911. Notes on British dinosaurs. Part IV: *Stegosaurus priscus*, sp. nov., *Geological Magazine* (5) 8, pp. 143–153.

—— 1917. Über Dinosaurier, *Centralblatt für Mineralogie, Geologie, und Paläontologie* 1917, pp. 332–351.

Norell, M. A. and Novacek, M. J. 1992. The fossil record and evolution: comparing cladistic and paleontologic evidence for vertebrate history. *Science* 255, pp. 1691–1693.

Novotny, V., Basset, Y., Miles, S. E., Weiblen, G. D., Bremer, B., Cizek, L. and Drozd, P. 2002. Low host specificity of herbivorous insects in a tropical forest. *Nature* 416, pp. 841–844.

Nützel, A. 2005. Recovery of gastropods in the Early Triassic. *Comptes Rendus Palevol* 4, pp.. 501–515.

Ochev, V. G. and Surkov, M. A. 2000. The history of excavation of Permo-Triassic vertebrates from eastern Europe. In Benton, M. J., Unwin, D. M., Shishkin, M. A. and Kurochkin, E. N. (eds), *The Age of Dinosaurs in Russia and Mongolia.* Cambridge: Cambridge University Press.

Officer, C. B., Hallam, A., Drake, C. L. and Devine, J. D. 1987. Late Cretaceous and paroxysmal Cretaceous-Tertiary extinctions. *Nature* 326, pp. 143–149.

Oldroyd, D. R. 1990. *The Highlands Controversy. Constructing Geological Knowledge through Field-work in Nineteenth-Century Britain.* Chicago: University of Chicago Press.

Olson, E. C. 1955. Parallelism in the evolution of the Permian reptilian faunas of the Old and New Worlds. *Fieldiana, Zoology* 37, pp. 385–401.

—— 1957. Catalogue of localities of Permian and Triassic vertebrates of the territories of the U.S.S.R. *Journal of Geology* 65, pp. 196–226.

—— 1990. *The Other Side of the Medal: A Paleobiologist Reflects on the Art and Serendipity of Science.* Blacksburg, Va.: McDonald & Woodward.

Osborne, R. 1998. *The Floating Egg. Episodes in the Making of Geology.* London: Jonathan Cape.

Outram, D. 1984. *Georges Cuvier. Vocation, Science and Authority in Post-Revolutionary France.* Manchester: Manchester University Press.

Owen, R. 1842. Report on British fossil reptiles. *Report of the British Association for the Advancement of Science 1841*, pp. 60–204.

—— 1845a. Description of certain fossil crania, discovered by A. G. Bain, Esq., in sandstone rocks at the south-eastern extremity of Africa, referable to different species of an extinct genus of Reptilia (*Dicynodon*), and indicative of a new tribe or sub-order of Sauria. *Quarterly Journal of the Geological Society of London* 1, pp. 318–322, and *Transactions of the Geological Society of London, Series 2* 7, pp. 59–84.

—— 1845b. Professor Owen upon certain saurians of the Permian rocks. Page 637 in Murchison et al. (1845).

—— 1876. Evidences of theriodonts elsewhere than in South Africa. *Quarterly Journal of the Geological Society of London* 32, pp. 352–363.

Payne, J. L., Lehrmann, D. J., Wei, J., Orchard, M. J., Schrag, D. P. and Knoll, A. H. 2004. Large perturbations of the carbon cycle during recovery from the end-Permian extinction. *Science* 305, pp. 506–509.

Phillips, J. 1838. Geology. *Penny Cyclopedia* 11, pp. 127–151.

—— 1840a. Organic remains. *Penny Cyclopedia* 16, pp. 487–491.

—— 1840b. Palaeozoic series. *Penny Cyclopedia* 17, pp. 153–154.

—— 1841. *Figures and descriptions of the Palaeozoic fossils of Cornwall, Devon and west Somerset; observed in the course of the Ordnance Geological Survey of that district.* London: Longman.

—— 1860. *Life on Earth. Its Origin and Succession.* Cambridge: Macmillan.

Pimm, S. L., Russell, G. J., Gittelman, J. L. and Brooks, T. M. 1995. The future of biodiversity. *Science* 269, pp. 347–350.

Pope, A. 1993. *Alexander Pope* [selections]. Edited by P. Rogers. Oxford: Oxford University Press.

Rampino, M. R. and Adler, A. C. 1998. Evidence for abrupt latest Permian mass extinction of foraminifera: results of tests for the Signor-Lipps effect. *Geology* 26, pp. 415–418.

Raup, D. M. 1979. Size of the Permo-Triassic bottleneck and its evolutionary implications. *Science* 206, pp. 217–218.

—— 1986. *The Nemesis Affair.* New York: Norton.

—— and Jablonski, D. 1993. Geography of end-Cretaceous bivalve extinctions. *Science* 260, pp. 971–973.

—— and Sepkoski, J. J., Jr. 1982. Mass extinctions in the marine fossil record. *Science* 215, pp. 1501–1503.

—— and —— 1984. Periodicities of extinctions in the geologic past. *Proceedings of the National Academy of Sciences, U.S.A.* 81, pp. 801–805.

Reichow, M. K., Saunders, A. D., White, R. V., Pringle, M. S., Al'Mukhamedov, A. I., Medvedev, A. I. and Kirda, N. P. 2002. $^{40}Ar/^{39}Ar$ dates from the West Siberian basin: Siberian flood basalt province doubled. *Science* 296, pp. 1846–1849.

Renne, P. R. and Basu, A. R. 1991. Rapid eruption of the Siberian Traps flood basalts at the Permo-Triassic boundary. *Science* 253, pp. 176–179.

—— Zhang, Z., Richardson, M. A., Black, M. T. and Basu, A. R. 1995. Synchrony and causal relations between Permo-Triassic boundary crises and Siberian flood volcanism. *Science* 269, pp. 1413–1416.

—— and 7 others. 2004. Is Bedout an impact structure? Take 2. *Science* 306, pp. 610–611.

Retallack, G. J. 1999. Postapocalyptic greenhouse paleoclimate revealed by earliest Triassic paleosols in the Sydney Basin, Australia. *Bulletin of the Geological Society of America* 111, pp. 52–70.

——, Seyedolali, A., Krull, E. S., Holser, W. T., Ambers, C. P. and Kyte, F. T. 1998. Search for evidence of impact at the Permian-Triassic boundary in Antarctica and Australia. *Geology* 26, pp. 979–982.

——, Veevers, J. J. and Morante, R. 1996. Global coal gap between Permian-Triassic extinction and Middle Triassic recovery of peat-forming plants. *Bulletin of the Geological Society of America* 108, pp. 195–207.

Rhodes, F. H. T. 1967. Permo-Triassic extinction. In Harland, W. B. *The Fossil Record* (ed.), pp. 57–76. Geological Society of London.

Rodland, D. L. and Bottjer, D. J. 2001. Biotic recovery from the end-Permian mass extinction: behavior of the inarticulate brachiopod *Lingula* as a disaster taxon. *Palaios* 16, pp. 95–101.

Romano, C., Koot, M. B., Kogan, I., Brayard, A., Minikh, A. V., Brinkmann, W., Bucher, H. and Kriwet, J. 2014. Permian-Triassic Osteichthyes (bony fishes): diversity dynamics and body size evolution. *Biological Reviews* online ahead of print (doi:10.1111/brv.12161).

Rubidge, B. S. (ed.) 1995. *Biostratigraphy of the Beaufort Group (Karoo Supergroup), South Africa.* Pretoria: Council of Geoscience.

Rudwick, M. J. S. 1969. The strategy of Lyell's *Principles of geology. Isis* 61, pp. 5–33.

—— 1975. Caricature as a source for the history of science: De la Beche's anti-Lyellian sketches of 1831. *Isis* 66, pp. 534–560.

—— 1976. *The Meaning of Fossils. Episodes in the History of Palaeontology.* 2nd edition. New York: Science History Publications.

—— 1985. *The Great Devonian Controversy: The Shaping of Scientific Knowledge Among Gentlemanly Specialists.* Chicago: Chicago University Press.

—— 1997. *Georges Cuvier, Fossil Bones, and Geological Catastrophes.* Chicago: Chicago University Press.

Rupke, N. A. 1994. *Richard Owen, Victorian Naturalist.* New Haven: Yale University Press.

Russell, D. A. and Tucker, W. 1971. Supernovae and the extinction of the dinosaurs. *Nature* 229, pp. 553–554.

Ruzhentsev, V. E. 1950. [Upper Carboniferous ammonoids from the Urals.] *Trudy Paleontologischeskogo Instituta AN SSSR* 29, pp. 1–223.

Salvador, A. (ed.) 1994. *International Stratigraphic Guide. A Guide to Stratigraphic Classification, Terminology, and Procedure.* 2nd edition. International Union of Geological Sciences.

Schindewolf, O. H. 1950. *Grundfragen der Paläontologie. Geologische Zeitmessung – Organische Stammesentwicklung – Biologische Systematik.* Stuttgart: Schweizerbart.

—— 1958. Zur Aussprache über die grossen erdgeschichtlichen Faunenschnitte und ihre Verursachung. *Neues Jahrbuch für Geologie und Paläontologie, Monatshefte* 1958, pp. 270–279.

—— 1963. Neokatastrophismus? *Zeitschrift der Deutschen Geologischen Gesellschaft* 114, pp. 430–445.

—— 1993. *Basic Questions in Paleontology. Geologic Time, Organic Evolution, and Biological Systematics.* (Judith Shaefer trans.). Chicago: University of Chicago Press.

Schulte, P. and 40 others. 2010. The Chicxulub asteroid impact and mass extinction at the Cretaceous-Paleogene boundary. *Science* 327, pp. 1214–1218.

Secord, J. A. 1986. *Controversy in Victorian Geology: The Cambrian-Silurian Dispute.* Princeton: Princeton University Press.

Sedgwick, A. 1831. Address to the Geological Society, delivered on the evening of the anniversary, Feb. 18, 1831. *Proceedings of the Geological Society of London*, 1, pp. 281–316.

—— and Murchison, R. I. 1839. Classification of the older stratified rocks of Devonshire and Cornwall. *Philosophical Magazine and Journal of Science*, Series 3, 14, pp. 241–260.

—— and —— 1840. On the classification and distribution of the older or Palaeozoic rocks of the north of Germany and of Belgium, as compared with formations of the same age in the British Isles. *Proceedings of the Geological Society of London* 3, pp. 300–311.

Seeley, H. G. 1894. Researches on the structure, organization, and classification of the fossil Reptilia. Part VIII. Further evidences of *Deuterosaurus* and *Rhopalodon* from the Permian rocks of Russia. *Philosophical Transactions of the Royal Society, Series B* 185, pp. 663–717.

Self, S., Schmidt, A. and Mather, T. A. 2014. Emplacement characteristics, time scales, and volcanic gas release rates of continental flood basalt eruptions on Earth. *Geological Society of America, Special Paper 505*, doi: 10.1130/2014.2505(16).

Sennikov, A. G. 1996. Evolution of the Permian and Triassic tetrapod communities of Eastern Europe. *Palaeogeography, Palaeoclimatology, Palaeoecology* 120, pp. 331–351.

Sepkoski, J. J., Jr. 1984. A kinetic model of Phanerozoic taxonomic diversity. III. Post-Paleozoic families and mass extinctions. *Paleobiology* 10, pp. 246–267.

—— 1993. Ten years in the library: how changes in taxonomic data bases affect perception of macroevolutionary pattern. *Paleobiology* 19, pp. 43–51.

—— 1996. Patterns of Phanerozoic extinction: a perspective from global data bases. In Walliser, O. H. (ed.) *Global Events and Event Stratigraphy*, pp. 35–52. Berlin: Springer-Verlag.

Sheehan, P. M., Fastovsky, D. E., Hoffman, R. G., Berghaus, C. B. and Gabriel, D. L. 1991. Sudden extinction of the dinosaurs: latest Cretaceous upper Great Plains, U.S.A. *Science* 254, pp. 835–839.

Shen, S. and 21 others. 2011. Calibrating the end Permian mass extinction. *Science* 334, pp. 1367–1372.

Shoemaker, E. M. and Chao, E. C. T. 1961. New evidence for the impact origin of the Ries Basin, Bavaria, Germany. *Journal of Geophysical Research* 66, pp. 3371–3378.

Signor, P. W. and Lipps, J. H. 1982. Sampling bias, gradual extinction patterns and catastrophes in the fossil record. *Special Paper of the Geological Society of America* 190, pp. 291–296.

Simpson, S. 2001. Deeper impact. *Scientific American* 276 (5), pp. 13–14.

Sloan, R. E., Rigby, I. K., Jr., Van Valen, L. M. and Gabriel, D. 1986. Gradual dinosaur extinction and simultaneous ungulate radiation in the Hell Creek Formation. *Science* 232, pp. 629–632.

Smit, J. and Hertogen, J. 1980. An extraterrestrial event at the Cretaceous-Tertiary boundary. *Nature* 285, pp. 198–200.

Smith, F. D. M., May, R. M., Pellew, R., Johnson, T. H. and Walter, K. R. 1993. How much do we know about the current extinction rate? *Trends in Ecology and Evolution* 8, pp. 375–378.

Smith, R. H. M. and Ward, P. D. 2001. Pattern of vertebrate extinctions across an event bed at the Permian-Triassic boundary in the Karoo Basin of South Africa. *Geology* 29, pp. 1147–1150.

—— and Botha, J. 2005. The recovery of terrestrial vertebrate diversity in the South African Karoo Basin after the end-Permian mass extinction. *Comptes Rendus Palevol* 4, pp. 555–568.

—— and Botha-Brink, J. 2014. Anatomy of a mass extinction: sedimentological and taphonomic evidence for drought-induced die-offs at the Permo-Triassic boundary in the main Karoo Basin, South Africa. *Palaeogeography, Palaeoclimatology, Palaeoecology* 396, pp. 99–118.

Solé, R., Saldaña, J., Montoya, J. M. and Erwin, D. H. 2010. Simple model of recovery dynamics after mass extinction. *Journal of Theoretical Biology* 267, pp. 193–200.

Song, H. J., Wignall, P. B., Tong, J. N. and Yin, H. F. 2013. Two pulses of extinction during the Permian-Triassic crisis. *Nature Geoscience* 6, pp. 52–56.

Stafford, R. A. 1989. *Scientist of Empire: Sir Roderick Murchison, Scientific Exploration and Victorian Imperialism.* Cambridge: Cambridge University Press.

Stanley, S. M. 1984. Temperature and biotic crises in the marine realm. *Geology* 12, pp. 205–208.

—— 1988. Paleozoic mass extinctions: shared patterns suggest global cooling as a common cause. *American Journal of Science* 288, pp. 334–352.

—— and Yang, X. 1994. A double mass extinction at the end of the Paleozoic era. *Science* 266, pp. 1340–1344.

—— 2008. *Earth System History.* New York: W. H. Freeman.

Stauffer, R. C. (ed.) 1975. *Charles Darwin's Natural Selection: Being the Second Part of his Big Species Book Written from 1856 to 1858.* Cambridge: Cambridge University Press.

Stöffler, D. and Ostertag, R. 1983. The Ries impact crater. *Fortschritte der Mineralogie* 61 (2), pp. 71–116.

Strangways, W. H. T. F. 1822. An outline of the geology of Russia. *Transactions of the Geological Society of London*, Series 2 1, pp. 1–39.

Sun, Y., Xu, D., Zhang, Q., Yang, Z., Sheng, J., Chen, C., Rui, L., Liang, X., Zhao, J. and He, J. 1984. The discovery of an iridium anomaly in the Permian-Triassic boundary clay in Changxing, Zhejiang, China and its significance. In *Developments in Geoscience, Contributions*, pp. 235–245. 27th International Geological Congress. Beijing: Science Press.

Sun, Y. D., Joachmiski, M. M., Wignall, P. B., Yan, C. B., Chen, Y. L., Jiang, H. S., Wang, L. and Lai, X. L. 2012. Lethally hot temperatures during the Early Triassic greenhouse. *Science* 338, pp. 366–370.

Surkov, M. V. 2002. *Lystrosaurus georgi* and the habits of the lystrosaurs. *Palaeontology*, in review.

Teichert, C. 1990. The Permian-Triassic boundary revisited. In Kauffman, E. G. and Walliser, O. H. (eds) *Extinction Events in Earth History*, pp. 199–238. Berlin: Springer Verlag.

—— and Kummel, B. 1976. Permian-Triassic boundary in the Kap Stosch area, East Greenland. *Meddelelser om Grønland* 597, pp. 1–54.

Terry, K. D. and Tucker, W. H. 1968. Biological effects of supernova. *Science* 159, pp. 421–423.

Thackray, J. C. 1978. R. I. Murchison's *Geology of Russia* (1845). *Journal of the Society for the Bibliography of Natural History* 8, pp. 421–433.

Tverdokhlebov, V. P. 1971. [On Early Triassic proluvial deposits of the Pre-Urals, and times of folding and mountain-building processes in the southern Urals.] *Izvestiya AN SSSR, Seriya Geologicheskaya* 1971 (4), pp. 42–50.

——, Tverdokhlebova, G. I., Benton, M. J. and Storrs, G. W. 1997. First record of footprints of terrestrial vertebrates from the Upper Permian of the Cis-Urals, Russia. *Palaeontology* 40, pp. 157–166.

——, ——, Surkov, M. V. and Benton, M. J. 2003. Tetrapod localities from the Triassic of the SE of European Russia. *Earth Science Reviews* 60, pp. 1–66.

——, ——, Minikh, A. V., Surkov, M. V. and Benton, M. J. 2005. Upper Permian vertebrates and their sedimentological context in the South Urals, Russia. *Earth-Science Reviews* 69, pp. 27–77.

Twelvetrees, W. H. 1880. On a new theriodont reptile (*Cliorhizodon orenburgensis*, Twelvetr.) from the Upper Permian cupriferous sandstones of Kargala, near Orenburg in south-eastern Russia. *Quarterly Journal of the Geological Society of London* 36, pp. 540–543.

—— 1882. On the organic remains from the Upper Permian strata of Kargala, in eastern Russia. *Quarterly Journal of the Geological Society of London* 38, pp. 490–501.

Twitchett, R. J. 1999. Palaeoenvironments and faunal recovery after the end-Permian mass extinction. *Palaeogeography, Palaeoclimatology, Palaeoecology* 154, pp. 27–37.

——, Looy, C. V., Morante, R., Visscher, H. and Wignall, P. B. 2001. Rapid and synchronous collapse of marine and terrestrial ecosystems during the end-Permian biotic crisis. *Geology* 29, pp. 351–354.

Valentine, J. W. and Moores, E. M. 1973. Provinciality and diversity across the Permian-Triassic boundary. In Logan, A. and Hills, L. V. (eds) *The Permian and Triassic Systems and their Mutual Boundary*, pp. 759–766. Canadian Society of Petroleum Geology, Memoir No. 2.

Van Valen, L. M. 1984. Catastrophes, expectations, and the evidence. *Paleobiology* 10, pp. 121–137.

Wang, X.-D. and Sugiyama, T. 2000. Diversity and extinction patterns of Permian coral faunas of China. *Lethaia* 33, pp. 285–294.

Ward, P. D. 1990. The Cretaceous/Tertiary extinctions in the marine realm: a 1990

perspective. *Geological Society of America Special Paper* 247, pp. 425–432.
——, Montgomery, D. R. and Smith, R. 2000. Altered river morphology in South Africa related to the Permian-Triassic extinction. *Science* 289, pp. 1740–1743.
Watson, D. M. S. 1952. Dr Robert Broom, F.R.S. *Obituaries of Fellows of the Royal Society* 8, pp. 37–70.
—— 1957. The two great breaks in the history of life. *Quarterly Journal of the Geological Society, London* 112, pp. 435–444.
Wieland, G. R. 1925. Dinosaur extinction. *American Naturalist* 59, pp. 557–565.
Wignall, P. B. 2001. Large igneous provinces and mass extinctions. *Earth-Science Reviews* 53, pp. 1–33.
—— and Hallam, A. 1992. Anoxia as a cause of the Permian/Triassic extinction: facies evidence from northern Italy and the western United States. *Palaeogeography, Palaeoclimatology, Palaeoecology* 93, pp. 21–46.
—— and —— 1993. Griesbachian (earliest Triassic) palaeoenvironmental changes in the Salt Range, Pakistan and southwest China and their bearing on the Permo-Triassic mass extinction. *Palaeogeography, Palaeoclimatology, Palaeoecology* 102, pp. 215–237.
——, Kozur, H., and Hallam, A. 1996. The timing of palaeoenvironmental changes at the Permo-Triassic (P/Tr) boundary using conodont biostratigraphy. *Historical Biology* 12, pp. 39–62.
—— and Twitchett, R. J. 1996. Oceanic anoxia and the end Permian mass extinction. *Science* 272, pp. 1155–1158.
——, —— 2002 Extent, duration and nature of the Permian-Triassic superanoxic event. *Geological Society of America Special Paper* 356, pp. 395–413.
Williams, J. S. 1938. Pre-congress Permian conference in the U.S.S.R. *Bulletin of the American Association of Petroleum Geologists* 22, pp. 771–776.
Wilson, E. O. 1992. *The Diversity of Life.* Cambridge, Mass.: Harvard University Press; London: Penguin.
—— 2002. *The Future of Life.* New York, Alfred A. Knopf; London: Little Brown.
—— and Peter, F. M. (eds) 1988. *Biodiversity.* Washington, D.C.: National Academy Press.
Winchester, S. 2001. *The Map that Changed the World. The Tale of William Smith and the Birth of a Science.* London: Penguin Viking; New York: HarperCollins.
Woodward, A. S. 1898, *Outlines of Vertebrate Paleontology for Students of Geology.* Cambridge: Cambridge University Press.
—— 1910, Presidential Address to Section C, *Report of the British Association for the Advancement of Science*, 1909, pp. 462–471.
Wooler, -. 1758. A description of the fossil skeleton of an animal found in the alum rock near Whitby. *Philosophical Transactions of the Royal Society of London* 50, pp. 786–791.
Xu, D., Na, L., Chi, Z., Mao, X., Su, Y., Zhang, Q. and Yong, Z. 1985. Abundance of iridium and trace metals at the Permian/Triassic boundary at Shangsi in China. *Nature* 314, pp. 154–156.
Yang, Z., Sheng, J. and Yin, H. 1995. The Permian-Triassic boundary: the global stratotype section and point. *Episodes* 18, pp. 49–53.
Yin, H., Huang, S., Zhang, K., Hansen, H. J., Yang, F., Ding, M. and Bie, X. 1992. The effects of volcanism on the Permo-Triassic mass extinction in South China. In Sweet, W. C., Yang, Z., Dickins, J. M. and Yin, H. (eds) *Permo-Triassic Events in Eastern Tethys*, pp. 146–157. Cambridge: Cambridge University Press.
Zhou, L. and Kyte, F. 1988. The Permian-Triassic boundary event: a geochemical study of three Chinese sections. *Earth and Planetary Science Letters* 90, pp. 411–421.
Zittel, K. A. von. 1901. *History of Geology and Palaeontology to the End of the Nineteenth Century.* London: Walter Scott.

ILLUSTRATION CREDITS

All chapter heading details are taken from the drawings in the book by John Sibbick
Frontispiece. Drawing by John Sibbick.

1 Drawing by John Sibbick
2 Drawing by John Sibbick
3 Drawing by John Sibbick
4 From Murchison 1841a
5 Drawing by John Sibbick, from various sources
6 From Phillips 1860
7 Drawing by John Sibbick
8 'Professor Ichthyosaurus' by Henry De la Beche, 1830
9 Drawing by John Sibbick
10 Drawing by John Sibbick
11 Based on Alvarez et al., 1980 and other sources
12 Drawing by John Sibbick
13 Based on various sources
14 Based on various sources
15 Based on various sources
16 Much modified from Benton, 1995
17 Based on data in Benton, 1995; redrawn and embellished by John Sibbick
18 Modified from Raup and Sepkoski, 1982
19 Modified from Raup and Sepkoski, 1984
20 Based on data in Keller and Barrera, 1990
21 Based on data in Maxwell and Benton 1990
22 Modified from Benton et al., 2000
23 Based on various sources
24 Based on the work of John Scotese, and other authors
25 Based on Wignall and Hallam, 1993
26 Courtesy of Tony Hallam
27 Based on Song et al., 2013
28 Drawing by John Sibbick
29 Drawing by John. Sibbick
30 Modified from Gould and Calloway, 1980
31 Drawing by John Sibbick, based on various sources
32 Drawing by John Sibbick, based on various sources
33 Based on Ochev and Surkov, 2001
34 M.J. Benton
35 Drawing by John Sibbick, based on various sources
36 Drawing by John Sibbick, based on Bystrov, 1957
37 Drawing by John Sibbick
38 Drawing by John Sibbick
39 Based on work during the 1995 expedition, from Newell et al., 1999
40 Based on the work of Valentin Tverdokhlebov
41 Courtesy of Paul Wignall
42 Drawing by John Sibbick
43 Courtesy of Tony Hallam and Paul Wignall
44 Based on Solé et al., 2010
45 Based on Payne et al., 2004
46 M.J. Benton
47 Based on data in Smith et al., 1993

INDEX

Page numbers in *italics* refer to illustrations